Catalytic Processes for Water and Wastewater Treatment

Catalytic Processes for Water and Wastewater Treatment

Editor

John Vakros

MDPI • Basel • Beijing • Wuhan • Barcelona • Belgrade • Manchester • Tokyo • Cluj • Tianjin

Editor
John Vakros
Chemical Engineering
University of Patras
Patras
Greece

Editorial Office
MDPI
St. Alban-Anlage 66
4052 Basel, Switzerland

This is a reprint of articles from the Special Issue published online in the open access journal *Catalysts* (ISSN 2073-4344) (available at: www.mdpi.com/journal/catalysts/special_issues/catalytic_wastewater_treatment).

For citation purposes, cite each article independently as indicated on the article page online and as indicated below:

LastName, A.A.; LastName, B.B.; LastName, C.C. Article Title. *Journal Name* **Year**, *Volume Number*, Page Range.

ISBN 978-3-0365-7299-4 (Hbk)
ISBN 978-3-0365-7298-7 (PDF)

© 2023 by the authors. Articles in this book are Open Access and distributed under the Creative Commons Attribution (CC BY) license, which allows users to download, copy and build upon published articles, as long as the author and publisher are properly credited, which ensures maximum dissemination and a wider impact of our publications.

The book as a whole is distributed by MDPI under the terms and conditions of the Creative Commons license CC BY-NC-ND.

Contents

About the Editor vii

Preface to "Catalytic Processes for Water and Wastewater Treatment" ix

John Vakros
Catalytic Processes for Water and Wastewater Treatment
Reprinted from: *Catalysts* 2023, 13, 677, doi:10.3390/catal13040677 1

Gururaj M. Neelgund, Sanjuana Fabiola Aguilar, Erica A. Jimenez and Ram L. Ray
Adsorption Efficiency and Photocatalytic Activity of Silver Sulfide Nanoparticles Deposited on Carbon Nanotubes
Reprinted from: *Catalysts* 2023, 13, 476, doi:10.3390/catal13030476 7

Shakeel Khan, Awal Noor, Idrees Khan, Mian Muhammad, Muhammad Sadiq and Niaz Muhammad
Photocatalytic Degradation of Organic Dyes Contaminated Aqueous Solution Using Binary $CdTiO_2$ and Ternary $NiCdTiO_2$ Nanocomposites
Reprinted from: *Catalysts* 2022, 13, 44, doi:10.3390/catal13010044 27

Cristina Pei Ying Kong, Nurul Amanina A. Suhaimi, Nurulizzatul Ningsheh M. Shahri, Jun-Wei Lim, Muhammad Nur and Jonathan Hobley et al.
Auramine O UV Photocatalytic Degradation on TiO_2 Nanoparticles in a Heterogeneous Aqueous Solution
Reprinted from: *Catalysts* 2022, 12, 975, doi:10.3390/catal12090975 47

Nurul Amanina A. Suhaimi, Cristina Pei Ying Kong, Nurulizzatul Ningsheh M. Shahri, Muhammad Nur, Jonathan Hobley and Anwar Usman
Dynamics of Diffusion- and Immobilization-Limited Photocatalytic Degradation of Dyes by Metal Oxide Nanoparticles in Binary or Ternary Solutions
Reprinted from: *Catalysts* 2022, 12, 1254, doi:10.3390/catal12101254 63

Jishu Rawal, Urooj Kamran, Mira Park, Bishweshwar Pant and Soo-Jin Park
Nitrogen and Sulfur Co-Doped Graphene Quantum Dots Anchored TiO_2 Nanocomposites for Enhanced Photocatalytic Activity
Reprinted from: *Catalysts* 2022, 12, 548, doi:10.3390/catal12050548 79

Souraya Goumri-Said and Mohammed Benali Kanoun
Insight into the Effect of Anionic–Anionic Co-Doping on $BaTiO_3$ for Visible Light Photocatalytic Water Splitting: A First-Principles Hybrid Computational Study
Reprinted from: *Catalysts* 2022, 12, 1672, doi:10.3390/catal12121672 99

Luminita Isac, Cristina Cazan, Luminita Andronic and Alexandru Enesca
CuS-Based Nanostructures as Catalysts for Organic Pollutants Photodegradation
Reprinted from: *Catalysts* 2022, 12, 1135, doi:10.3390/catal12101135 111

Marie Le Pivert, Aurélie Piebourg, Stéphane Bastide, Myriam Duc and Yamin Leprince-Wang
Direct One-Step Seedless Hydrothermal Growth of ZnO Nanostructures on Zinc: Primary Study for Photocatalytic Roof Development for Rainwater Purification
Reprinted from: *Catalysts* 2022, 12, 1231, doi:10.3390/catal12101231 129

Akbar Mehdizadeh, Zahra Derakhshan, Fariba Abbasi, Mohammad Reza Samaei, Mohammad Ali Baghapour and Mohammad Hoseini et al.
The Effect of Arsenic on the Photocatalytic Removal of Methyl Tet Butyl Ether (MTBE) Using Fe_2O_3/MgO Catalyst, Modeling, and Process Optimization
Reprinted from: *Catalysts* **2022**, *12*, 927, doi:10.3390/catal12080927 143

Maria Antonopoulou, Maria Papadaki, Ilaeira Rapti and Ioannis Konstantinou
Photocatalytic Degradation of Pharmaceutical Amisulpride Using g-C_3N_4 Catalyst and UV-A Irradiation
Reprinted from: *Catalysts* **2023**, *13*, 226, doi:10.3390/catal13020226 155

Ilaeira Rapti, Vasiliki Boti, Triantafyllos Albanis and Ioannis Konstantinou
Photocatalytic Degradation of Psychiatric Pharmaceuticals in Hospital WWTP Secondary Effluents Using g-C_3N_4 and g-C_3N_4/MoS_2 Catalysts in Laboratory-Scale Pilot
Reprinted from: *Catalysts* **2023**, *13*, 252, doi:10.3390/catal13020252 169

Lalit Goswami, Anamika Kushwaha, Saroj Raj Kafle and Beom-Soo Kim
Surface Modification of Biochar for Dye Removal from Wastewater
Reprinted from: *Catalysts* **2022**, *12*, 817, doi:10.3390/catal12080817 183

Spyridon Giannakopoulos, John Vakros, Zacharias Frontistis, Ioannis D. Manariotis, Danae Venieri and Stavros G. Poulopoulos et al.
Biochar from Lemon Stalks: A Highly Active and Selective Carbocatalyst for the Oxidation of Sulfamethoxazole with Persulfate
Reprinted from: *Catalysts* **2023**, *13*, 233, doi:10.3390/catal13020233 211

Konstantinos Kouvelis, Adamantia A. Kampioti, Athanasia Petala and Zacharias Frontistis
Degradation of Sulfamethoxazole Using a Hybrid CuO_x–$BiVO_4$/SPS/Solar System
Reprinted from: *Catalysts* **2022**, *12*, 882, doi:10.3390/catal12080882 231

About the Editor

John Vakros

John Vakros serves as researcher in the Department of Chemical Engineering at University of Patras. His teaching activities cover general chemistry, physical chemistry, catalysis and instrumental chemical analysis. His research activities are in the field of heterogeneous catalysis, preparation of catalysts, preparation and characterization of biochars and application of novel materials for environmental protection.

Preface to "Catalytic Processes for Water and Wastewater Treatment"

Water and wastewater treatment are still facing significant challenges today. Due to climate change, water shortages are already noticeable in many regions of the world. To overcome these problems, alternative water sources must be generated and used; for example, treated wastewater, rainwater, surface water, etc. These water sources have to be treated before use since the development of highly sensitive methods reveals the presence of hazardous compounds in low concentrations. These pollutants should be removed, and many processes have been proposed for their degradation/removal, among them catalytic methods.

Those where catalysts are in solid form in particular exhibit significant advantages.

Catalytic oxidation or reduction processes are among the most efficient processes. These include photocatalysis, sulfate and hydroxyl-radical-based advanced oxidation processes.

This Special Issue contains the contributions of different research groups and discusses the recent progress and advances in catalytic processes in water and wastewater treatment.

John Vakros
Editor

Editorial

Catalytic Processes for Water and Wastewater Treatment

John Vakros [1,2]

[1] Department of Chemical Engineering, University of Patras, Caratheodory 1, University Campus, GR-26504 Patras, Greece; vakros@chemistry.upatras.gr
[2] School of Sciences and Engineering, University of Nicosia, 2417 Nicosia, Cyprus

Water and wastewater treatment still face significant challenges today. Due to climate change, water shortages are already noticeable in many regions of the world. To overcome these problems, alternative water sources must be generated and used, for example, treated wastewater, rainwater, surface water, etc. Since the development of highly sensitive methods reveals the presence of hazardous compounds in low concentrations, these water sources must be treated before use. The detected compounds include pharmaceuticals, which have a serious influence on human health. Generally, the concentrations of these compounds is low. However, recent statistics have revealed that the per capita consumption of medicines within the European Union for the period of 2000–2014 has almost tripled, reflecting the excessive and perhaps reckless use of similar products. In particular, there are numerous different ways that remedies may permeate into surface recipients, such as landfill leachates, excessive consumption by people themselves, high volumes of organic hospital effluents, and the undeniably incorrect rejection of medicines that have expired or were not used. Due to the incomplete biodegradation of complex organic molecules in conventional wastewater treatment plants, the accumulation of organic pollutants on surface waters endangers biodiversity through its potential toxicity. Thus, these pollutants should be removed. Many processes have been proposed for their degradation/removal, including catalytic methods. These methods, in which catalysts are in a solid form in particular, exhibit significant advantages such as easy separation, better stability, higher activity, and a lower cost of the process. Catalytic oxidation or reduction processes are among the most efficient processes. These include photocatalysis and sulfate- and hydroxyl-radical-based advanced oxidation processes.

The application of a solid catalyst in a removal experiment combines not only the oxidation reaction but also the adsorption of the contaminant on the catalyst surface. The adsorption is a necessary step for the catalytic process, and hydrophilicity of the surface is required in such cases. By itself, adsorption very useful for removing heavy metal ions. Heavy metals are highly soluble in water and are toxic, carcinogenic, and non-degradable. In the study by Neelgund et al. [1], a nanocomposite CNTs-Ag_2S was prepared through the facile deposition of Ag_2S nanoparticles onto oxidized carbon nanotubes (CNTs), using the hydrothermal method. At first, the hydrophobic CNTs were treated with concentrated nitric and sulfuric acids under reflux. The modification increased the hydrophilic nature of the CNTs by introducing oxygen species onto the surface. The Ag_2S nanoparticles were then deposited onto the modified CNTs. The CNTs-Ag_2S is more efficient than CNTs and Ag_2S and can completely adsorb Cd (II) within 80 min. The adsorption follows pseudo-second-order kinetics, while intraparticle diffusion and the boundary layer effect contribute to the removal of Cd (II) [1]. The CNTs-Ag_2S can also act as a photocatalyst for the degradation of alizarin yellow R (AYR), a dye used in the textile industry. With the irradiation of the material with natural sunlight, complete degradation of AYR is achieved. The degradation follows pseudo-first-order kinetics. Different reactive species such as electrons, holes, hydrogen peroxide, and ROS are the active species for the degradation, and the CNTs-Ag_2S was found to be stable for three cycles [1].

Photocatalysis is probably the most common method used for the degradation of organic compounds in water matrices, and titania-based materials are widely used as photocatalysts. In the study by Khan et al., [2] binary $CdTiO_2$ and ternary $NiCdTiO_2$ were compared with bare TiO_2 for the photocatalytic degradation of organic dyes such as methylene blue (MB) and methyl green (MG). It was found that the photocatalytic degradation of TiO_2, $CdTiO_2$, and $NiCdTiO_2$ for MB was 77, 82, and 86%, while for MG the binary and ternary materials were even better, achieving a degradation of 63.5, 88, and 97.5%, respectively. In all three materials, the TiO_2 was in the form of anatase, while the deposition of Cd and Ni caused a red shift in the absorbance spectra by coupling with Ni and Cd. This shift can be considered responsible for the enhanced photocatalytic activity.

Kong et. al. [3] used anatase TiO_2 to degrade Auramine O (AO) under 365 nm light irradiation. The titania nanoparticles had a mean diameter of 100nm. The degradation process followed a pseudo-first order reaction, with k = 0.048 ± 0.002 min^{-1}. This value is significantly lower than the corresponding value for the degradation of methylene blue (0.173 ± 0.019 min^{-1}); this is due to the nonplanar structure of AO. The higher photocatalytic efficiency was found to be 96%, and it decreased nonlinearly with an increase in the initial concentration and catalyst dosage. Finally, the authors confirmed the superiority of titania nanoparticles compared to the use of micro-sized powders.

The degradation process is more complicated when more than one dye is present in the solution; however, this scenario is also more realistic. Therefore, there are numerous studies in which binary or ternary solutions of dyes are used in photocatalytic degradation. In this recent review [4], the photocatalytic behavior of methylene, rhodamine B, and methyl orange in their binary or ternary solutions was summarized. In this review, the importance of diffusion was discussed. It was revealed that smaller dyes with a planar conformation are easier to degrade and dominate in photocatalytic degradation in their binary or ternary solutions.

Although titania is likely the better photocatalyst, it has some limitations, among which is its limited optical adsorption in the visible light spectrum. Many methods have been used to further expand the optical UV–Vis absorption region of TiO_2 and to enhance the efficient light-induced charge separation. The best option reported is to combine TiO_2 with typical narrow-bandgap quantum dots (QDs) of different semiconductors and carbon quantum dots (CQDs). In the work of Rawal et. al., [5] nitrogen (N) and sulfur (S) co-doped graphene quantum dots (GQDs) were used. This material has a narrowed Eg value that improves photogenerated electron transfer due to π-conjugation. The activity was found to be 2.3–3 times higher than that of TiO_2. The photocatalyst was stable for at least three cycles, and the formation of C-O-Ti bonds provided a charge transfer pathway.

$BaTiO_3$ is another promising material, especially when it is combined with the double-hole coupling of anion–anion combinations. Barium titanate has a perovskite structure and large band gap of 3.0–3.3 eV; it has the appropriate band positions to split water into hydrogen and oxygen, although the band gap should be regulated in lower values to provide the ability to absorb more visible light. It has also been reported that by substituting a metal dopant at a Ba or Ti site, a direct effect on the band gap energy (Eg), the stability, and the formation of oxygen vacancies can be achieved. On this basis, the electronic properties and optical absorption characteristics of $BaTiO_3$, with the double-hole coupling of anion–anion combinations using first-principles methods, were studied. It was found that the N–N co-doped $BaTiO_3$ exhibited a more favorable formation energy under an O-poor condition, and all the co-doping configurations reduced the value of Eg. The tuning of the Eg makes the photocatalyst a promising candidate for visible-light water splitting [6].

Due to the importance of photocatalysis, as it is one of the most innovative advanced oxidation techniques used, many materials have been tested for their photocatalytic activity. A photocatalytic process should present a high efficiency, low cost, and avoid the production of intermediate products with a high toxicity. The selection of a photocatalyst is very important, as it must be cheap, stable, environmentally friendly, and have high activity. Usually, a photocatalyst is a semiconductor with a high degree of absorption of light, preferably solar light, and high rate of photogenerated charge carriers. One review summarized the application of CuS as a photocatalyst [7]. CuS is a p-type semiconductor with different nanostructures, such as nanoparticles and quantum dots or heterojunctions with carbon materials, organic materials, or metal oxides. The recent developments in organic pollutant photodegradation with a CuS as catalyst are discussed on the basis of different synthesis parameters (Cu:S molar ratios, surfactant concentration etc.) and properties (particle size, morphology, bandgap energy, and surface properties)

In the study by Le Pivert et. al., [8], ZnO nanoparticles were synthesized onto Zn and used as roof for the photocatalytic removal of organic contaminants in rainwater. The authors prepared the ZnO nanoparticles on the Zn roof with a hydrothermal growth method in 2 h. The presence of Zn (II) ions and the native oxide film on the Zn surface acted as active sites for the growth of ZnO nanoparticles. The photocatalytic activity was determined using Methylene Blue under UV irradiation and Acid Red 14 with solar light.

The removal of Methyl Tert Butyl Ether (MTBE) using Fe_2O_3/MgO as a catalyst was studied in the presence of arsenic under UV irradiation [9]. The catalyst contained 33.06% Fe_2O_3 and 45.06% MgO. The main parameters of the degradation process, including the effect of arsenic and MTBE concentrations, catalyst mass, pH, etc., were studied. It was found that the concentration of MTBE influenced the most the photocatalytic process. The higher performance in the removal of MTBE was achieved at pH = 5 and an initial concentration of MTBe equal to 37.5 mg/L, with a catalyst dose of 1.58 mg/L without the presence of As. The presence of arsenic decreased the removal efficiency remarkably. Therefore, pretreatment for the removal of arsenic and more details of this interference effect are suggested.

Psychiatric drugs, including amisulpride, a typical antipsychotic drug, are a class of pharmaceuticals that are often prescribed and used in a wide range of mental health problems. Their use is increasing worldwide. Graphitic carbon nitride (g-C_3N_4) can be used as photocatalyst for the degradation of this type of drug. In the work of Antonopoulou et. al. [10], the application of g-C_3N_4 catalysts for the degradation of amisulpride was studied. Particularly, in the degradation process of amisulpride under UV-A irradiation with the application of g-C_3N_4, the transformation products and the ecotoxicological assessment were investigated. It was found that the degradation pathways of amisulpride included mainly oxidation, dealkylation, and the cleavage of the methoxy group. The main reactive species were found to be h^+ and $O_2^{\bullet-}$. In ultrapure water, the transformation products had a low toxicity, while an increased toxicity was observed when wastewater was used as a matrix, especially at the beginning of the process.

g-C_3N_4 is an interesting material with a graphene-like structure, chemical and thermal stability, and a low cost. Is a good visible light absorber and can be used in a wide variety of photocatalytic applications. It has a narrow band gap of ~2.7 eV, but it exhibits a high recombination process and thus limited activity. This drawback can be overcome with the combination of g-C_3N_4 and MoS_2, since heterostructuring can decrease the recombination effects between electrons and holes. The photocatalytic removal of psychiatric drugs in secondary effluents of hospital wastewater with g-C_3N_4 or 1% MoS_2/g-C_3N_4 was studied [11] in a laboratory-scale pilot plant and in a solar simulator apparatus. The results showed that the 1% MoS_2/g-C_3N_4 was more active than the g-C_3N_4, with 54% and 30% removal, respectively. The degradation followed first order kinetics, and six transformation products were generated during degradation; these were totally degraded at the end of the treatment.

As previously noted, research on new materials is always challenging. These materials should be active, selective, and of a low cost. It is also desirable that they are ecofriendly, and, if possible, they should originate from the valorization of waste. Biochar is such a materials. It can be produced from any kind of biomass through a pyrolysis process under a limited- or no-oxygen atmosphere. Generally, biochar has a moderate or high specific surface area, a hierarchical pore structure, and active surface groups. The utilization of biochar and biochar-based nanocomposites for dye removal from wastewater was reviewed [12]. The removal of dyes from wastewater via natural and modified biochar follows numerous mechanisms, such as precipitation, surface complexation, ion exchange, cation–π interactions, and electrostatic attraction. The preparation parameters and the post treatment of biochar can alter the surface speciation, and the modified biochar can exhibit better performance in the removal of dyes. A framework for artificial neural networking and machine learning to model the dye removal efficiency of biochar from wastewater was also proposed.

In addition to the excellent adsorption properties of the biochar, it can also act as a catalyst for the activation of persulfate ions and the oxidation of organic compounds. Biochar from lemon stalks, pyrolyzed at 850 °C under a limited oxygen atmosphere, produced a highly active and selective biochar for the oxidation of sulfamethoxazole (SMX), using sodium persulfate as oxidant [13]. The biochar had a significant amount of carbonates and could completely degrade 0.5 mg L^{-1} SMX within 20 min using 500 mg L^{-1} sodium persulfate (SPS) and 100 mg L^{-1} biochar in ultrapure water (UPW). The degradation process is favored at a low pH and in the presence of chlorides. The complexity of the water matrices usually has a negative impact on the degradation. The mechanism follows radicals with hydroxyl radicals as the main active species and non-radical pathways; electron transfer pathway was proven with electrochemical characterization. The biochar from lemon stalks is stable and active.

The combination of more than one advanced oxidation process often leads to synergistic effects and has proved to be a more efficient strategy. In this respect, the degradation of SMX was studied using a $CuOx$–$BiVO_4$ catalyst as either a photocatalyst or persulfate (sodium persulfate, SPS) activator [14]. This hybrid process (catalyst/SPS/Solar System) was found to be synergetic for copper-promoted $BiVO_4$ photocatalysts. Specifically, different loadings of CuOx (0.75–10% wt.) were deposited on $BiVO_4$ and tested under solar irradiation for the degradation of SMX. The most active catalyst was found to be the 0.75 $CuOx$–$BiVO_4$, while a higher loading of CuOx delayed the degradation process. The complexity of the water matrix (from ultrapure to bottled, BW, and wastewater, WW) also has a negative effect on the degradation process. The addition of SPS is beneficial for the degradation process, and the two processes exhibit a synergistic effect. Specifically, in a BW matrix, the application of SPS results in complete elimination after 60 min, while in the absence of SPS, an SMX degradation of only 40% took place in 120 min under solar light. For the WW, ~37%, 45%, and 66% degrees of symmetry were determined using 0.75, 3.0, and 10.0 $CuOx$–$BiVO_4$, respectively.

Catalytic processes for water and wastewater treatment are a promising approach to improving the quality of the water. Although photocatalysis is the most popular and attractive process due to its simplicity and utilization of the abundant solar light, the combination of other advanced oxidation processes can lead to effective systems. Interestingly, the presented experimental results highlight that it is difficult to decide which catalyst or process is better, and the determination of the activity must be performed in a wide window of operating parameters.

Conflicts of Interest: The author declares no conflict of interest.

References

1. Neelgund, G.M.; Aguilar, S.F.; Jimenez, E.A.; Ray, R.L. Adsorption Efficiency and Photocatalytic Activity of Silver Sulfide Nanoparticles Deposited on Carbon Nanotubes. *Catalysts* **2023**, *13*, 476. [CrossRef]
2. Khan, S.; Noor, A.; Khan, I.; Muhammad, M.; Sadiq, M.; Muhammad, N. Photocatalytic Degradation of Organic Dyes Contaminated Aqueous Solution Using Binary $CdTiO_2$ and Ternary $NiCdTiO_2$ Nanocomposites. *Catalysts* **2023**, *13*, 44. [CrossRef]
3. Kong, C.P.Y.; Suhaimi, N.A.A.; Shahri, N.N.M.; Lim, J.-W.; Nur, M.; Hobley, J.; Usman, A. Auramine O UV Photocatalytic Degradation on TiO_2 Nanoparticles in a Heterogeneous Aqueous Solution. *Catalysts* **2022**, *12*, 975. [CrossRef]
4. Suhaimi, N.A.A.; Kong, C.P.Y.; Shahri, N.N.M.; Nur, M.; Hobley, J.; Usman, A. Dynamics of Diffusion- and Immobilization-Limited Photocatalytic Degradation of Dyes by Metal Oxide Nanoparticles in Binary or Ternary Solutions. *Catalysts* **2022**, *12*, 1254. [CrossRef]
5. Rawal, J.; Kamran, U.; Park, M.; Pant, B.; Park, S.-J. Nitrogen and Sulfur Co-Doped Graphene Quantum Dots Anchored TiO_2 Nanocomposites for Enhanced Photocatalytic Activity. *Catalysts* **2022**, *12*, 548. [CrossRef]

6. Goumri-Said, S.; Kanoun, M.B. Insight into the Effect of Anionic–Anionic Co-Doping on BaTiO$_3$ for Visible Light Photocatalytic Water Splitting: A First-Principles Hybrid Computational Study. *Catalysts* **2022**, *12*, 1672. [CrossRef]
7. Isac, L.; Cazan, C.; Andronic, L.; Enesca, A. CuS-Based Nanostructures as Catalysts for Organic Pollutants Photodegradation. *Catalysts* **2022**, *12*, 1135. [CrossRef]
8. Le Pivert, M.; Piebourg, A.; Bastide, S.; Duc, M.; Leprince-Wang, Y. Direct One-Step Seedless Hydrothermal Growth of ZnO Nanostructures on Zinc: Primary Study for Photocatalytic Roof Development for Rainwater Purification. *Catalysts* **2022**, *12*, 1231. [CrossRef]
9. Mehdizadeh, A.; Derakhshan, Z.; Abbasi, F.; Samaei, M.R.; Baghapour, M.A.; Hoseini, M.; Lima, E.C.; Bilal, M. The Effect of Arsenic on the Photocatalytic Removal of Methyl Tet Butyl Ether (MTBE) Using Fe$_2$O$_3$/MgO Catalyst, Modeling, and Process Optimization. *Catalysts* **2022**, *12*, 927. [CrossRef]
10. Antonopoulou, M.; Papadaki, M.; Rapti, I.; Konstantinou, I. Photocatalytic Degradation of Pharmaceutical Amisulpride Using g-C$_3$N$_4$ Catalyst and UV-A Irradiation. *Catalysts* **2023**, *13*, 226. [CrossRef]
11. Rapti, I.; Boti, V.; Albanis, T.; Konstantinou, I. Photocatalytic Degradation of Psychiatric Pharmaceuticals in Hospital WWTP Secondary Effluents Using g-C$_3$N$_4$ and g-C$_3$N$_4$/MoS$_2$ Catalysts in Laboratory-Scale Pilot. *Catalysts* **2023**, *13*, 252. [CrossRef]
12. Goswami, L.; Kushwaha, A.; Kafle, S.R.; Kim, B.-S. Surface Modification of Biochar for Dye Removal from Wastewater. *Catalysts* **2022**, *12*, 817. [CrossRef]
13. Giannakopoulos, S.; Vakros, J.; Frontistis, Z.; Manariotis, I.D.; Venieri, D.; Poulopoulos, S.G.; Mantzavinos, D. Biochar from Lemon Stalks: A Highly Active and Selective Carbocatalyst for the Oxidation of Sulfamethoxazole with Persulfate. *Catalysts* **2023**, *13*, 233. [CrossRef]
14. Kouvelis, K.; Kampioti, A.A.; Petala, A.; Frontistis, Z. Degradation of Sulfamethoxazole Using a Hybrid CuO$_x$–BiVO$_4$/SPS/Solar System. *Catalysts* **2022**, *12*, 882. [CrossRef]

Disclaimer/Publisher's Note: The statements, opinions and data contained in all publications are solely those of the individual author(s) and contributor(s) and not of MDPI and/or the editor(s). MDPI and/or the editor(s) disclaim responsibility for any injury to people or property resulting from any ideas, methods, instructions or products referred to in the content.

Article

Adsorption Efficiency and Photocatalytic Activity of Silver Sulfide Nanoparticles Deposited on Carbon Nanotubes

Gururaj M. Neelgund [1,*], Sanjuana Fabiola Aguilar [1], Erica A. Jimenez [1] and Ram L. Ray [2]

1. Department of Chemistry, Prairie View A&M University, Prairie View, TX 77446, USA
2. College of Agriculture and Human Sciences, Prairie View A&M University, Prairie View, TX 77446, USA
* Correspondence: gmneelgund@pvamu.edu

Abstract: A multimode, dual functional nanomaterial, CNTs-Ag$_2$S, comprised of carbon nanotubes (CNTs) and silver sulfide (Ag$_2$S) nanoparticles, was prepared through the facile hydrothermal process. Before the deposition of Ag$_2$S nanoparticles, hydrophobic CNTs were modified to become hydrophilic through refluxing with a mixture of concentrated nitric and sulfuric acids. The oxidized CNTs were employed to deposit the Ag$_2$S nanoparticles for their efficient immobilization and homogenous distribution. The CNTs-Ag$_2$S could adsorb toxic Cd(II) and completely degrade the hazardous Alizarin yellow R present in water. The adsorption efficiency of CNTs-Ag$_2$S was evaluated by estimating the Cd(II) adsorption at different concentrations and contact times. The CNTs-Ag$_2$S could adsorb Cd(II) entirely within 80 min of the contact time, while CNTs and Ag$_2$S could not pursue it. The Cd(II) adsorption followed the pseudo-second-order, and chemisorption was the rate-determining step in the adsorption process. The Weber–Morris intraparticle pore diffusion model revealed that intraparticle diffusion was not the sole rate-controlling step in the Cd(II) adsorption. Instead, it was contributed by the boundary layer effect. In addition, CNTs-Ag$_2$S could completely degrade alizarin yellow R in water under the illumination of natural sunlight. The Langmuir-Hinshelwood (L-H) model showed that the degradation of alizarin yellow R proceeded with pseudo-first-order kinetics. Overall, CNTs-Ag$_2$S performed as an efficient adsorbent and a competent photocatalyst.

Keywords: CNTs; Ag$_2$S; cadmium; adsorption; alizarin yellow R

1. Introduction

Water pollution engrossed by heavy metals is a primary environmental, ecological, and public health concern [1]. Heavy metals are highly soluble in water and are toxic, carcinogenic, and non-degradable [2,3]. Rapid industrialization and technological development have resulted in the discharge of heavy metal-containing effluents into surface and groundwater [4]. The discharged heavy metals are adsorbed by soil and enter the human body through the food chain. Heavy metals accumulate in various organs and body tissues and can cause irreparable damage, including death [5]. Among the heavy metals, cadmium is enormously used and highly toxic [6]. The source for discharging the cadmium into the environment is industrial activities, such as electroplating, smelting, alloy manufacturing, pigments, plastic, battery, fertilizers, pesticides, pigments, dyes, textile operations, and refining [7]. In water, cadmium exists as bivalent, Cd(II), which is responsible for several adverse effects, such as kidney dysfunction, nephritis, hypertension, renal dysfunction, nervous system dysfunction, bone lesions, digestive system dysfunction, cancer, and reproductive organ damage [8–10]. The itai-itai disease, caused by Cd(II)-contaminated water from the Jinzu river in Japan, has instigated severe pain, bone fractures, proteinuria, and osteomalacia [11]. The Cd^{2+} ions have a high affinity for binding to sulfhydryl (-SH) groups of proteins in biological systems [12]. The presence of Cd(II) in water can lead to severe health and environmental problems. Therefore, its elimination is critically needed. The Cd(II) present in water could be eradicated through chemical precipitation, ion exchange,

solvent extraction, membrane separation, electrochemical removal, coagulation, and adsorption [13–21]. In these techniques, adsorption is superior because of its high efficiency, relative simplicity in design, easy operation, and low operational cost [20,21]. Because of this, many adsorbents have been developed and tested for their efficiency [20–22].

Another class of pollutants that also causes major environmental problems, such as heavy metals, is azo dyes. Azo dyes have attractive colors and are enormously used in industries due to their availability and stability [23]. These dyes contain azo bonds (-N=N-) and substituted aromatic rings [24]. The complex molecular structure of azo dyes has made them recalcitrant and resistant to biodegradation [25]. Azo dyes are mutagenic, teratogenic, and carcinogenic [25]. Moreover, azo dyes can decompose into potentially carcinogenic polychlorinated naphthalenes, benzidine, and amines [26,27]. The toxicity of azo dyes can result in lung cancer, heart diseases, chromosomal aberrations, neurotoxicity, skin disease, and respiratory problems in humans [25]. These dyes can cause disorders of the central nervous system and the inactivation of enzymatic activities [28]. Beyond toxicity, azo dyes are abundantly used in industries because of their cost and advancement in colors that cover the entire spectrum. After their use in industries, about 10-15% of the azo dyes are discharged into the environment through water effluents [29,30]. Releasing azo dyes content water can result in several environmental problems including reduced light penetration in water bodies, which leads to diminished photosynthetic activities, lessened growth and reproduction of aquatic creatures, and aesthetic damage [30,31]. The consequences of the xenobiotic and recalcitrant azo dyes impact the ecosystem's structure and functioning. Considering the adverse effects, it is essential to completely obliterate azo dyes to prevent severe damage and protect health and the environment. Different techniques have been developed to remove the azo dyes in water, viz., coagulation/flocculation, ultrafiltration and membrane processing, chemical precipitation, electrochemical degradation, and ozonation [32–37]. In comparison, photocatalysis is an excellent method for eradicating azo dyes from water in an environmentally friendly approach as the end products of this process are harmless [38,39]. Due to its efficiency in degrading azo dyes, many photocatalysts have been developed to eliminate dyes [37–45]. Among azo dyes, alizarin yellow R (AYR) is a prominently used dye in industries. It is a highly water-soluble anionic dye that contains polycyclic aromatic hydrocarbons [25]. AYR is a derivative of salicylic acid and was prepared in 1887 by Rudolf Nietzki through the reaction of m-nitroaniline and salicylic acid [46]. AYR is an industrially important dye and is vastly used in the textile, leather, plastics, paints, and lacquer industries [24]. Furthermore, it is used as an acid-base indicator and employed in histology, stains, and nutrient media preparations [47].

Considering the health and environmental problems associated with Cd(II) and AYR, we designed the dual applicable nanocomposite, CNTs-Ag$_2$S for efficient Cd(II) adsorption and photodegradation of AYR. The multimode CNTs-Ag$_2$S was produced through the facile deposition process of Ag$_2$S nanoparticles over oxidized carbon nanotubes (CNTs) using the hydrothermal method. The adsorption efficiency of CNTs-Ag$_2$S was estimated by evaluating the adsorption rate of Cd(II) from water. The dynamics and controlling mechanisms of Cd(II) adsorption were assessed by pseudo-first- and second-order kinetic models. The reaction pathways and the rate-controlling step underlying Cd(II) adsorption were evaluated using the Weber−Morris intraparticle pore diffusion model. The adsorption equilibrium was determined by fitting the experimental results with Langmuir, Freundlich, and Temkin isotherm models. Furthermore, the catalytic activity of CNTs-Ag$_2$S was determined through the degradation of AYR under exposure to natural sunlight. The photocatalytic activity of CNTs-Ag$_2$S was quantified using the Langmuir-Hinshelwood (L-H) model.

2. Results and Discussion

The ATR-FTIR spectrum of CNTs-COOH, shown in Figure 1a, demonstrated a band at 3427 cm^{-1} corresponding to the O-H bond. The band for the C=O bond of the -COOH groups appeared at 1699 cm^{-1}, and the band for the C=C bonds of CNTs was found at

1554 cm^{-1} [48–50]. The spectrum of Ag$_2$S (Figure 1b) displayed the peak of the Ag-S bond at 552 cm^{-1} [51]. The spectrum of CNTs-Ag$_2$S (Figure 1c) revealed the characteristic absorption bands related to CNTs and Ag$_2$S. The prominent peaks observed at 1066 and 1096 cm^{-1} in Figure 1c were attributed to C-O stretching. The peak at 3574 cm^{-1}, was due to the −OH stretching vibrations of adsorbed water molecules. The band due to C=O was observed at 1693 cm^{-1}, and the peak, at 1517 cm^{-1}, was attributed to C-H. The peak related to Ag-S was situated around 550 cm^{-1}. The XRD pattern of CNTs-Ag$_2$S (Figure 2) showed a characteristic (0 0 2) reflection of hexagonal graphite of CNTs at 26.1° [52]. It revealed the reflections related to Ag$_2$S at 22.7 (−1 0 1), 25.2 (−1 1 1), 26.5 (0 1 2), 29.2 (1 1 1), 31.7 (−1 1 2), 33.8 (1 2 0), 34.6 (−1 2 1), 34.9 (0 2 2), 36.8 (1 1 1), 37.0 (1 2 1), 37.3 (0 1 3), 37.9 (−1 0 3), 40.9 (0 3 1), 43.6 (0 2 3), 44.4 (−1 3 1), 46.4 (−1 2 3), 48.0 (−2 1 2), 48.9 (0 1 4), 53.0 (0 4 0), 53.5 (−2 1 3), 58.3 (−1 4 1), 58.4 (−2 2 3), 60.2 (−1 0 5), 61.4 (0 1 5), 62.8 (2 3 1), 63.4 (2 1 3), 63.9 (−1, 3 4), and 68.0° (2 3 2). These values agree with the values found in the standard pattern of acanthite Ag$_2$S (JCPDS file 14-0072) [52]. The TEM images of CNTs-Ag$_2$S presented in Figure 3 explored the deposition of spherical-resembling Ag$_2$S nanoparticles over the surface of tubular-structured CNTs. The proportion of Ag$_2$S nanoparticles was low, possibly due to the low quantity of precursor AgNO$_3$ used in the preparation compared to the ratio of CNTs. The presence of Ag$_2$S nanoparticles over the surface of CNTs is perceptible in Figure 3a–f. It is noticeable in Figure 3a that the few Ag$_2$S nanoparticles have combined and formed clusters. Figure 3b,d reveal the ruptured surface of some CNTs. It could have happened through the harsh treatment of CNTs with the mixture of concentrated HNO$_3$ and H$_2$SO$_4$. This process was required to make the CNTs hydrophilic and provide a suitable environment for the effective deposition of Ag$_2$S nanoparticles. However, the entire surface of the CNTs was not distracted and the smooth surface of the CNT is perceptible in Figure 3e,f. Overall, a few defective sites were formed in CNTs. Figure 3d-f reveals the strong adherence of Ag$_2$S nanoparticles to the smooth-surfaced CNT. The absence of leached or free-standing nanoparticles in the void area shows the strong adherence of Ag$_2$S nanoparticles to the surface of CNTs. The size of the Ag$_2$S nanoparticle present in Figure 3d,e, was around 38 and 15 nm, respectively. Therefore, Ag$_2$S nanoparticles have broad size distribution. The average size of Ag$_2$S nanoparticles, calculated from XRD, was around 29 nm, which agrees with the size determined from the TEM images. The core of the Ag$_2$S nanoparticles (Figure 3d) appears denser than the edge. The CNTs have a width of around 110 nm and a length of several micrometers. Overall, CNTs have effectively exfoliated. Further, the EDS of CNTs-Ag$_2$S (Figure S1, Supplementary Materials) demonstrated the presence of O, Ag, and S in CNTs-Ag$_2$S. The peaks for Cu and Ca were found in Figure S1, those are by the copper grid used in the TEM measurement.

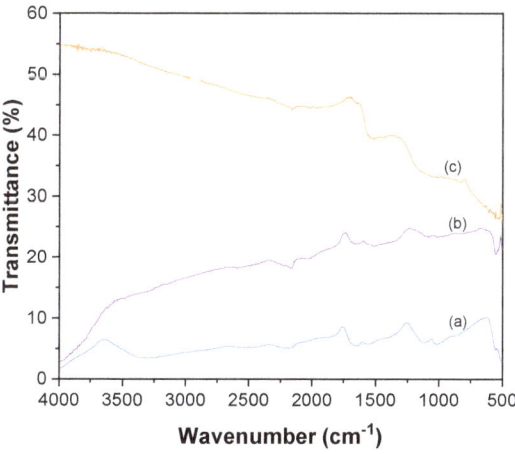

Figure 1. ATR-FTIR spectra of (**a**) CNTs-COOH, (**b**) Ag$_2$S, and (**c**) CNTs-Ag$_2$S.

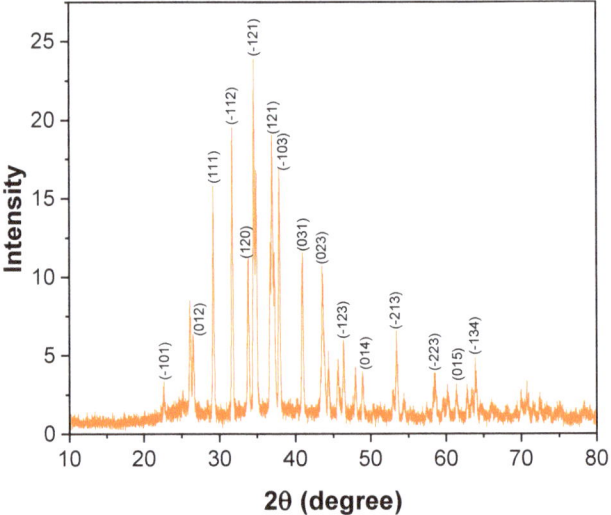

Figure 2. XRD of CNTs-Ag$_2$S.

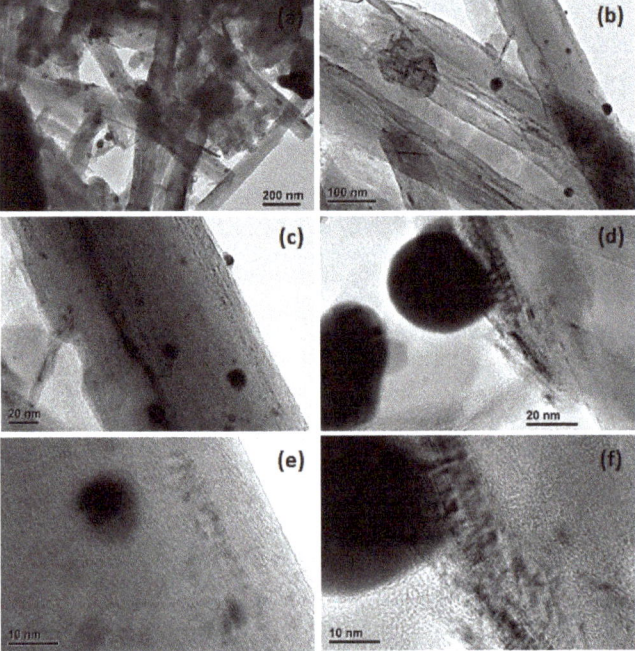

Figure 3. (**a–f**) TEM images of CNTs-Ag$_2$S.

The TGA pattern of CNTs-Ag$_2$S (Figure 4) displayed four weight loss steps. The initial weight loss of 6.5% that occurred before 100 °C was attributed to the desorption of physically adsorbed water molecules over the surface of CNTs-Ag$_2$S. The following weight loss of 15% ensued within the range of 100–490 °C was due to the detachment and decomposition of oxygen-containing functional groups existing on the surface of the CNTs. The subsequent weight loss of 10.5% transpired within the range of 500–540 °C, was owing to the breaking of the bond between silver and sulfide in Ag_2S nanoparticles and releasing sulfur. The successive sharp and significant weight loss of 31.5% taken place within the range of 540–550 °C was by the decomposition of sulfur and the formation of metallic silver and silver oxide. The residual weight remained was about 64%. The broad endothermic peak found in the DTA curve around 80 °C was because of the dehydration of the sample. The small endothermic peak found at 170 °C accounted for the removal of functional groups from the surface of the CNTs. The intense exothermic peak at 545 °C was assigned to release sulfur by breaking the bond between silver and sulfide. The UV-vis absorption spectrum of CNTs-Ag_2S, shown in Figure 5, exhibited a characteristic absorption band of the C=C bonds of CNTs at 232 nm [52]. In addition, the broad absorption tail corresponding to Ag_2S nanoparticles was observed in the visible region [52,53]. If semiconductor nanoparticles are conjugated to CNTs, it shows the charge-transfer band [52,54]. However, no such band was observed in Figure 5, illustrating that the conjugation of Ag_2S nanoparticles to CNTs has not altered their energy states [52,54].

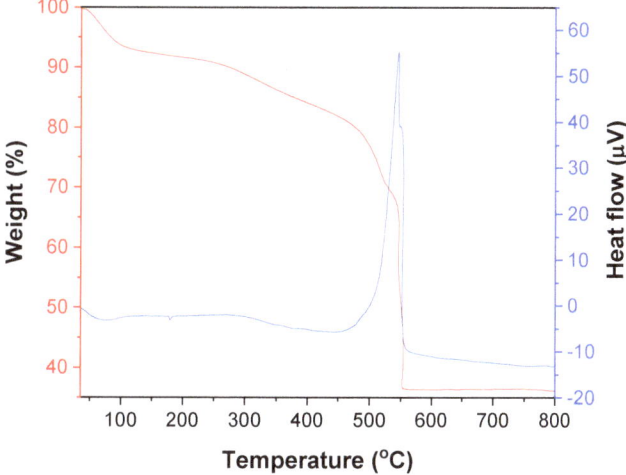

Figure 4. TG/DTA of CNTs-Ag_2S.

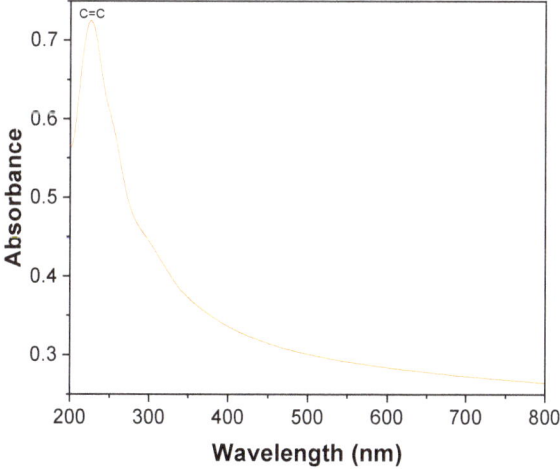

Figure 5. UV-vis absorption spectrum of CNTs-Ag_2S.

The XPS survey spectrum of CNTs-Ag$_2$S, shown in Figure 6a, confirmed the presence of C, O, Ag, and S. The high-resolution spectrum of C1s (Figure 6b) divulged four distinct peaks by Gaussian fitting situated at 284.5, 285.5, 286.4, and 287.8 eV. Among these, peaks at 284.5 and 285.5 eV were assigned to the C-C bonds of the sp^2 and sp^3 hybridized carbon atoms of CNTs, respectively [55]. The peaks at 286.4 and 287.8 eV were due to the C-O and C=O bonds of CNTs, respectively [55]. The deconvoluted O 1s peak (Figure 6c) exhibited a peak at 531.3 eV for the lattice oxygen and a peak at 532.4 eV corresponding to the carbonyl (=C-O) functional groups of the CNTs [56]. The core-electron binding energy of Ag 3d3/2 was found at 373.9 eV, and that of Ag 3d5/2 was at 367.9 eV (Figure 6d). The presence of Ag 3d3/2 and Ag 3d5/2 peaks and their positions confirm the Ag$^+$ state of silver [57]. The difference in the position of Ag 3d5/2 and Ag 3d3/2 peaks was 6 eV, and no shoulder or satellite peaks were observed between them. The high-resolution S 2p spectrum (Figure 6e) for the spin-orbit splitting of S^{2+} was deconvoluted into the 2p3/2 level peak at 161.2 eV, and the 2p1/2 peak at 162.3 eV. These peaks occurred by the Ag-S bonds of the Ag$_2$S nanoparticles [58]. The additional peak at 163.9 eV is related to the presence of sulfur in CNTs-Ag$_2$S [58].

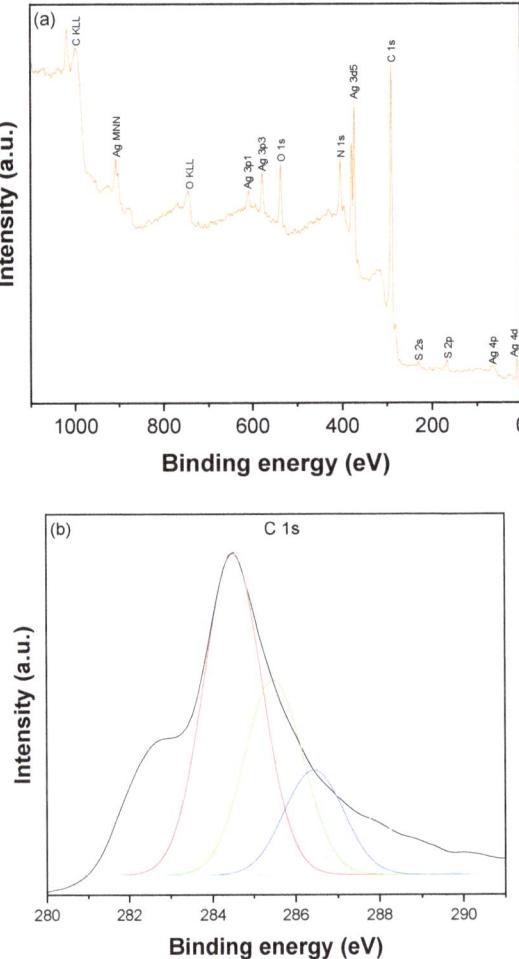

Figure 6. Cont.

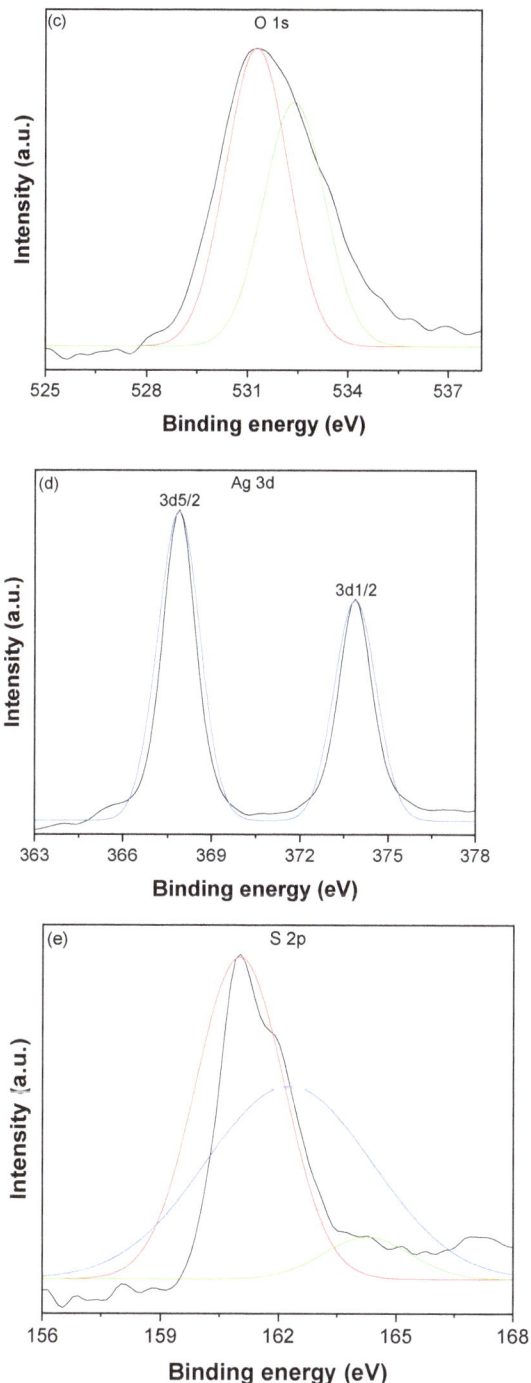

Figure 6. (**a**) XPS survey spectrum of CNTs-Ag$_2$S. (**b**) High-resolution spectrum of C1s. (**c**) High-resolution spectrum of O1s. (**d**) High-resolution spectrum of Ag3d. (**e**) High-resolution spectrum of S2p.

The adsorption ability of CNTs-Ag$_2$S was determined by estimating the Cd(II) adsorption present in water. The contact time between adsorbent and adsorbate is critical and it controls adsorption. Therefore, the Cd(II) adsorption was determined by allowing for contact with CNTs-Ag$_2$S at different times and compared with that of CNTs and Ag$_2$S (Figure 7). The plot of the adsorption capacity, q_t versus t, is presented in Figure S2. Cd(II) adsorption is time dependent and occurs as a gradient function of time. Hence, contact time is crucial in controlling the Cd(II) adsorption. The tendency in Cd(II) adsorption by CNTs, Ag$_2$S, and CNTs-Ag$_2$S was identical like the adsorption was rapid, then gradually reduced, and finally attained equilibrium. The initial rapid Cd(II) adsorption could have occurred due to the difference in the concentration of Cd^{2+} ions in the solution and over the surface of the CNTs-Ag$_2$S, which facilitated the movement of Cd^{2+} ions from the solution to the surface of the CNTs-Ag$_2$S [8]. Also, in the beginning, many active sites were available on the surface of CNTs-Ag$_2$S to occupy by Cd^{2+} ions [8]. With the elapsed contact time, the active sites of the CNTs-Ag$_2$S were predominantly occupied by Cd^{2+} ions. Apart, there is a possibility of repulsion between Cd^{2+} ions in the solution and Cd^{2+} ions exist over the surface of CNTs-Ag$_2$S. Owing to these reasons, the adsorption rate could gradually decrease and eventually attain equilibrium [8]. Within 80 min of the contact time, CNTs-Ag$_2$S could adsorb the Cd(II) completely; however, under identical conditions, the adsorption rates of CNTs and Ag$_2$S were 79 and 53%, respectively. Thus, the adsorption efficiency of CNTs and Ag$_2$S was significantly improved by their conjugation in CNTs-Ag$_2$S. Further, to identify the adsorption kinetics of CNTs-Ag$_2$S, the data were simulated with the pseudo-first-order kinetic model using the expression presented in Equation (1).

$$ln(q_e - q_t) = lnq_e - k_1 t \qquad (1)$$

where q_e and q_t are the quantity of adsorbate (mg/g) at equilibrium and particular time, t (min), respectively, and k_1 (min^{-1}) is the pseudo-first-order rate constant.

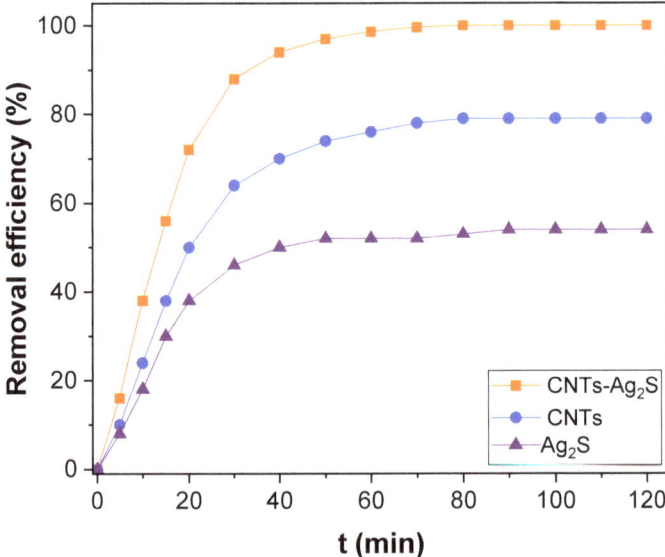

Figure 7. Adsorption kinetics of Cd(II) over CNTs, Ag$_2$S, and CNTs-Ag$_2$S.

The pseudo-first-order plot obtained for Cd(II) adsorption over CNTs, Ag$_2$S, and CNTs-Ag$_2$S is shown in Figure 8. The correlation coefficients (R^2) perceived for CNTs, Ag$_2$S, and CNTs-Ag$_2$S were 0.9906, 0.9540, and 0.9930, respectively. The k_1 and q_e were computed using the slope and intercept values of the straight lines acquired in Figure 8. The estimated values of k_1 and q_e is summarized in Table 1. The calculated q_e(cal) could not agree with the experimental value q_e(exp) for all samples. Thus, the experimental data for the Cd(II) adsorption could not be projected by the pseudo-first-order kinetic model. Alternatively, it was analyzed using the pseudo-second-order kinetic model using Equation (2).

$$\frac{t}{q_t} = \frac{1}{k_2 \, q_e^2} + \frac{t}{q_e} \qquad (2)$$

where k_2 [g/(mg.min)] is the pseudo-second-order rate constant.

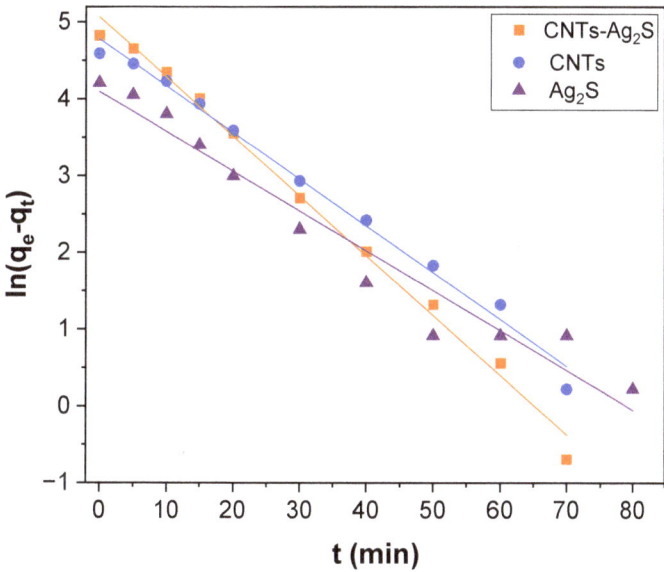

Figure 8. Pseudo-first-order kinetics for Cd(II) adsorption over CNTs, Ag$_2$S, and CNTs-Ag$_2$S.

Table 1. Parameters calculated for the Cd(II) adsorption using adsorption kinetic models.

Adsorbent	q_e (exp) mg g^{-1}	Pseudo-First-Order Kinetic Model			Pseudo-Second-Order Kinetic Model		
		q_e (cal) mg g^{-1}	k_1 (min^{-1})	R^2	q_e (cal) mg g^{-1}	$k_2 \times 10^{-3}$ (g mg^{-1} min^{-1})	R^2
CNTs-Ag$_2$S	125.0	5.0633	0.0778	0.9930	133.2	1.2140	0.9980
CNTs	98.75	4.7785	0.0608	0.9906	109.8	0.8360	0.9963
Ag$_2$S	67.50	4.0970	0.0519	0.9540	72.62	1.7750	0.9985

The pseudo-second-order model plot is illustrated in Figure 9, and the determined parameters are presented in Table 1. The R^2 values derived from the pseudo-second-order plots for CNTs, Ag$_2$S, and CNTs-Ag$_2$S were 0.9963, 0.9985, and 0.9980, respectively. The R^2 value of the second-order plot was relatively higher than that found for the first-order plot, which was close to 1. The q_e(cal) of the second-order model matched the q_e(exp) for all samples. The linearity of the pseudo-second-order plot and the agreeing values of q_e(exp) and q_e(cal) show that the Cd(II) adsorption could be analyzed using the pseudo-second-order kinetics rather than the pseudo-first-order kinetics. The agreement of the second-order kinetics demonstrates that chemisorption was the rate-determining step in the Cd(II) adsorption [9]. Moreover, it represents the rapid transfer of Cd(II) from a solution with a lower initial concentration to the surface of the CNTs-Ag$_2$S due to the concentration gradient [9]. Further, the experimental data were explored using the Weber–Morris intraparticle pore diffusion model to evaluate the diffusion mechanism using Equation (3).

$$q_t = k_{id} \, t^{0.5} + C \quad (3)$$

where k_{id} (mg/g.min) is the intraparticle diffusion rate constant, which can be calculated from the linear plot of q_t versus $t^{0.5}$. c (mg/g) is the intraparticle diffusion constant, estimated from the intercept and directly proportional to the boundary layer thickness. It is assumed that the higher the intercept value, the more significant the contribution of the surface adsorption in the rate-controlling step. If the regression of the q_t versus $t^{0.5}$ plot is linear and passes through the origin, intraparticle diffusion plays a significant role in controlling the kinetics of the adsorption process. If it does not pass through the origin, intraparticle diffusion is not the only rate-limiting step. Instead, it is contributed by the boundary layer effect [59]. The intraparticle diffusion plot of q_t versus $t^{0.5}$ acquired for the Cd(II) adsorption on CNTs-Ag$_2$S is demonstrated in Figure 10. The plot in Figure 10 does not pass through the origin of the coordinates (Figure S3). So, intraparticle diffusion is not the sole rate-limiting step in Cd(II) adsorption over CNTs-Ag$_2$S. The association of the two straight lines in the intraparticle diffusion plot (Figure S3) and the intercept values (Table S1) support that the intraparticle diffusion was not only a rate-controlling step in the Cd(II) adsorption; it also contributed through the boundary layer effect [59]. The multilinearity of the plot in Figure 10 (Figure S3) shows that the Cd(II) adsorption transpires through multiple phases [60]. Among these, the initial stage occurred through the rapid adsorption of Cd^{2+} ions, the second phase was owing to the diffusion of Cd^{2+} ions into the pores of CNTs-Ag$_2$S, and the third phase was due to equilibrium of the adsorption that caused a chemical reaction/bonding [60].

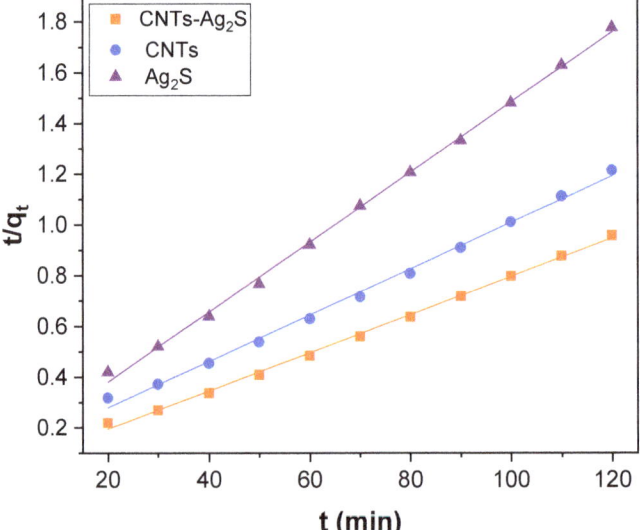

Figure 9. Pseudo-second-order kinetics for Cd(II) adsorption over CNTs, Ag$_2$S, and CNTs-Ag$_2$S.

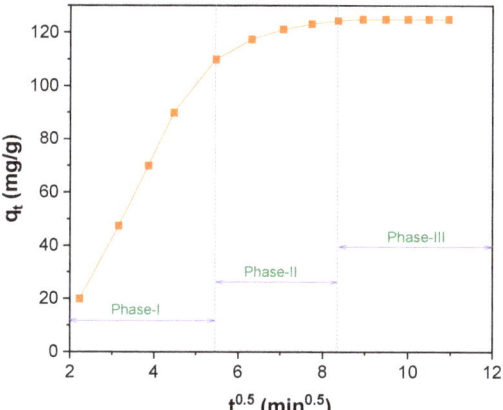

Figure 10. Weber-Morris intraparticle diffusion plot for the Cd(II) adsorption over CNTs-Ag$_2$S.

The experimental equilibrium parameters for Cd(II) adsorption were determined by applying three isotherm models: Langmuir, Freundlich, and Temkin. The Langmuir model predicts that the adsorbed molecules form a monolayer and adsorption emerges at a static number of adsorption sites, each of which is equivalent in efficiency. Moreover, each molecule has a constant enthalpy and adsorption activation energy, it means that all molecules have an affinity equal to entire adsorption sites [61]. With the help of Langmuir isotherm model, it could be find the value of the maximum adsorption capacity of the adsorbent using Equation (4) [62]:

$$\frac{C_e}{q_e} = \frac{C_e}{q_m} + \frac{1}{K_L q_m} \quad (4)$$

where q_e (mg/g) is the amount of adsorbed Cd(II) per unit mass of CNTs-Ag$_2$S; C_e (mg/L) is the concentration of Cd(II) at equilibrium; q_m is the maximum amount of the Cd(II) adsorbed per unit mass of CNTs-Ag$_2$S to form a complete monolayer on the surface-bound at high C_e. K_L is the Langmuir adsorption constant related to the free energy of adsorption. The Langmuir plot for Cd(II) adsorption over CNTs-Ag$_2$S is presented in Figure 11. The estimated parameters are shown in Table 2. The maximum adsorption capacity (q_m) calculated for Cd(II) adsorption over CNTs-Ag$_2$S was 256.4 mg/g.

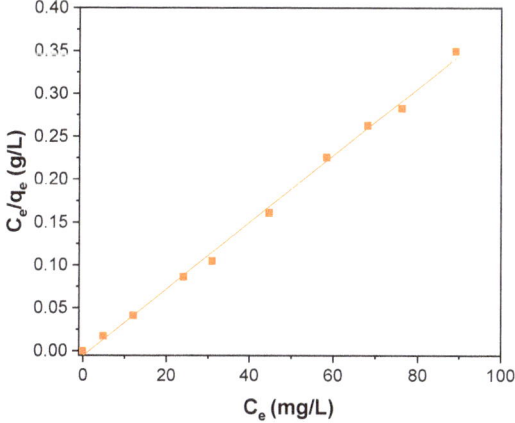

Figure 11. Langmuir isotherm plot for Cd(II) adsorption over CNTs-Ag$_2$S.

Table 2. Parameters calculated for Cd(II) adsorption over CNTs-Ag$_2$S using Langmuir, Freundlich, and Temkin adsorption isotherm models.

Langmuir Isotherm			Freundlich Isotherm			Temkin Isotherm		
q_m (mg g^{-1})	K_L (L mg^{-1})	R^2_{Lan}	K_F (mg g^{-1})	n	R^2_{Fre}	A	B	R^2_{Tem}
256.4	−0.6830	0.9972	305.0	−31.21	0.3562	1.1×10^{256}	0.4565	0.0019

The Freundlich isotherm is developed based on the assumption that the adsorption sites are distributed exponentially concerning the heat of adsorption [62,63]. It provides the relationship between the equilibrium of liquid and solid phase capacities with multilayer adsorption properties consisting of the heterogeneous surface of the adsorbent. The Freundlich isotherm that supports multilayer adsorption agrees with the Langmuir model over moderate ranges of concentrations but differs at low and high concentrations. The linear form of the Freundlich isotherm can be represented b Equation (5):

$$ln\, q_e = ln\, K_F + \frac{ln\, C_e}{n} \qquad (5)$$

where q_e (mg/g) is the amount of Cd(II) adsorbed at equilibrium. K_F and n are Freundlich constants. K_F symbolizes the affinity of the adsorbent, and n signifies the adsorption intensity. C_e (mg/L) is the Cd(II) concentration at equilibrium. The Freundlich isotherm plot attained for Cd(II) adsorption is shown in Figure S4, and the calculated parameters are included in Table 2.

The Temkin isotherm model is based on the fact that, during adsorption, the heat of all molecules decreases linearly with an increase in the coverage of the adsorbent surface and that adsorption is characterized by the uniform distribution of binding energies up to the maximum binding energy [62,64]. The linear form of the Temkin isotherm model is shown in Equation (6):

$$q_e = B\, lnA + B\, ln\, C_e \qquad (6)$$

where B = RT/K_T, K_T is the Temkin constant related to the heat of adsorption (J/mol); A is the Temkin isotherm constant (L/g), R is the gas constant (8.314 J/mol K), and T is the absolute temperature (K). The Temkin isotherm fitting plot for the Cd(II) adsorption is shown in Figure S5. The values estimated from Figure S5 are illustrated in Table 2.

The R^2_{Lan} value found for the Langmuir isotherm plot was 0.9972; for Freundlich and Temkin isotherm plots, the R^2_{Fre} and R^2_{Tem} values were 0.3562 and 0.0019, respectively. The Freundlich and Temkin isotherm models provided scattered points and failed to accomplish linearity. Therefore, the Freundlich and Temkin isotherm models were unsuitable for explaining Cd(II) adsorption over CNTs-Ag$_2$S. However, the Langmuir isotherm model produced linearity. Therefore, the Langmuir model is appropriate for elucidating the Cd(II) adsorption. The validation of the Langmuir model indicates that the Cd(II) adsorption over CNTs-Ag$_2$S ensued with monolayer molecular covering and chemisorption, together [65].

Further, the photocatalytic activity of CNTs-Ag$_2$S was investigated through the degradation of AYR under exposure to sunlight. The degradation of AYR by CNTs-Ag$_2$S is presented in Figure 12. The activity of CNTs-Ag$_2$S was compared with that of CNTs and Ag$_2$S. The CNTs-Ag$_2$S could degrade the AYR completely within 120 min of illumination. However, CNTs and Ag$_2$S were capable to degrade 77.3 and 41% of AYR, respectively, in 120 min. Hence, the conjugation of CNTs and Ag$_2$S was substantially improved the photocatalytic activity in CNTs-Ag$_2$S. The degradation of AYR was quantified by the reduction in the electronic absorption band located at 373 nm (Figure 13). The hybrid nano-architecture of CNTs-Ag$_2$S was proficient in adsorbing higher number of AYR molecules and capable in absorption of sunlight. In addition, the effectual hindrance of recombining of electrons and holes during photocatalysis facilitated the degradation process. The photocatalytic activity of CNTs-Ag$_2$S was further explored by finding the apparent rate constants using the Langmuir-Hinshelwood (L-H) model and with the help of Equation (7) that could be applicable for low concentrations of dyes [38,39,44,45].

$$\ln \frac{C_0}{C} = k_{app}\, t \qquad (7)$$

where C_0 is the initial concentration of AYR and C is the concentration at a particular time of irradiation. k_{app} is the apparent rate constant of the reaction, and t is the irradiation time.

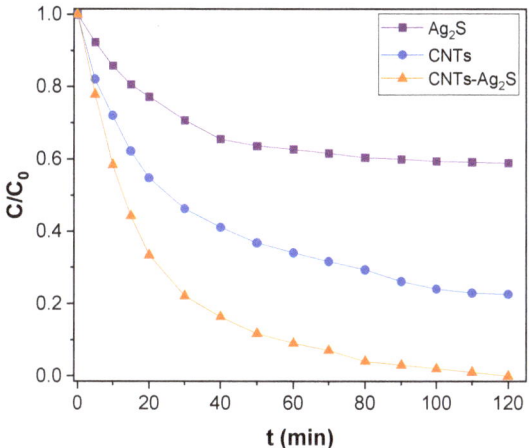

Figure 12. The degradation of alizarin yellow R in the presence of CNTs, Ag_2S, and CNTs-Ag_2S under the illumination to sunlight.

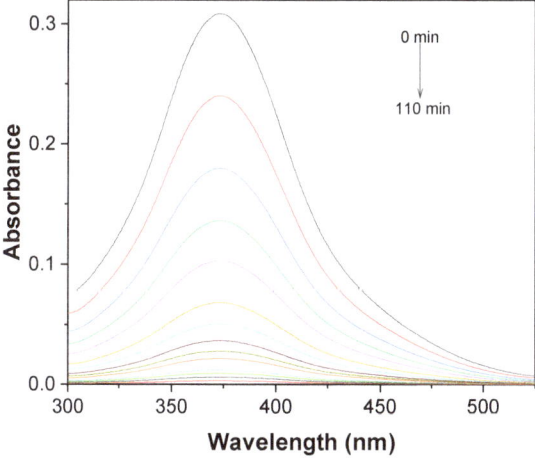

Figure 13. Reduction in the electronic absorption of alizarin yellow R in the presence of CNTs-Ag_2S under the irradiation to sunlight.

The L-H plots perceived for the degradation of AYR were linear, as depicted in Figure 14. The linearity of the L-H plots reveals that the degradation of AYR ensue with pseudo-first-order kinetics. The apparent rate constants determined by the L-H plots for CNTs, Ag_2S, and CNTs-Ag_2S were 0.0108, 0.0035, and 0.0378 min^{-1}, respectively. The value determined for CNTs-Ag_2S was about four-fold higher than CNTs and eleven-fold greater than Ag$_2$S. Therefore, the conjugation of CNTs and Ag2S has magnificently improved the photocatalytic activity in CNTs-Ag$_2$S. Due to the conjugated structure, containing aromatic, carbonyl, and azo groups, the AYR solution possessed an intense color in water. Accordingly, AYR degradation could happen by breaking the conjugated system in the presence of CNTs-Ag$_2$S under sunlight irradiation. With this assumption, the possible mechanism for the rapid degradation of AYR by CNTs-Ag$_2$S could be explained as follows (Figure 15).

CNTs-Ag$_2$S + hν → e$^-$ + h$^+$

h$^+$ + H$_2$O → HO$^-$ + H$^+$

e$^-$ + O$_2$ → O$_2^{\bullet-}$

O$_2^{\bullet-}$ + H$^+$ → HO$_2^{\bullet}$

HO$_2^{\bullet}$ + H$_2$O → H$_2$O$_2$ + HO$^{\bullet}$

AYR + HO$^{\bullet}$ → H$_2$O + CO$_2$ + nontoxic products

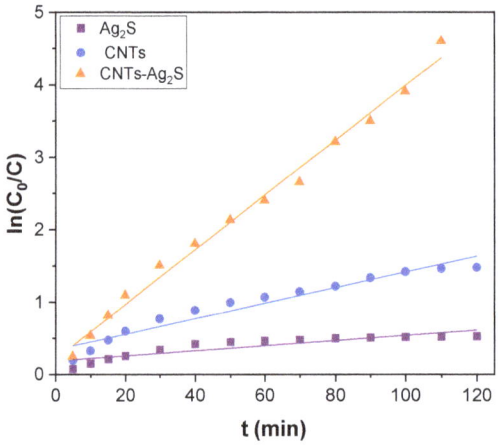

Figure 14. Langmuir-Hinshelwood plot for the degradation of alizarin yellow R in the presence of CNTs-Ag$_2$S under illumination to sunlight.

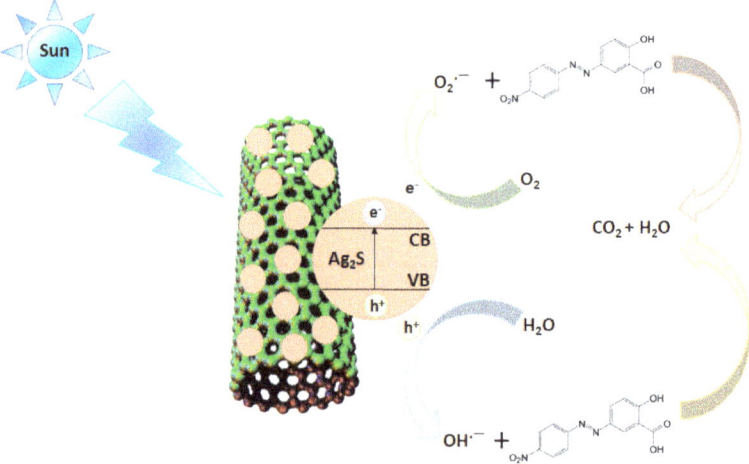

Figure 15. A possible mechanism for the degradation of alizarin yellow R in the presence of CNTs-Ag$_2$S under exposure to sunlight.

Further, to find the stability of CNTs-Ag$_2$S, it was recovered by centrifugation after photocatalysis. It was washed with DI water, dried and used in further two cycles of the photocatalysis. The reused CNTs-Ag$_2$S was able to degrade AYR completely in further two cycles without any significant reduction in activity (Figure S6). The XRD of CNTs-Ag$_2$S recorded before and after using in the three degradation cycles of AYR could not show any structural modification or rapture (Figure S7). Consequently, CNTs-Ag$_2$S is table and suitable for reuse.

The CNT/Ag$_2$S nanocomposite prepared by Di et al. and applied it to the degradation of rhodamine B revealed its higher activity than bare Ag$_2$S nanoparticles under illumination to visible and near-infrared (NIR) light [57]. In addition, the recycled CNT/Ag$_2$S nanocomposite could not lost the activity [57]. The Ag$_2$S-CNT nanocomposite, reported by Meng et al. [66], efficiently degraded texbrite BA-L in presence of visible light. In this study, the Ag$_2$S-CNT nanocomposite was for four degradation cycles of texbrite BA-L [66]. Further, the photocatalytic activity of CNTs-Ag$_2$S in the degradation of AYR was compared with different photocatalysts and presented in Table 3 [24,25,46,67–69].

Table 3. Comparison of the degradation rate of AYR by CNTs-Ag$_2$S with reported photocatalysts.

Photocatalyst	Source of Irradiation	Degradation Rate (%)	Ref
CNTs-Ag$_2$S	Sunlight	100	This work
Fe nanoparticles	Sunlight	93.7	25
H$_2$O$_2$	UV	100	24
Zinc oxide	UV	92.5	46
β-MnO$_2$ nanowires	Mercury lamp	98.0	69
Mn$_3$O$_4$	Mercury lamp	62.0	69
MnO(OH) nanorods	Mercury lamp	54.0	69
Bi$_2$O$_3$@RGO	Sunlight	41.5	70
ZnO nanoparticles	UV	95.0	71
ZnO nanoparticles	Sunlight	13.2	71
ZnO nanoparticles	Visible	06.2	71

3. Experimental

3.1. Materials

The chemicals and CNTs were purchased from Millipore Sigma and used as received. The aqueous solutions were prepared using ultrapure water obtained by the Milli-Q Plus system (Millipore; Burlington, MA, USA).

3.2. Preparation of CNTs-Ag$_2$S

The hydrophobic pristine CNTs were modified to hydrophilic through the oxidization by refluxing in a mixture of 1:3 (v/v) concentrated HNO$_3$ and H$_2$SO$_4$ at 70 °C for 24 h. The resulting oxidized CNTs were collected through centrifugation and purified by washing them with DI water [70,71]. The resulting oxidized CNTs were dried under a vacuum and used to deposit Ag$_2$S nanoparticles. About 40 mg of oxidized CNTs (CNTs-COOH) were dispersed in 40 mL of DI water using sonicator, and a solution of 0.04 mol/L of AgNO$_3$ in 40 mL of DI water was added. The mixture was allowed to stir in room temperature for 30 min, and a freshly prepared 40 mL aqueous solution of 0.02 mol/L sodium sulfide was added slowly. This suspension was stirred for 10 min and transferred to an autoclave. The autoclave was heated at 180 °C for 6 h to yield CNTs-Ag$_2$S. Thus, the formed CNTs-Ag$_2$S was collected by centrifugation and purified with washing with DI water.

3.3. Preparation of Ag$_2$S

For control experiments, Ag$_2$S nanoparticles were prepared using the procedures used for CNTs-Ag$_2$S without CNTs.

3.4. Adsorption Experiments

The stock solution of Cd(II), with a concentration of 1 g/L, was prepared in DI water using cadmium chloride and was diluted to the desired concentrations. The kinetic Cd(II) adsorption experiments were conducted to find the contact time needed to attain equilibrium. In the typical experiment, 100 mg of CNTs-Ag$_2$S was dispersed into 500 mL of a Cd(II) solution with a concentration of 0.5 mg/L and stirred at room temperature. After the required contact time, an adequate sample was collected, and the dispersed CNTs-Ag$_2$S was separated by centrifugation. The concentration of residual Cd(II) in the collected samples was estimated by atomic absorption spectrometer. The amount of Cd(II), adsorbed by CNTs-Ag$_2$S was monitored as a function of time for 120 min. The quantity of Cd(II) adsorbed was calculated using Equation (8).

$$q_t = \frac{(C_0 - C_t)\,V}{M} \tag{8}$$

where q_t is the amount of adsorbed Cd(II) (mg/g) at time t; C_0 is the initial concentration of the Cd(II) (mg/L), and C_t is the concentration of the Cd(II) (mg/L) at time t; V is the volume of the solution (L), and M is the amount of adsorbent (g).

Further, the efficiency of CNTs-Ag$_2$S in Cd(II) adsorption was estimated using Equation (9).

$$Removal\ efficiency\ (\%) = \frac{(C_0 - C_t)\,V}{C_0} \times 100 \tag{9}$$

For adsorption isotherm experiments, 10 mg of CNTs-Ag$_2$S was mixed with 50 mL of the Cd(II) solution and stirred at room temperature for 24 h to reach equilibrium in the concentration between 50 and 140 mg/L. After separating the dispersed CNTs-Ag$_2$S, the concentration of Cd(II) in the solution was measured using an atomic absorption spectrometer. The amount of Cd(II) adsorbed at equilibrium, q_e (mg/g), was determined by Equation (10):

$$q_e = \frac{(C_0 - C_e)\,V}{M} \tag{10}$$

where q_e is the amount of Cd(II) adsorbed (mg/g) at equilibrium.

3.5. Photocatalytic Activity

To evaluate the photocatalytic activity, 10 mg of CNTs-Ag$_2$S was added to the 100 mL aqueous solution of AYR with a concentration of 10 mg/L. It was allowed to stir in the dark for 30 min to reach adsorption/desorption equilibrium of the AYR molecules over the surface of CNTs-Ag$_2$S. This suspension was transferred to a photocatalytic reactor having a water jacket with a water circulation system to maintain a constant temperature, and the suspension was exposed to sunlight. At the required time, 5 mL of the reaction mixture was withdrawn, and the suspended CNTs-Ag$_2$S was separated using centrifugation. The concentration of AYR after photocatalysis was assessed with UV-vis spectrophotometer by recording the absorbance at 373 nm. The normalized concentration of AYR after photocatalysis was calculated as C/C_0, where C_0 is the initial concentration of AYR and C is its concentration after photocatalysis. All photocatalytic experiments were conducted in the month of June, between 1 pm and 4 pm. The intensity of sunlight measured during photocatalysis was 800–900 W/m^2.

3.6. Characterization

The UV-vis absorption spectra were recorded using a Jasco V-770 UV-vis-NIR spectrophotometer (Easton, MD, USA), and ATR-FTIR spectra were collected with a Smiths ChemID diamond attenuated total reflection (DATR) spectrometer (Smiths Detection, Inc., London, United Kingdom). The XRD was obtained by a Scintag X-ray diffractometer (Cupertino, CA, USA), model PAD X, equipped with a Cu-Kα photon source (45 kV, 40 mA), at a scanning rate of 3°/min. The thermogravimetry differential thermal analysis

(TG/DTA) was performed using a Perkin Elmer Diamond TG/DTA instrument (Waltham, MA, USA) at a 10 °C/min heating rate. Transmission electron microscopy (TEM) images and X-ray energy-dispersive spectroscopy (EDS) were perceived by a Hitachi H-8100 microscope (Tokyo, Japan). The X-ray photoelectron spectra (XPS) were acquired by a Perkin Elmer PHI 5600 ci X-ray photoelectron spectrometer (Waltham, MA, USA). The Cd(II) concentration was estimated using a Varian SpectrAA 220FS atomic absorption spectrometer (Lake Forest, CA, USA).

4. Conclusions

The facile hydrothermal process produced an efficient adsorbent and photocatalyst, CNTs-Ag_2S. The ATR-FTIR, XRD, EDS, and XPS confirmed the formation of CNTs-Ag_2S in the right structure and phase. The TEM explored the deposition of Ag_2S nanoparticles over the surface of CNTs. The TG/DTA revealed the high thermal stability of CNTs-Ag_2S. The dual-tasking CNTs-Ag_2S could accomplish the complete Cd(II) adsorption and the degradation of AYR in water. The agreement of second-order kinetics for Cd(II) adsorption reveals that chemisorption is the rate-determining step of the adsorption process. The Weber−Morris intraparticle pore diffusion model represented that intraparticle diffusion could not be the sole rate-limiting step in Cd(II) adsorption, instead, it occurred through multiple phases. The validation of the Langmuir model illustrates that the Cd(II) adsorption takes place with monolayer molecular covering and chemisorption. Not limiting to Cd(II) adsorption, the CNTs-Ag_2S could also be an excellent adsorbent for adsorption of other toxic heavy metals. Apart from excellent adsorbent, CNTs-Ag_2S could also be an exceptional photocatalyst as reveled by degradation of AYR. The elevated degradation of AYR demonstrated that CNTs-Ag_2S is the strong sunlight-active photocatalyst that could be applied in the degradation of other toxic dyes. The linearity of the L-H plots depicted that the degradation of AYR occurred through pseudo-first-order kinetics. The CNTs-Ag_2S could be easily recovered and used for several times without losing its activity. Overall, CNTs-Ag_2S is a robust adsorbent as well as photocatalyst that could be employed in the adsorption of different heavy metals and the photodegradation of alternative dyes.

Supplementary Materials: The following supporting information can be downloaded at: https://www.mdpi.com/article/10.3390/catal13030476/s1. Table S1: Parameters calculated for Cd(II) adsorption over CNTs-Ag2S from the intra-particle diffusion plot; Figure S1: EDS spectrum of CNTs-Ag_2S; Figure S2: Plot perceived qt as a function of time for Cd(II) adsorption over CNTs- Ag_2S; Figure S3: Intraparticle diffusion model for Cd(II) adsorption over CNTs- Ag_2S; Figure S4: Freundlich isotherm plot for Cd(II) adsorption over CNTs- Ag_2S; Figure S5: Temkin isotherm plot for Cd(II) adsorption over CNTs- Ag_2S; Figure S6: Degradation of alizarin yellow R in presence of CNTs-Ag_2S for successive three cycles under illumination to sunlight; Figure S7: XRD of CNTs- Ag_2S (a) before and (b) after using in successive three cycles of photodegradation of alizarin yellow R under illumination to sunlight.

Author Contributions: G.M.N.: data curation, writing original draft, review, and editing funding acquisition. S.F.A.: investigation, data curation. E.A.J.: investigation, data curation. R.L.R.: writing, review, and editing. All authors have read and agreed to the published version of the manuscript.

Funding: The author (GMN) acknowledges the support of the National Academy of Sciences of the U.S.-Egypt Science and Technology Joint Fund and the Welch Foundation, Texas, United States, for departmental grant L-0002-20181021.

Institutional Review Board Statement: Not applicable.

Informed Consent Statement: Not applicable.

Data Availability Statement: Not applicable.

Conflicts of Interest: The authors declare no conflict of interest.

References

1. Xie, K.; Fang, J.; Li, L.; Deng, J.; Chen, F. Progress of graphite carbon nitride with different dimensions in the photocatalytic degradation of dyes: A review. *J. Alloy. Compd.* **2022**, *901*, 163589. [CrossRef]
2. Zhan, S.; Wu, Y.; Wang, L.; Zhan, X.; Zhou, P. A mini-review on functional nucleic acids-based heavy metal ion detection. *Biosens. Bioelectron.* **2016**, *86*, 353–368. [CrossRef] [PubMed]
3. Pang, Y.; Zhao, C.; Li, Y.; Li, Q.; Bayongzhong, X.; Peng, D.; Huang, T. Cadmium adsorption performance and mechanism from aqueous solution using red mud modified with amorphous MnO_2. *Sci. Rep.* **2022**, *12*, 1–18. [CrossRef] [PubMed]
4. Wang, J.; Chen, C. Biosorbents for heavy metals removal and their future. *Biotechnol. Adv.* **2009**, *27*, 195–226. [CrossRef]
5. Bailey, S.E.; Olin, T.J.; Bricka, R.; Adrian, D. A review of potentially low-cost sorbents for heavy metals. *Water Res.* **1999**, *33*, 2469–2479. [CrossRef]
6. Drush, G. Increase of cadmium body burden for this century. *Sci. Total Environ.* **1993**, *67*, 75–89.
7. Elkhatib, E.; Mahdy, A.; Sherif, F.; Elshemy, W. Competitive adsorption of cadmium (II) from aqueous solutions onto nano-particles of water treatment residual. *J. Nanomater.* **2016**, *2016*, 8496798. [CrossRef]
8. Lei, T.; Li, S.-J.; Jiang, F.; Ren, Z.-X.; Wang, L.-L.; Yang, X.-J.; Tang, L.-H.; Wang, S.-X. Adsorption of Cadmium Ions from an Aqueous Solution on a Highly Stable Dopamine-Modified Magnetic Nano-Adsorbent. *Nanoscale Res. Lett.* **2019**, *14*, 1–17. [CrossRef]
9. Yusuff, A.S.; Popoola, L.T.; Babatunde, E.O. Adsorption of cadmium ion from aqueous solutions by copper-based metal organic framework: Equilibrium modeling and kinetic studies. *Appl. Water Sci.* **2019**, *9*, 106. [CrossRef]
10. Masoudi, R.; Moghimi, H.; Azin, E.; Taheri, R. Adsorption of cadmium from aqueous solutions by novel Fe_3O_4- newly isolated Actinomucor sp. bionanoadsorbent: Functional group study. *Artif. Cells Nanomed. Biotechnol.* **2018**, *46*, S1092–S1101. [CrossRef]
11. Tran, H.N.; You, S.-J.; Chao, H.-P. Effect of pyrolysis temperatures and times on the adsorption of cadmium onto orange peel derived biochar. *Waste Manag. Res. J. Sustain. Circ. Econ.* **2015**, *34*, 129–138. [CrossRef]
12. Asuquo, E.; Martin, A.; Nzerem, P.; Siperstein, F.; Fan, X. Cd(II) adsorption and Pb(II) ions from aqueous solutions using mesoporous activated carbon adsorbent: Equilibrium, kinetics and characterization studies. *J. Environ. Chem. Eng.* **2017**, *5*, 679–698. [CrossRef]
13. Bulut, V.; Demirci, H.; Ozdes, D.; Gundogdu, A.; Bekircan, O.; Soylak, M.; Duran, C. A novel carrier element-free coprecipi-tation method for separation/preconcentration of lead and cadmium ions from environmental matrices. *Environ. Prog. Sustain. Energy* **2016**, *35*, 1709–1715. [CrossRef]
14. Wong, C.-W.; Barford, J.P.; Chen, G.; McKay, G. Kinetics and equilibrium studies for the removal of cadmium ions by ion exchange resin. *J. Environ. Chem. Eng.* **2014**, *2*, 698–707. [CrossRef]
15. Kumar, P.R.; Swathanthr, P.A.; Rao, V.B.; Rao, S.R.M. Adsorption of Cadmium and Zinc Ions from Aqueous Solution Using Low Cost Adsorbents. *J. Appl. Sci.* **2014**, *14*, 1372–1378. [CrossRef]
16. Li, Y.; Yang, L.; Xu, Z.; Sun, Q. Separation and recovery of heavy metals from waste water using synergistic solvent extraction. In Proceedings of the IOP Conference Series: Materials Science and Engineering, Sanya, China, 19–21 November 2016; Volume 167, p. 12005. [CrossRef]
17. Meng, Q.; Nan, J.; Wang, Z.; Ji, X.; Wu, F.; Liu, B.; Xiao, Q. Study on the efficiency of ultrafiltration technology in dealing with sudden cadmium pollution in surface water and ultrafiltration membrane fouling. *Environ. Sci. Pollut. Res.* **2019**, *26*, 16641–16651. [CrossRef]
18. Bhadrinarayana, N.S.; Basha, C.A.; Anantharaman, N. Electrochemical Oxidation of Cyanide and Simultaneous Cathodic Removal of Cadmium Present in the Plating Rinse Water. *Ind. Eng. Chem. Res.* **2007**, *46*, 6417–6424. [CrossRef]
19. Kim, H.S.; Seo, B.-H.; Kuppusamy, S.; Lee, Y.B.; Lee, J.-H.; Yang, J.-E.; Owens, G.; Kim, K.-R. A DOC coagulant, gypsum treatment can simultaneously reduce As, Cd and Pb uptake by medicinal plants grown in contaminated soil. *Ecotoxicol. Environ. Saf.* **2018**, *148*, 615–619. [CrossRef]
20. Lan, Z.; Lin, Y.; Yang, C. Lanthanum-iron incorporated chitosan beads for adsorption of phosphate and cadmium from aqueous solutions. *Chem. Eng. J.* **2022**, *448*. [CrossRef]
21. Foroutan, R.; Peighambardoust, S.; Mohammadi, R.; Peighambardoust, S.; Ramavandi, B. Cadmium ion removal from aqueous media using banana peel biochar/ Fe_3O_4/ZIF-67. *Envir. Res.* **2022**, *211*, 113020. [CrossRef]
22. Dou, D.; Wei, D.; Guan, X.; Liang, Z.; Lan, L.; Lan, X.; Liu, P.; Mo, H.; Lan, P. Adsorption of copper (II) and cadmium (II) ions by in situ doped nano-calcium carbonate high-intensity chitin hydrogels. *J. Hazard. Mater.* **2021**, *423*, 127137. [CrossRef] [PubMed]
23. Alsantali, R.I.; Alam Raja, Q.; Alzahrani, A.Y.; Sadiq, A.; Naeem, N.; Mughal, E.U.; Al-Rooqi, M.M.; El Guesmi, N.; Moussa, Z.; Ahmed, S.A. Miscellaneous azo dyes: A comprehensive review on recent advancements in biological and industrial applications. *Dye. Pigment.* **2022**, *199*, 110050. [CrossRef]
24. Narayanasamy, L.; Murugesan, T. Degradation of Alizarin Yellow R using UV/H_2O_2 Advanced Oxidation Process. *Environ. Prog. Sustain. Energy* **2014**, *33*, 482–489. [CrossRef]
25. Ahmed, A.; Usman, M.; Yu, B.; Ding, X.; Peng, Q.; Shen, Y.; Cong, H. Efficient photocatalytic degradation of toxic Alizarin yellow R dye from industrial wastewater using biosynthesized Fe nanoparticle and study of factors affecting the degradation rate. *J. Photochem. Photobiol. B Biol.* **2020**, *202*, 111682. [CrossRef] [PubMed]
26. Suteu, D.; Zaharia, C.; Muresan, A.; Muresan, R.; Popescu, A. Textile wastewater treatment by homogenous oxidation with hydrogen peroxide. *Environ. Eng. Manage. J.* **2009**, *8*, 1359–1369.

27. Jiao, X.; Yu, H.; Kong, Q.; Luo, Y.; Chen, Q.; Qu, J. Theoretical mechanistic studies on the degradation of alizarin yellow R initiated by hydroxyl radical. *J. Phys. Org. Chem.* **2014**, *27*, 519–526. [CrossRef]
28. Lellis, B.; Fávaro-Polonio, C.Z.; Pamphile, J.A.; Polonio, J.C. Effects of textile dyes on health and the environment and bioremediation potential of living organisms. *Biotechnol. Res. Innov.* **2019**, *3*, 275–290. [CrossRef]
29. Al Nafiey, A.; Addad, A.; Sieber, B.; Chastanet, G.; Barras, A.; Szunerits, S.; Boukherroub, R. Reduced graphene oxide deco-rated with Co3O4 nanoparticles (rGO-Co3O4) nanocomposite: A reusable catalyst for highly efficient reduction of 4-nitrophenol, and Cr(VI) and dye removal from aqueous solutions. *Chem. Eng. J.* **2017**, *322*, 375–384. [CrossRef]
30. Zhu, C.; Mahmood, Z.; Siddique, M.S.; Wang, H.; Anqi, H.; Sillanpää, M. Structure-Based Long-Term Biodegradation of the Azo Dye: Insights from the Bacterial Community Succession and Efficiency Comparison. *Water* **2021**, *13*, 3017. [CrossRef]
31. Zhang, Y.; He, P.; Jia, L.; Li, C.; Liu, H.; Wang, S.; Zhou, S., F. Dong Ti/PbO2-Sm2O3 composite based electrode for highly effi-cient electrocatalytic degradation of alizarin yellow R. *J. Colloid Interface Sci.* **2019**, *533*, 750–761. [CrossRef]
32. Mcyotto, F.; Wei, Q.; Macharia, D.K.; Huang, M.; Shen, C.; Chow, C.W. Effect of dye structure on color removal efficiency by coagulation. *Chem. Eng. J.* **2020**, *405*, 126674. [CrossRef]
33. Lee, J.-W.; Choi, S.-P.; Thiruvenkatachari, R.; Shim, W.-G.; Moon, H. Evaluation of the performance of adsorption and coagulation processes for the maximum removal of reactive dyes. *Dye. Pigment.* **2006**, *69*, 196–203. [CrossRef]
34. Unsal, Y.E.; Tuzen, M.; Soylak, M. Separation and Preconcentration of Sudan Blue II Using Membrane Filtration and UV-Visible Spectrophotometric Determination in River Water and Industrial Wastewater Samples. *J. AOAC Int.* **2015**, *98*, 213–217. [CrossRef]
35. Kadirvelu, K.; Kavipriya, M.; Karthika, C.; Radhika, M.; Vennilamani, N.; Pattabhi, S. Utilization of various agricultural wastes for activated carbon preparation and application for the removal of dyes and metal ions from aqueous solutions. *Bioresour. Technol.* **2002**, *87*, 129–132. [CrossRef]
36. Rajkumar, D.; Song, B.J.; Kim, J.G. Electrochemical degradation of Reactive Blue 19 in chloride medium for the treatment of textile dyeing wastewater with identification of intermediate compounds. *Dye. Pigment.* **2007**, *72*, 1–7. [CrossRef]
37. Mahmoodi, N. Photocatalytic ozonation of dyes using multiwalled carbon nanotube. *J. Mol. Catal. A* **2013**, *366*, 254–260. [CrossRef]
38. Neelgund, G.; Bliznyuk, V.; Oki, A. Photocatalytic activity and NIR laser response of polyanilineconjugated graphene nanocomposite prepared by a novel acid-less method. *Appl. Catal. B* **2016**, *187*, 357–366. [CrossRef]
39. Neelgund, G.M.; Oki, A. Folic acid and CuS conjugated graphene oxide: An efficient photocatalyst for explicit degradation of toxic dyes. *Appl. Surf. Sci.* **2021**, *566*, 150648. [CrossRef]
40. Singh, S.; Rao, C.; Nandi, C.K.; Mukherjee, T.K. Quantum Dot-Embedded Hybrid Photocatalytic Nanoreactors for Visible Light Photocatalysis and Dye Degradation. *ACS Appl. Nano Mater.* **2022**, *5*, 7427–7439. [CrossRef]
41. Chouke, P.; Dadure, K.; Potbhare, A.; Bhusari, G.; Mondal, A.; Chaudhary, K.; Singh, V.; Desimone, M.; Chaudhary, R.; Masram, D. Biosynthesized δ-Bi2O3 nanoparticles from crinum viviparum flower extract for photocatalytic dye degradation and molecular docking. *ACS Omega* **2022**, *7*, 20983–20993. [CrossRef]
42. Neelgund, G.M.; Oki, A. Photocatalytic activity of hydroxyapatite deposited graphene nanosheets under illumination to sunlight. *Mater. Res. Bull.* **2021**, *146*, 111593. [CrossRef]
43. Hamd, A.; Shaban, M.; AlMohamadi, H.; Dryaz, A.R.; Ahmed, S.A.; Abu Al-Ola, K.A.; El-Mageed, H.R.A.; Soliman, N.K. Novel Wastewater Treatment by Using Newly Prepared Green Seaweed–Zeolite Nanocomposite. *ACS Omega* **2022**, *7*, 11044–11056. [CrossRef] [PubMed]
44. Neelgund, G.M.; Oki, A. ZnO conjugated graphene: An efficient sunlight driven photocatalyst for degradation of organic dyes. *Mater. Res. Bull.* **2020**, *129*, 110911. [CrossRef]
45. Neelgund, G.; Oki, A. Photothermal effect: An important aspect for the enhancement of photocatalytic activity under illumi-nation by NIR radiation. *Mater. Chem. Front.* **2018**, *2*, 64–75. [CrossRef]
46. Abass, A.K.; Raoof, S.D. Photocatalytic Removal of Alizarin Yellow R from Water using Modified Zinc Oxide Catalyst. *Asian J. Chem.* **2016**, *28*, 312–316. [CrossRef]
47. Ramamoorthy, S.; Das, S.; Balan, R.; Lekshmi, I. TiO2-ZrO2 nanocomposite with tetragonal zirconia phase and photocatalytic degradation of Alizarin Yellow GG azo dye under natural sunlight. *Mater. Today: Proc.* **2021**, *47*, 4641–4646. [CrossRef]
48. Mawhinney, D.B.; Naumenko, V.; Kuznetsova, A.; Yates, J.J.T.; Liu, J.; Smalley, R.E. Infrared Spectral Evidence for the Etching of Carbon Nanotubes: Ozone Oxidation at 298 K. *J. Am. Chem. Soc.* **2000**, *122*, 2383–2384. [CrossRef]
49. Chingombe, P.; Saha, B.; Wakeman, R. Surface modification and characterization of a coal-based activated carbon. *Carbon* **2005**, *43*, 3132–3143. [CrossRef]
50. Neelgund, G.M.; Oki, A.R. Influence of carbon nanotubes and graphene nanosheets on photothermal effect of hydroxyapatite. *J. Colloid Interface Sci.* **2016**, *484*, 135–145. [CrossRef]
51. Zamiri, R.; Ahangar, H.A.; Zakaria, A.; Zamiri, G.; Shabani, M.; Singh, B.K.; Ferreira, J.M.F. The structural and optical constants of Ag2S semiconductor nanostructure in the Far-Infrared. *Chem. Central J.* **2015**, *9*, 1–6. [CrossRef]
52. Neelgund, G.M.; Oki, A. Photocatalytic activity of CdS and Ag2S quantum dots deposited on poly(amidoamine) functionalized carbon nanotubes. *Appl. Catal. B: Environ.* **2011**, *110*, 99–107. [CrossRef]
53. Hota, G.; Jain, S.; Khila, K. Synthesis of CdS-Ag2S core-shell/composite nanoparticles using AOT/n-heptane/water microemulsions. *Colloids Surf. A* **2004**, *232*, 119–127. [CrossRef]

54. Wu, H.; Cao, W.; Chen, Q.; Liu, M.; Qian, S.; Jia, N.; Yang, H.; Yang, S. Metal sulfide coated multiwalled carbon nanotubes synthesized by an in situ method and their optical limiting properties. *Nanotechnology* **2009**, *20*, 195604–195613. [CrossRef]
55. Kim, B.-J.; Kim, J.-P.; Park, J.-S. Effects of Al interlayer coating and thermal treatment on electron emission characteristics of carbon nanotubes deposited by electrophoretic method. *Nanoscale Res. Lett.* **2014**, *9*, 236. [CrossRef]
56. Lin, Y.; Wu, S.; Yang, C.; Chen, M.; Li, X. Preparation of size-controlled silver phosphate catalysts and their enhanced photocatalysis performance via synergetic effect with MWCNTs and PANI. *Appl. Catal. B* **2019**, *245*, 71–86. [CrossRef]
57. Di, L.; Xian, T.; Sun, X.; Li, H.; Zhou, Y.; Ma, J.; Yang, H. Facile Preparation of CNT/Ag_2S Nanocomposites with Improved Visible and NIR Light Photocatalytic Degradation Activity and Their Catalytic Mechanism. *Micromachines* **2019**, *10*, 503. [CrossRef]
58. Zhang, Y.; Xia, J.; Li, C.; Zhou, G.; Yang, W.; Wang, D.; Zheng, H.; Du, Y.; Li, X.; Li, Q. Near-infrared-emitting colloidal Ag2S quantum dots excited by an 808 nm diode laser. *J. Mater. Sci.* **2017**, *52*, 9424–9429. [CrossRef]
59. Hameed, B.; Salman, J.; Ahmad, A. Adsorption isotherm and kinetic modeling of 2,4-D pesticide on activated carbon derived from date stones. *J. Hazard. Mater.* **2009**, *163*, 121–126. [CrossRef]
60. Zeng, H.; Zhai, L.; Qiao, T.; Yu, Y.; Zhang, J.; Li, D. Efficient removal of As(V) from aqueous media by magnetic nanoparticles prepared with Iron-containing water treatment residuals. *Sci. Rep.* **2020**, *10*, 1–12. [CrossRef]
61. Langmuir, I. The constitution and fundamental properties of solids and liquids. Part I. Solids. *J. Am. Chem. Soc.* **1916**, *38*, 2221–2295. [CrossRef]
62. Neelgund, G.M.; Aguilar, S.F.; Kurkuri, M.D.; Rodrigues, D.F.; Ray, R.L. Elevated Adsorption of Lead and Arsenic over Silver Nanoparticles Deposited on Poly(amidoamine) Grafted Carbon Nanotubes. *Nanomaterials* **2022**, *12*, 3852. [CrossRef] [PubMed]
63. Freundlich, H. Over the adsorption in solution. *J. Phys. Chem.* **1906**, *57*, 385–471.
64. Temkin, M.; Pyzhev, V. Recent modifications to Langmuir isotherms. *Acta Physiochim. USSR* **1940**, *12*, 217–222.
65. Tabuchi, A.; Ogata, F.; Uematsu, Y.; Toda, M.; Otani, M.; Saenjum, C.; Nakamura, T.; Kawasaki, N. Granulation of nick-el-aluminum-zirconium complex hydroxide using colloidal silica for adsorption of chromium(VI) ions from the liquid phase. *Molecules* **2022**, *27*, 2392. [CrossRef]
66. Meng, Z.; Sarkar, S.; Zhu, L.; Ullah, K.; Ye, S.; Oh, W. Facile preparation of Ag2S-CNT nanocomposites with enhanced pho-to-catalytic activity. *J. Korean Ceram. Soc.* **2014**, *51*, 1–6. [CrossRef]
67. Ahmed, K.A.M.; Peng, H.; Wu, K.; Huang, K. Hydrothermal preparation of nanostructured manganese oxides (MnOx) and their electrochemical and photocatalytic properties. *Chem. Eng. J.* **2011**, *172*, 531–539. [CrossRef]
68. Iqbal, S.; Iqbal, M.; Sibtain, A.; Iqbal, A.; Farooqi, Z.; Ahmad, S.; Mustafa, K.; Musaddiq, S. Solar driven photocatalytic degradation of organic pollutants via Bi2O3@reduced graphene oxide nanocomposite. *Desalination Water Treat.* **2021**, *216*, 140–150. [CrossRef]
69. Mashentseva, A.A.; Aimanova, N.A.; Parmanbek, N.; Temirgaziyev, B.S.; Barsbay, M.; Zdorovets, M.V. *Serratula coronata* L. Mediated Synthesis of ZnO Nanoparticles and Their Application for the Removal of Alizarin Yellow R by Photocatalytic Degradation and Adsorption. *Nanomaterials* **2022**, *12*, 3293. [CrossRef]
70. Neelgund, G.M.; Oki, A. Deposition of silver nanoparticles on dendrimer functionalized multiwalled carbon nanotubes: Synthesis, characterization and antimicrobial activity. *J. Nanosci. Nanotechnol.* **2011**, *11*, 3621–3629. [CrossRef]
71. Neelgund, G.M.; Oki, A. Pd nanoparticles deposited on poly(lactic acid) grafted carbon nanotubes: Synthesis, characterization and application in Heck C–C coupling reaction. *Appl. Catal. A Gen.* **2011**, *399*, 154–160. [CrossRef]

Disclaimer/Publisher's Note: The statements, opinions and data contained in all publications are solely those of the individual author(s) and contributor(s) and not of MDPI and/or the editor(s). MDPI and/or the editor(s) disclaim responsibility for any injury to people or property resulting from any ideas, methods, instructions or products referred to in the content.

Article

Photocatalytic Degradation of Organic Dyes Contaminated Aqueous Solution Using Binary CdTiO$_2$ and Ternary NiCdTiO$_2$ Nanocomposites

Shakeel Khan [1], Awal Noor [2,*], Idrees Khan [3], Mian Muhammad [4], Muhammad Sadiq [4] and Niaz Muhammad [1,*]

[1] Department of Chemistry, Abdul Wali Khan University Mardan, Mardan 23200, Pakistan
[2] Department of Basic Sciences, Preparatory Year Deanship, King Faisal University, Al-Hassa 31982, Saudi Arabia
[3] School of Chemistry and Chemical Engineering, Northwestern Polytechnical University, Xi'an 710072, China
[4] Department of Chemistry, University of Malakand, Chakdara 18800, Dir (Lower), Pakistan
* Correspondence: anoor@kfu.edu.sa (A.N.); drniaz@awkum.edu.pk (N.M.)

Abstract: The synergistic effect of binary CdTiO$_2$ and ternary NiCdTiO$_2$ on the photocatalytic efficiency of TiO$_2$ nanoparticles was investigated. The SEM analysis demonstrates spherical TiO$_2$ NPs of different sizes present in agglomerated form. The structural analysis of the nanocomposites reveals a porous structure for TiO$_2$ with well deposited Cd and Ni NPs. TEM images show NiCdTiO$_2$ nanocomposites as highly crystalline particles having spherical and cubical geometry with an average particle size of 20 nm. The EDX and XRD analysis confirm the purity and anatase phase of TiO$_2$, respectively. Physical features of NiCdTiO$_2$ nanocomposite were determined via BET analysis which shows that the surface area, pore size and pore volume are 61.2 m^2/g, 10.6 nm and 0.1 cm^3/g, respectively. The absorbance wavelengths of the CdTiO$_2$ and NiCdTiO$_2$ nanocomposites have shown red shift as compared to the neat TiO$_2$ due to coupling with Ni and Cd that results in the enhanced photocatalytic activity. The photocatalytic activity demonstrated that TiO$_2$, CdTiO$_2$ and NiCdTiO$_2$ degrade methylene blue (MB) and methyl green (MG) about 76.59, 82, 86% and 63.5, 88, 97.5%, respectively, at optimum reaction conditions.

Keywords: TiO$_2$; nanocomposites; photocatalysts; photodegradation; methylene blue; methyl green

1. Introduction

Industrial effluents containing synthetic dyes is a formidable challenge for water remedy processes [1,2]. Dyes are used for the coloration of several materials such as textile fibers, cosmetics, paper, tannery, food, leather and pharmaceutical products [3]. These synthetic dyes are major water pollutants and cause serious environmental problems due to their high aromaticity, low biodegradability, toxicity, chemical stability and carcinogenic nature [4]. These dyes also reduce the light penetration which reduces the photosynthetic activity that causes a deficiency in dissolved O$_2$ content of the water [5]. Various approaches are applied for the remediation of these pollutants such as adsorption [6], nanofiltration [7], ozonation [8], coagulation [9], biodegradation [10] and phytoremediation [11] etc. These conventional approaches are expensive, destructive, difficult and transform pollutants into sludge [12].

Advanced Oxidation Processes (AOPs) generate and use powerful transitory species such as hydroxyl radicals [13] to eliminate the organic pollutants by final conversion into small and stable molecules such as H$_2$O and CO$_2$ [14]. Among the AOPs, photocatalytic degradation is believed to be the most appropriate low-cost approach to treat organic pollutants [15]. Photodegradation has advantages over other conventional approaches owing to its simplicity, complete pollutants mineralization, cost-effectiveness, no harmful byproducts formation, ambient pressure and temperature operation [16]. Various semiconducting photocatalysts are used for the photodegradation of dyes such as ZnO [17], Fe_3O_4 [12], SnO_2 [18], TiO_2 [19] etc. Among these photocatalysts, titanium (IV) oxide has been the most investigated material for the environmental photocatalysis owing to its abundance, high specific surface area, nontoxicity, photostability, strong oxidation capability, low price, high photoactivity and chemical stability [20–22]. Titanium dioxide (TiO_2) as an intrinsically n-type semiconductor material with a band gap of around 3 eV [23,24] is extensively suggested for diverse applications such as lithium-ion batteries [25], supercapacitors [26], solar cells [27], sensors [28] and photocatalysts [29–31]. However, TiO_2 as photocatalyst represents low photocatalytic activity due to its high electron–hole pair recombination rate, wide band gap and its excitation only under UV light [32]. In order to retard these deficiencies, various approaches are developed such as doping [33], sensitization [34], supporting on a medium [35] and coupling with semiconductors [36–38]. Among these, coupling of TiO_2 with other semiconducting material having lower band gap energy forming a heterojunction is a strategic option. The semiconductor having lower gap energy plays the role of sensitizer by being excited first, and then inducing the excitation of TiO_2 by passing photoelectrons from its conduction band to that of TiO_2 [39].

In the present work, TiO_2 nanoparticles were prepared by precipitation technique and then coupled with Cd and Ni to obtain $CdTiO_2$ and $NiCdTiO_2$ nanocomposites through co-precipitation method. The $CdTiO_2$ and $NiCdTiO_2$ nanocomposites are not reported in literature nor utilized as photocatalysts in the photodegradation of dyes to the best of our knowledge. These photocatalysts were prepared from economical materials and simple approach. The photocatalysts are very efficient toward the photodegradation of both dyes. The photocatalytic efficacy of the TiO_2, $CdTiO_2$ and $NiCdTiO_2$ was assessed by degradation of methylene blue (MB) and methyl green (MG) dyes in aqueous solution under UV-light irradiation. MB and MG are selected as model dyes because these are recalcitrant organic pollutants with carcinogenic and mutagenic nature with LD50 = 1180 mg/kg [40]. At higher concentration these dyes cause great damage to the human body and environment [41,42]. In the photodegradation of MB and MG dyes, the effect of irradiation time, catalyst dosage and pH were assessed.

2. Results and Discussion
2.1. Morphological and Elemental Analysis

The surface morphology of TiO_2, $CdTiO_2$ and $NiCdTiO_2$ was studied via SEM analysis and the images at different magnifications are shown in Figures 1–3, respectively. The images show spherical TiO_2 NPs of different shapes and sizes and present mostly in agglomerated form. The particles are also present and dispersed when highly magnified. The SEM analysis of $CdTiO_2$ shows that Cd NPs are deposited on the surface and embedded in the porous structure of TiO_2. The morphology of $CdTiO_2$ displayed that TiO_2 are present in porous nanotubes form and Cd NPs are dispersed on its surface, and inserted in the nanochannels. The $CdNiTiO_2$ nanocomposites are mostly agglomerated, and the Cd and Ni NPs significantly cover the pores and surface of TiO_2. The particles have different shapes and morphology.

The ternary $NiCdTiO_2$ nanocomposites were also examined via TEM analysis and the results are as shown in Figure 4 at different magnifications which support the SEM analysis. Images shows that Cd and Ni NPs are uniformly mixed with TiO_2 NPs. Highly crystalline $NiCdTiO_2$ nanocomposites of cubical as well as spherical geometry with an average particle size of 20 nm are confirmed by TEM images.

Figure 1. (**a**–**d**) SEM images of TiO$_2$ nanoparticles.

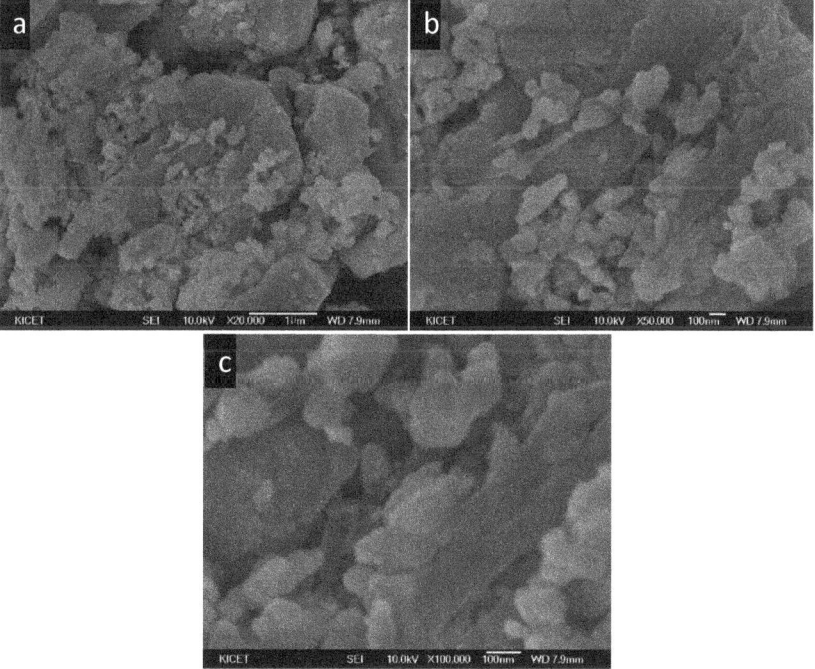

Figure 2. (**a**–**c**) SEM images of CdTiO$_2$ nanocomposite.

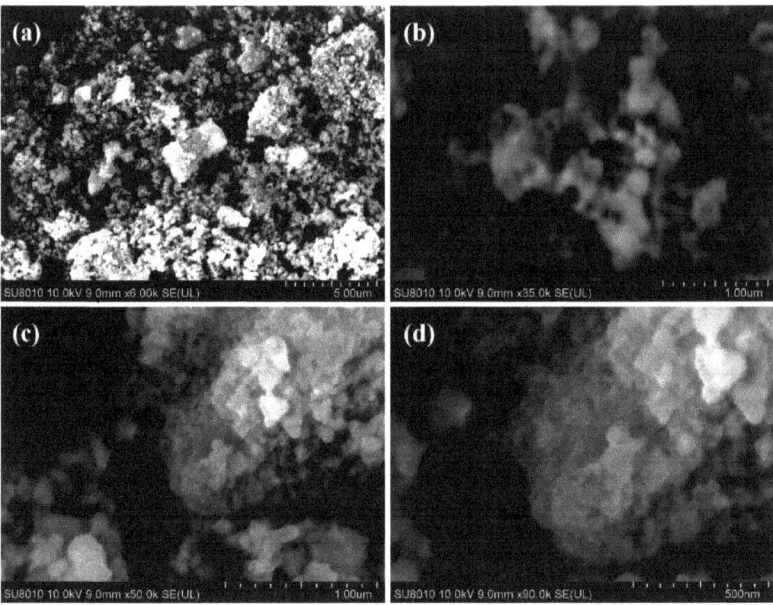

Figure 3. (**a**–**d**) SEM images of NiCdTiO$_2$ nanocomposite.

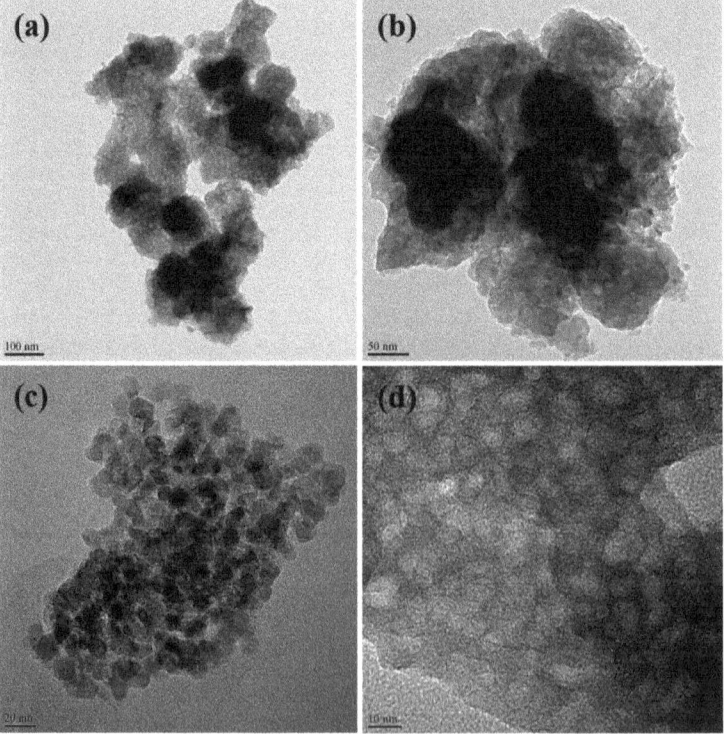

Figure 4. (**a**–**d**) TEM images of NiCdTiO$_2$ nanocomposite at different magnifications.

The composition of the ternary NiCdTiO$_2$ nanocomposite was ascertained via EDX and the spectrum along with %composition in tabulated form is presented in Figure 5. The EDX spectrum shows signals for the constituent elements (Ti, Ni, Cd and O). The carbon and silver signals are present due to their coating on samples prior to EDX analysis for obtaining good quality images.

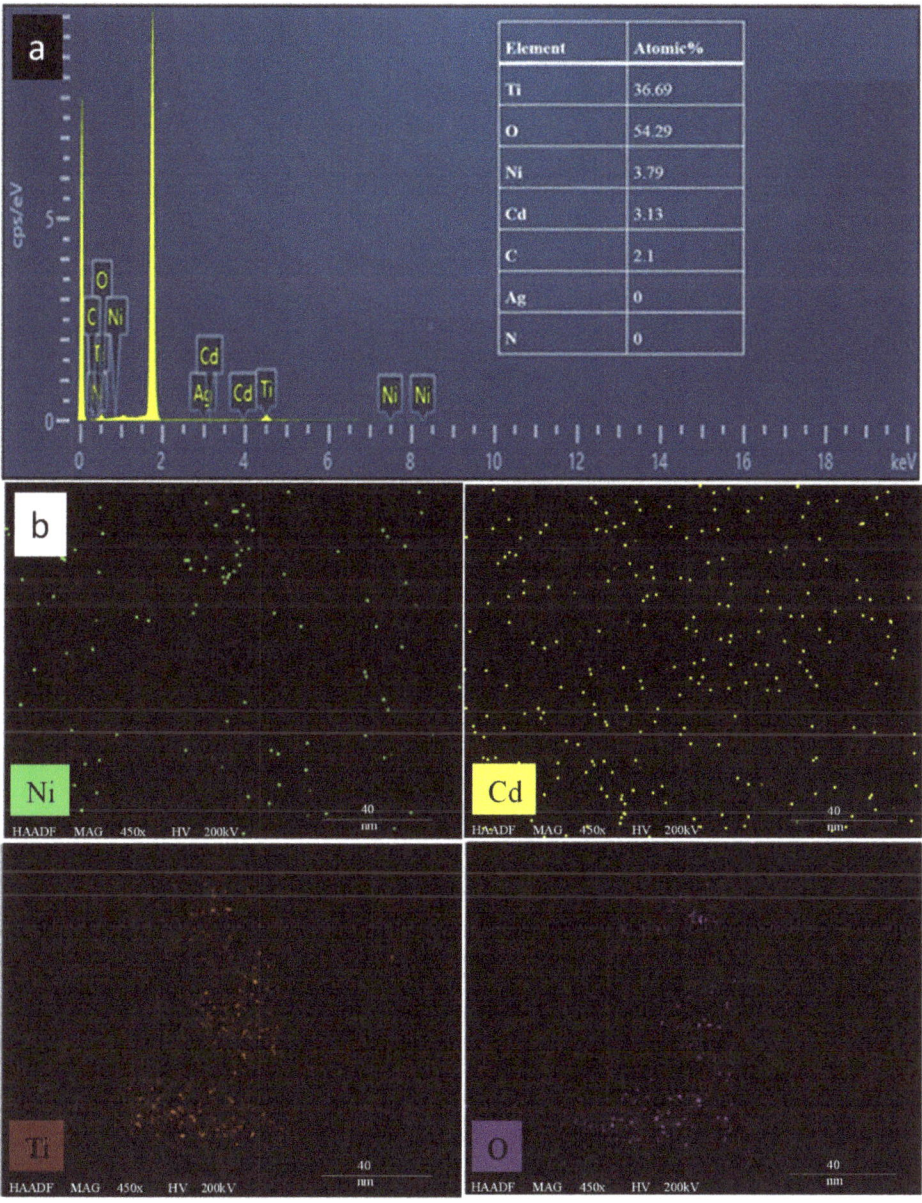

Figure 5. (a) EDX spectra and (b) mapping of the ternary NiCdTiO$_2$ nanocomposite.

2.2. XRD Analysis

The XRD patterns of the photocatalysts are displayed in Figure 6. The observed signals can be related to the corresponding (101), (004), (200), (105), (211), (204) and (116) crystal planes. The identified diffraction signals can be allocated to the anatase TiO_2 (JCPDS-21-1272). Peaks for the Cd and Ni NPs are not observed owing to their minute quantity. However, (105) and (211) crystal plane peaks observed in the TiO_2 patterns are replaced by single broadened peaks in the $CdTiO_2$ and $NiCdTiO_2$ nanocomposites patterns.

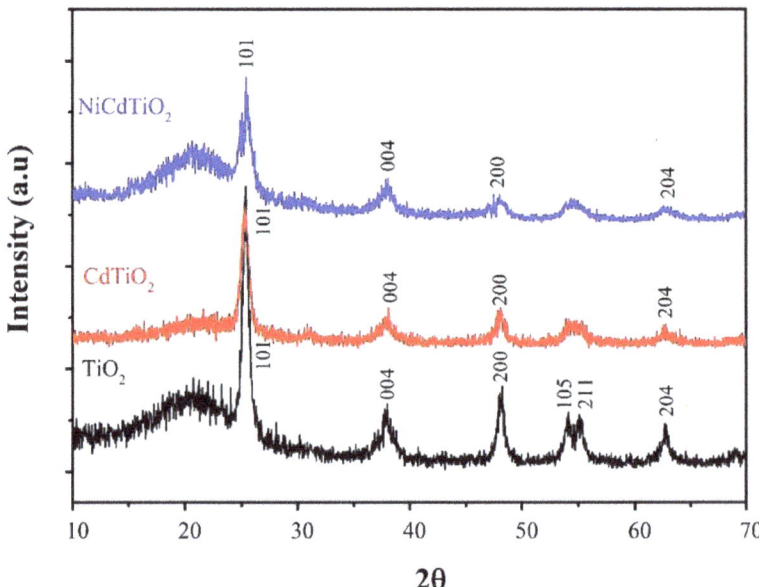

Figure 6. XRD patterns of the TiO_2, $CdTiO_2$ and $NiCdTiO_2$.

2.3. BET Analysis

The optimum porosity and high specific surface area are considered as essential parameters for the efficiency of nanocomposite materials used in photocatalysis. Figure 7a presents the BET and adsorption-desorption plot for $NiCdTiO_2$ nanocomposite. The adsorption/desorption of N_2 is important for investigating the surface area, average pore size and pore volume of the $NiCdTiO_2$ photocatalyst. The study revealed that the $NiCdTiO_2$ nanocomposite exhibit type IV isotherm with a sharp increase of the adsorbed volume starting from P/P_0 = 0.84, confirming the mesoporous nanosized nature of the nanocomposite. When the relative pressure approaches 1, the hysteresis loop shifts higher and shows that the microporous particles are also present. Figure 7b shows BJH plot for the porosity investigation of $NiCdTiO_2$ nanocomposite. Different surface parameters like BET surface area, pore size, volume and BJH average pore width of $NiCdTiO_2$ nanocomposite are represented in the Table 1.

Table 1. The specific surface area, pore size, pore volume and BJH average pore width of $NiCdTiO_2$ nanocomposite.

BET (m²/g)	Pore Volume (cm³/g)	Pore Size (nm)	BJH Average Pore Width Ads/Des (nm)
61.2	0.1	10.6	9.6/8.7

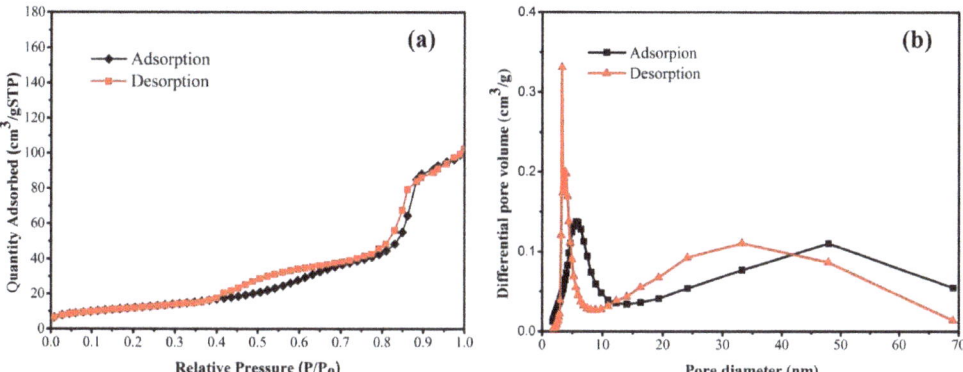

Figure 7. (a) BET and adsorption desorption plot for NiCdTiO$_2$ (b) BJH plot showing porosity evaluation of NiCdTiO$_2$.

2.4. UV-Visible Analysis

The absorbance wavelength and band gap energy of the TiO$_2$, CdTiO$_2$ and NiCdTiO$_2$ were recorded using UV–VIS spectroscopy. Figure 8 displays the UV–VIS absorption spectra of TiO$_2$, CdTiO$_2$ and NiCdTiO$_2$. The UV–Visible absorption spectrum of TiO$_2$ shows absorption band at 265 nm. The maximum absorbance wavelengths of the CdTiO$_2$ and NiCdTiO$_2$ have shown slight red shifts. This shift can make nanocomposites better photocatalysts compared to pure TiO$_2$.

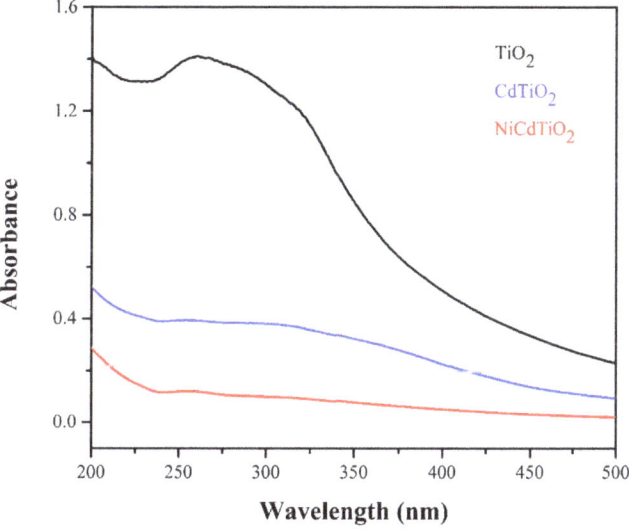

Figure 8. UV–VIS absorption spectra of TiO$_2$, CdTiO$_2$ and NiCdTiO$_2$.

Figure 9a–c displays the Tauc plots: $(\alpha h\nu)^2$ versus energy of TiO$_2$, CdTiO$_2$ and NiCdTiO$_2$, respectively. The band gap was calculated applying Tauc plots, which represents the relation between the sample absorption edge with the energy of the incident photon.

$$\alpha h\nu = A(h\nu - E_g)^n \tag{1}$$

where α = molar extinction coefficient, h = Planck constant, ν = photon's frequency, A = constant, E_g = band gap energy, and n = parameter associated with the electronic transition ($\frac{1}{2}$ in the present case). The results demonstrate 2.7 eV band gap energy for TiO_2. The band gap energy values for $CdTiO_2$ and $NiCdTiO_2$ are 2.64 eV and 2.52 eV, respectively, as shown in the Figure 9c inset. The results clearly show the effect of Cd doping and Ni, Cd co-doping on the band gap energy of TiO_2 which can further be correlated with the photocatalytic activity of the catalysts.

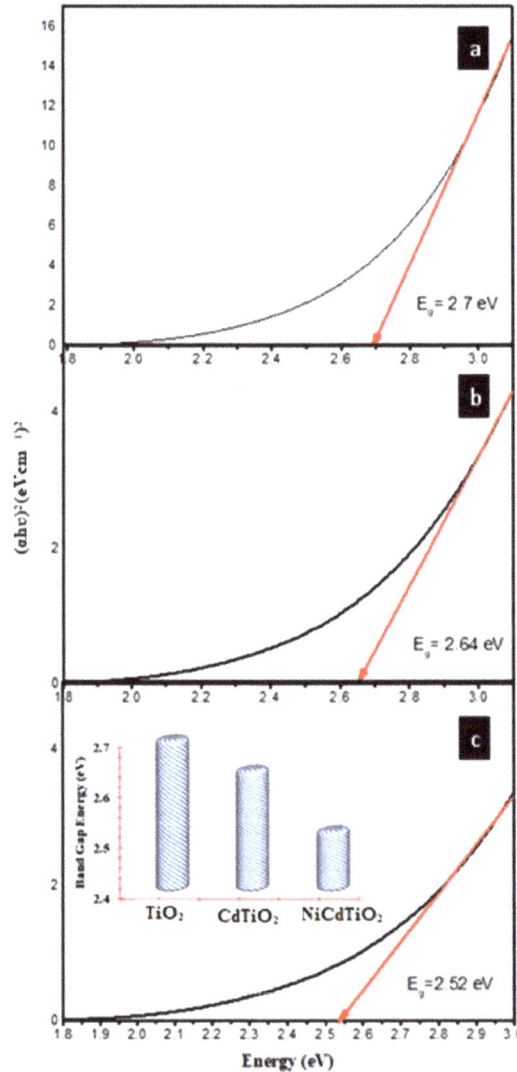

Figure 9. Tauc plots: $(\alpha h\nu)^2$ versus energy for the (**a**) TiO_2 (**b**) $CdTiO_2$ (**c**) $NiCdTiO_2$, inset; comparison of band gap energy.

2.5. Photodegradation of Methylene Blue (MB) Dye

2.5.1. Effect of Irradiation Time

The photocatalysts were utilized for the photodegradation of MB dye in aqueous medium under UV–light irradiation. Figure 10a–c displays the UV–VIS spectra of MB dye before reaction and after varying UV light-irradiation times in the presence of TiO_2, $CdTiO_2$ and $NiCdTiO_2$, respectively. The MB dye has shown maximum absorbance in the absence of catalysts. A sharp decrease in the absorbance was observed in the presence of catalysts due to photodegradation of the dye. The absorbance shows a regular decrease in the presence of catalysts with increasing irradiation time. The %degradation of MB dye by the synthesized catalysts as presented in Figure 10d shows higher photocatalytic activity for the ternary $NiCdTiO_2$ as compared to the binary $CdTiO_2$ and neat TiO_2. A 40, 48 and 65% degradation of the dye was observed within 20 min irradiation time with TiO_2, $CdTiO_2$ and $CdNiTiO_2$, respectively. The degradation was increased to 76.59, 82 and 86% in the presence of TiO_2, $CdTiO_2$ and $CdNiTiO_2$, respectively, by increasing irradiation time to 100 min. The increase in photodegradation of dye with an increase in irradiation time is due to the availability of more and more time to generate more hydroxyl radicals, which is a key species in dye degradation. The photocatalytic degradation experiments of MB using 30 ppm initial concentration were carried out under optimal reaction conditions and a pseudo first-order kinetics was observed. The results were in good agreement with the reported literature [42–50].

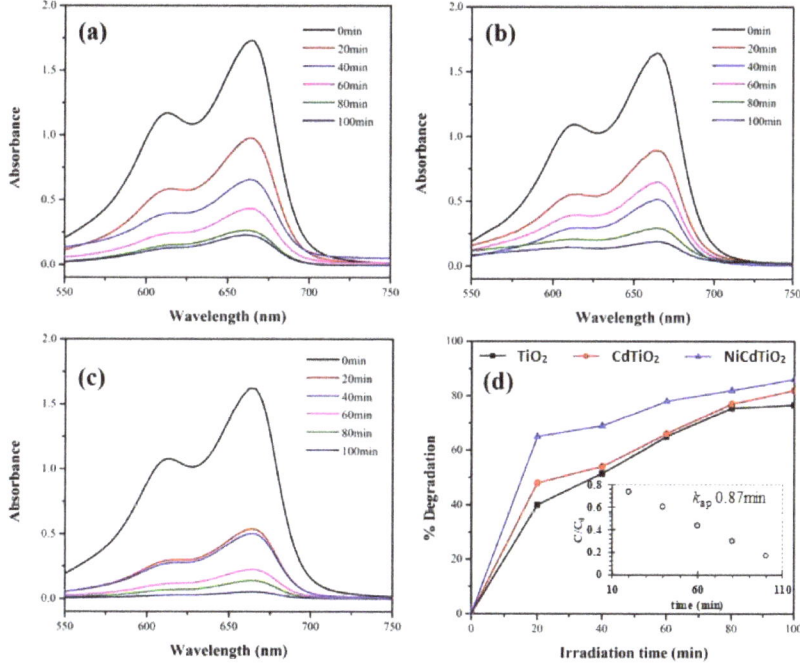

Figure 10. UV–Visible absorption spectra of MB photodegraded by (**a**) TiO_2 (**b**) $CdTiO_2$ (**c**) $NiCdTiO_2$ (**d**) Comparison of %degradation of MB dye inset kinetic model of degradation.

The photocatalytic degradation of dye depends upon the light-harvesting efficiency, the efficiency of the reaction of the photogenerated electron/hole charges and the reaction of photogenerated charges with substrate molecules. Photodegradation of the dye is achieved when UV-light interacts with the photocatalyst. Photons having energy equal to or greater than the band gap of the catalyst, excite the electrons from their valence band (VB) to the conduction band (CB) and produce positive holes (h^+) in the VB. The h^+ of the VB react with the water molecules and produce hydroxyl radicals (·OH) while the excited electrons present in the CB react with oxygen molecules and generate superoxide anion radicals ($O_2^{·-}$) [43,44]. These radicals are highly reactive species and effectively degrade dye molecules into simple and small species such as H_2O and CO_2. In the case of pristine TiO_2, a major portion of the separated electron–hole pairs recombine and reduces the photocatalytic activity. However, in $CdTiO_2$ and $NiCdTiO_2$ nanocomposites, the electrons present in the VB of TiO_2 get captured by the coupled Cd and Ni so the electron–hole pairs recombination rate decreases. This makes the nanocomposites more efficient photocatalysts compared to the neat TiO_2. The suggested mechanism for the photodegradation of MB dye by ternary $NiCdTiO_2$ nanocomposite is presented in Figure 11.

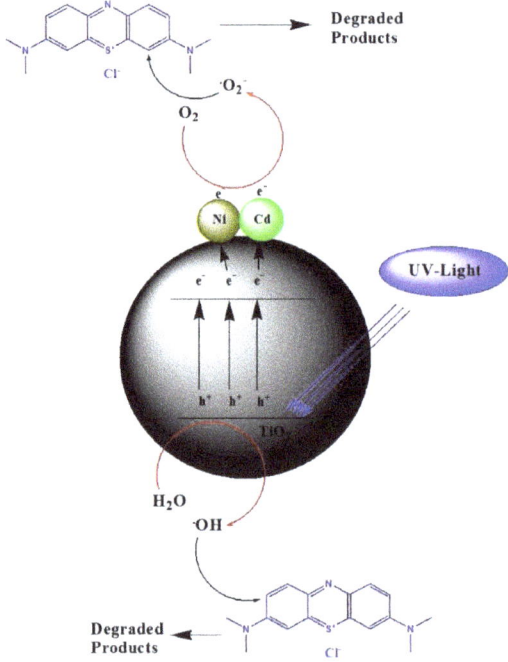

Figure 11. Proposed mechanism of photodegradation of MB dye by $NiCdTiO_2$.

2.5.2. Effect of Photocatalyst Dosage

As photocatalytic activity is greatly affected by the available active sites, the effect of catalyst dosage on the dye degradation was, therefore, also investigated. Different amounts of catalysts (0.010, 015, 0.020, 0.025 and 0.030 g) were taken with a fixed irradiation time (30 min) and dye amount (10 mL) and the results for the degradation process as monitored by UV-Vis spectroscopy are shown in Figure 12a–c for TiO_2, $CdTiO_2$ and $NiCdTiO_2$ NPs, respectively. The %degradation of MB dye by TiO_2, $CdTiO_2$ and $NiCdTiO_2$ is compared in the Figure 12d. With an increase in catalyst dosage, the %photodegradation was also increased. Maximum photodegradation was achieved with 0.030 g of the catalyst. With this amount, 76.5, 83.5 and 86% dye was degraded by TiO_2, $CdTiO_2$ and $NiCdTiO_2$, respectively. Further increases in catalyst dosage beyond the limit (0.030 g) had no significance on the enhancement of the photocatalytic activity for the degradation process. No further increase in the catalytic activity could be attributed to the agglomeration of photocatalysts beyond the optimum dosage [16].

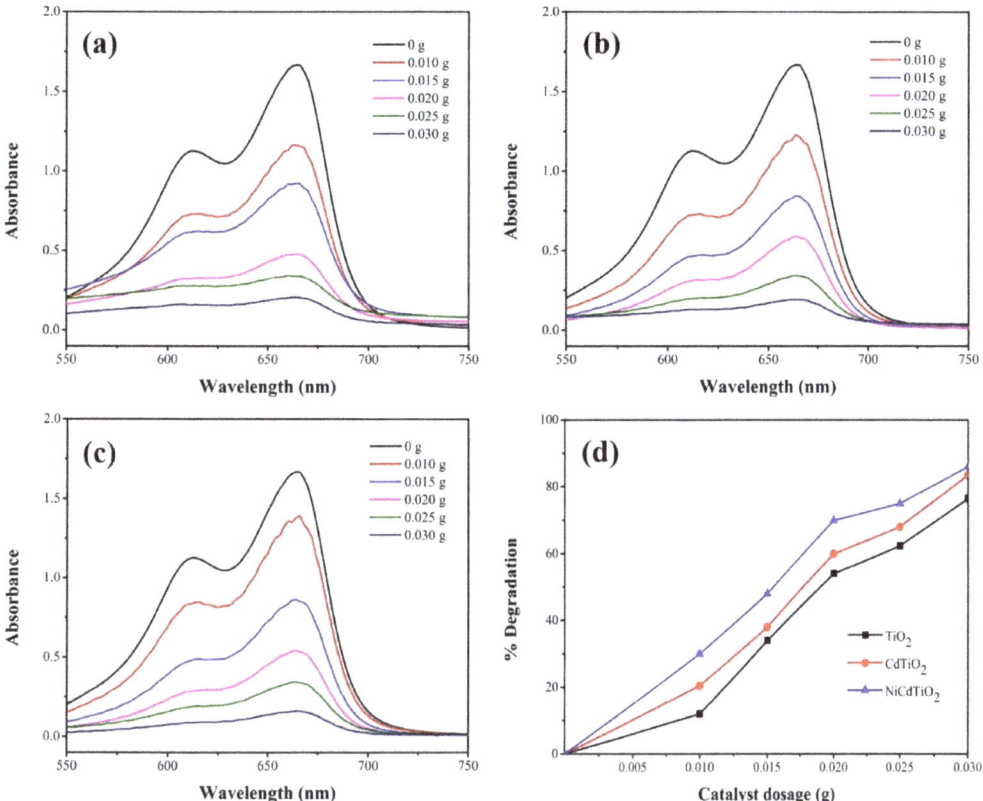

Figure 12. UV-VIS absorption spectra of MB dye before and after different dosage of (**a**) TiO$_2$ (**b**) CdTiO$_2$ (**c**) NiCdTiO$_2$ (**d**) Comparison of %degradation of MB dye photodegraded by different dosage of photocatalysts.

2.5.3. Effect of pH of the Medium

pH has an important role in the production of hydroxyl radicals responsible for the photodegradation process. Therefore, the effect of pH variation (2, 4, 6, 8, and 10) for a constant photocatalyst dosage (0.02 g) and irradiation time (30 min) on the photodegradation process was investigated and the results are shown in Figure 13a–c. The %degradation of the MB dye in different pH media are compared for the catalysts in the Figure 13d. The degradation process was low in acidic media and at pH 2 only 15, 22.07 and 26% MB dye was degraded by TiO$_2$, CdTiO$_2$ and NiCdTiO$_2$, respectively. However, in basic media the degradation percentages of the MB dye were quite high and at pH 10 about 79, 80.1 and 85.71% dye was degraded by TiO$_2$, CdTiO$_2$ and NiCdTiO$_2$, respectively. The cationic MB dye favors a high pH value for the adsorption process, which leads to an improved photocatalytic degradation [16].

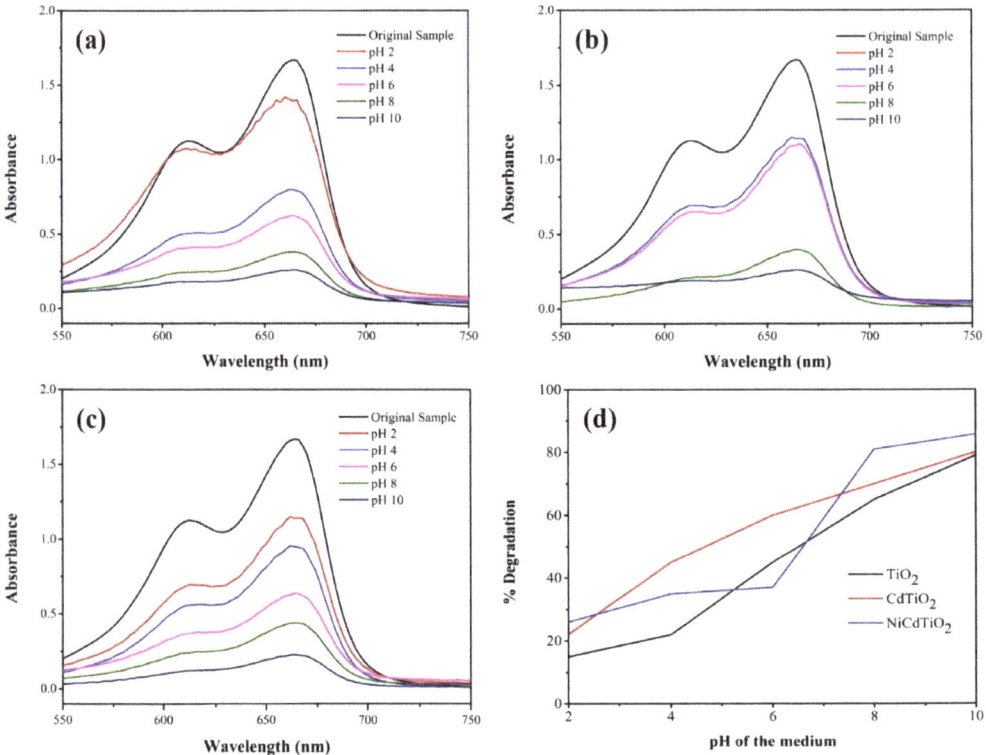

Figure 13. UV–VIS spectra of MB dye photodegraded in different pH medium in the presence of (**a**) TiO$_2$ (**b**) CdTiO$_2$ (**c**) NiCdTiO$_2$ (**d**) Comparison of %degradation of MB dye.

2.6. Photodegradation of Methyl Green (MG) Dye

2.6.1. Effect of Irradiation Time

The photodegradation efficacy of the synthesized catalysts was also investigated against methyl green (MG) dye in aqueous solutions in the presence of UV light. Figure 14a–c demonstrates the UV–VIS spectra of MG dye before reaction and after different UV light irradiation times in the presence of TiO$_2$, CdTiO$_2$ and NiCdTiO$_2$, respectively. Figure 14d represents the %degradation of MG dye at varying irradiation times in the presence of catalysts. The graph clearly demonstrates that MG photodegradation increases effectively with increasing UV irradiation time. The %degradation results show that about 28, 45 and 59.5% of the MG dye was photodegraded by TiO$_2$, NiTiO$_2$ and CdNiTiO$_2$, respectively, within 20 min. The %degradation was increased to 63.3, 88 and 97.5% by TiO$_2$, NiTiO$_2$ and CdNiTiO$_2$, respectively, when irradiation time was increased to 100 min. The results clearly demonstrate that an increase in irradiation time results in an increased dye degradation due to availability of more and more time for dye adsorption followed by photodegradation.

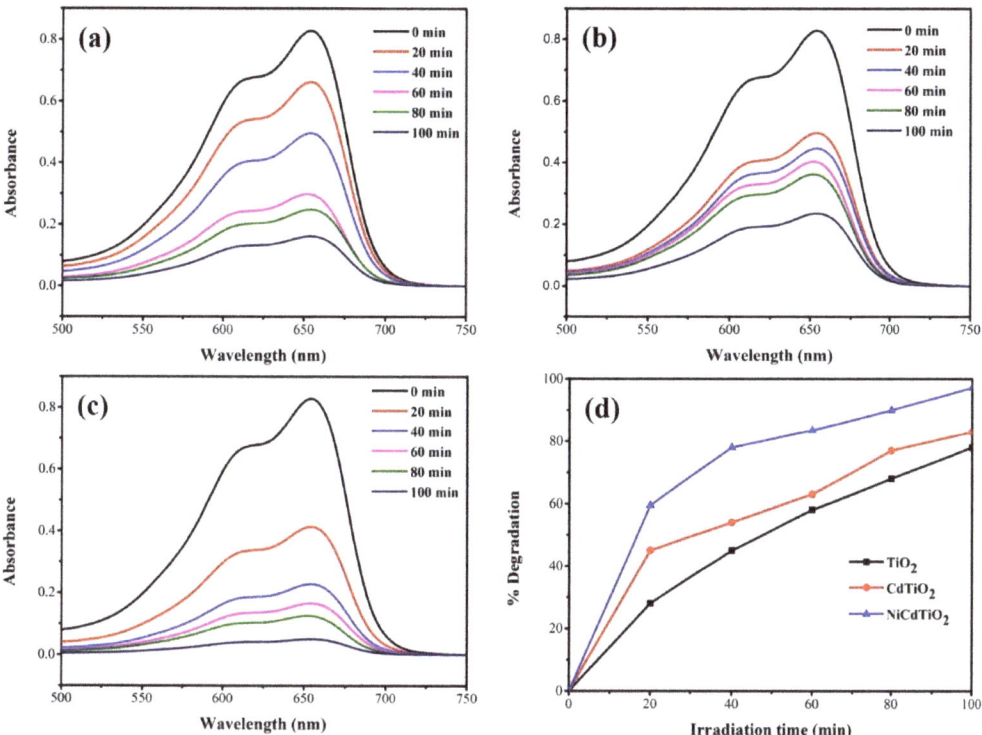

Figure 14. UV–VIS absorption spectra of MG degraded in the presence of (**a**) TiO$_2$ (**b**) CdTiO$_2$ (**c**) NiCdTiO$_2$ (**d**) Comparison of %degradation of MG dye.

2.6.2. Effect of Photocatalyst Dosage and Medium pH

The effect of catalyst dosage was evaluated by applying different dosages of photocatalysts keeping another parameters constant. Figure 15a represents the %degradation of MG dye photodegraded by different dosages of the catalysts The results show increased percent degradation of the dye with increased photocatalyst dosage. A 68, 87 and 98% degradation was achieved by 0.030 g (maximum dosage) of the TiO$_2$, CdTiO$_2$ and NiCdTiO$_2$, respectively, within 30 min. The effect of pH on Mg dye degradation was also evaluated by degrading dye in different pH solutions keeping other parameters constant. Figure 15b represents the %photocatalytic degradation of MG dye in different pH media. At pH 2, the TiO$_2$, NiTiO$_2$ and CdNiTiO$_2$ degraded about 12, 20 and 26% MG dye, respectively. The efficiency of MG dye degradation increases and about 67, 85 and 96.5% dye degraded at pH 10 by TiO$_2$, NiTiO$_2$ and CdNiTiO$_2$, respectively, within 30 min. The increased degradation of MG dye at higher pH is due to the production of more hydroxyl radicals in the basic medium.

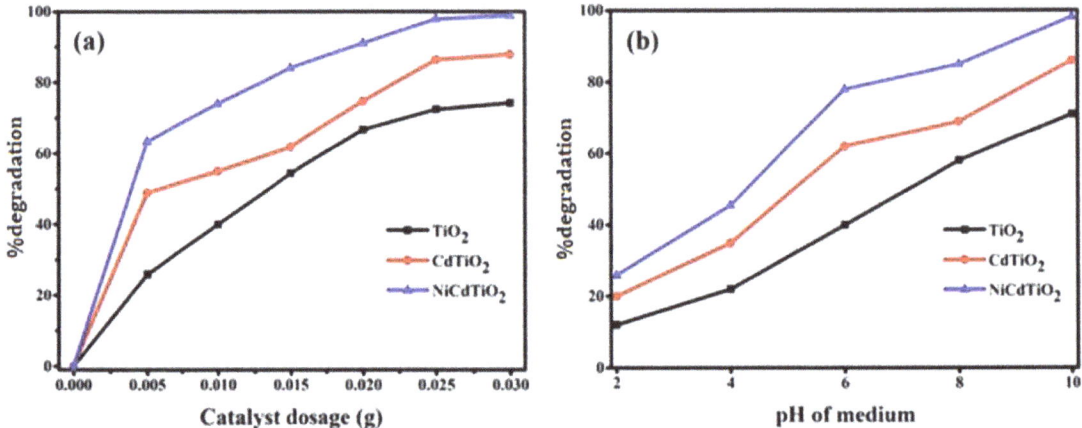

Figure 15. %degradation of MG dye photodegradation (**a**) by different dosage of photocatalysts and (**b**) at different pH.

2.7. Photocatalytic Activity Comparison

The comparative photocatalytic efficiency of the catalysts in the photodegradation of MB and MG dyes is shown as %degradation in the Figure 16. The dyes were irradiated for 100 min in the presence of catalysts. The data shows that neat TiO_2 is more efficient in degrading MB compared to MG dye. However, the nanocomposites are more effective in degrading MG dye compared to the MB dye. The ternary $NiCdTiO_2$ has shown the highest photocatalytic efficiency and degrades about 97.5% and 86% of MG and MB dye, respectively. The present study shows supremacy over the reported results [45–51] due to the photodegradation capability in correlation with band gap energy as presented in Table 2.

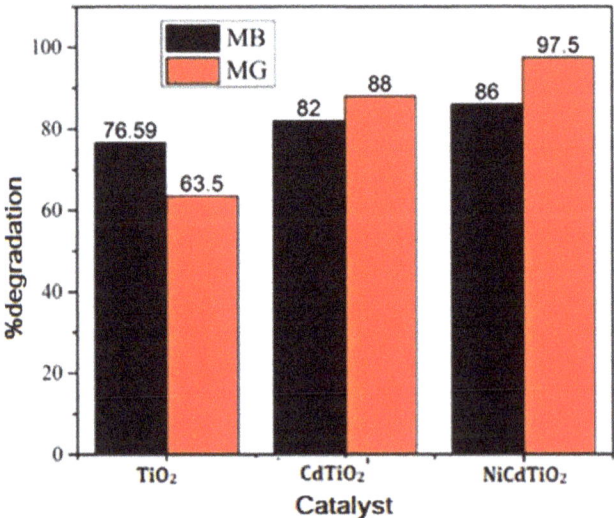

Figure 16. Comparison of %degradation of MB and MG dyes photodegraded by all the photocatalysts.

Table 2. Comparative analysis of the synthesized catalysts with the reported studies.

No.	Catalysts	Dye	Band Gap (eV)	Irradiation (nm)	Time (min)	Conversion (%)	Ref
1	Ni-Cr/TiO_2	methylene B	2.45	Sunlight	90	95.6	1
2	B-La/TiO_2	methyl O	-	Sunlight	90	98	2
3	Ag-MoO_3/TiO_2	methyl O	2.89	UV	330	95.6	3
4	Sn-La/TiO_2	rhodamine B	3.17	UV	180	98	4
5	Ce-B/TiO_2	malachite G	3.42	UV	120	90	5
6	CdS/TiO_2	Acid B	2.55	UV	90	84	6
7	Ni/TiO_2	methylene B	2.92	UV	240	99	7
8	Ni-Cd/TiO_2	methylene G	2.52	UV	100	97.5	Work

2.8. Lifespan of the Catalyst

The catalyst was regenerated from the reaction medium by simple centrifugation process and washed and dried in an oven overnight. The regenerated catalysts were utilized subsequently for five experimental runs under the optimum reaction conditions with no significant loss of activity as shown in Figure 17. The extended lifespan of the catalysts revealed the industrial scale applicability of the catalysts.

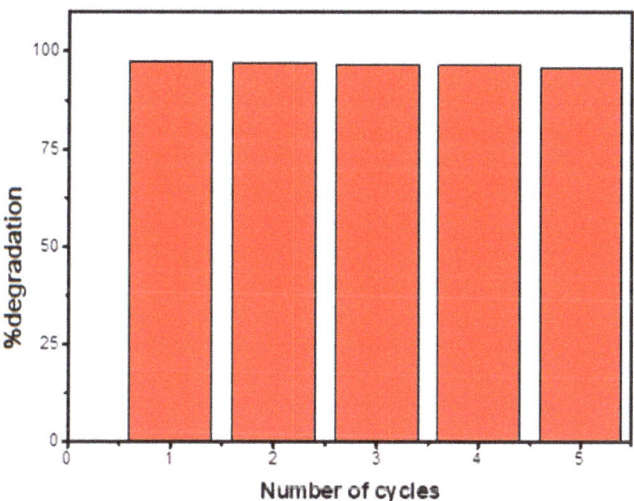

Figure 17. Recycling of catalyst for five experimental runs.

3. Experimental Work

3.1. Materials

Titanium (IV) isopropoxide ($C_{12}H_{28}O_4Ti$) (97%), cadmium (II) chloride hemi(pentahydrate) ($CdCl_2 \cdot 2.5H_2O$) (98%), and nickel (II) chloride hexahydrate ($NiCl_2 \cdot 6H_2O$) (98%) were purchased from Riedel-de Haen, Germany. Analytical grade sodium hydroxide (NaOH) (98%), methylene blue (98%), and methyl green (97.5%) were obtained from Sigma Aldrich, USA.

3.2. Synthesis of TiO_2 Nanoparticles

The precipitation procedure was utilized for the preparation of TiO_2 NPs by titanium (IV) isopropoxide precursor. A careful addition of 5 mL of titanium (IV) isopropoxide to 8 mL deionized water at 45 °C was followed by constant stirring for 1 h and resulted in white precipitate formation. The precipitates were centrifuged and washed three times with deionized water followed by methanol washings. The resultant powder was dried in an oven at 80 °C for 8 h followed by calcination at about 450 °C for 5 h.

3.3. Synthesis of Binary CdTiO$_2$ Nanocomposites

CdTiO$_2$ nanocomposites were synthesized via co-precipitation method. Cadmium (II) chloride hemi(pentahydrate) (0.030 mol, 6.85 g) was added to the deionized water (250 mL) and the mixture was stirred for 30 min at room temperature. The pH (9–10) of the mixture was adjusted by dropwise addition of 1 M NaOH solution. The precipitates formed during the process were centrifuged, separated and washed with deionized water followed by ethanol washing. The precipitates were dried in an oven at 80 °C for 5 h and annealed at 500 °C for 4 h. The resultant CdO NPs (0.01 M) were dispersed in methanol (5 mL) via sonication for 30 min. The dispersion was added to the TiO$_2$ (1 M) dispersion made in 5 mL methanol. The mixture was stirred at 80 °C for 30 min and the precipitates formed were separated via centrifuge. The CdTiO$_2$ binary nanocomposites obtained were kept overnight and oven dried at 100 °C for 2 h.

3.4. Synthesis of Ternary NiCdTiO$_2$ Nanocomposites

Ternary NiCdTiO$_2$ nanocomposites were prepared via co-precipitation method. Nickel (II) chloride hexahydrate (0.030 mol, 7.13 g) and cadmium (II) chloride hemi(pentahydrate) (0.030 mol, 6.85 g) were dissolved separately in deionized water (250 mL) and stirred for 30 min at ambient temperature. The pH (9–10) of the solutions were adjusted via dropwise addition of 1 M NaOH solution. The precipitates formed during the process were centrifuged and washed with deionized water followed by methanol three times. The obtained precipitates were oven dried at 80 °C for 6 h and annealed at 550 °C for 4 h. The synthesized Ni (0.001 M) and Cd (0.01 M) NPs were dispersed separately in 5 mL aqueous methanol via sonication for 30 min. The Ni and Cd dispersions were mixed with TiO$_2$ (1 M) solution made in aqueous methanol. The mixture was sonicated continuously at 80 °C for 30 min. The precipitates were centrifuged and the product NiCdTiO$_2$ was kept overnight and then dried at 100 °C for 5 h.

3.5. Characterization

The morphology and surface analysis were achieved through SEM (Model No. JEOL-5910; company Japan) and TEM (Tecnai F-20 FEI, USA). The elemental analysis was carried out via EDX (model INCA 200/oxford instrument, UK), while the size and phase of crystals were evaluated through XRD measurements (Model JEOL-300). The surface area and the porosity of the materials were evaluated applying a N$_2$ adsorption instrument (Micrometrics ASAP 2020). The photocatalytic degradation of the dyes was monitored using UV-VIS spectrophotometer (UV-1800, Shimadzu, Japan).

3.6. Dyes Photodegradation

The photocatalytic activity of the prepared catalysts was evaluated for the photodegradation (PD) of methylene blue (MB) and methyl green (MG) dyes in aqueous media under UV-light. The photocatalysts were individually added to 5 mL deionized water and sonicated constantly for 30 min, followed by an addition of 10 mL dye solutions (30 ppm) to each beaker. The beakers were covered with transparent plastic sheets. The mixtures were again sonicated for 30 min, kept in the dark to attain the adsorption/desorption equilibria and then kept under UV light (UV lamp (254 nm, 15 W)) with constant stirring for specific times. The catalysts were carefully removed through centrifugation (1200 rpm, 10 min). The study was completed by evaluating the effect of irradiation time (20, 40, 60, 80 and 100 min with a 0.03 g of catalyst), photocatalyst dosage (0.010, 0.015, 0.020, 0.025 and 0.03 g at 30 min irradiation time) and medium pH (2, 4, 6, 8 and 10 with 0.03 g catalyst and 30 min irradiation time) on dyes degradation. The degradation study was followed by UV-Vis spectrophotometer and the %degradation was found by applying the equation:

$$\text{Degradation rate (\%)} = \left(\frac{A_o - A}{A_o}\right) \times 100 \quad (2)$$

where A_o and A show dye absorbance before and after UV light irradiation, respectively. While for photocatalytic degradation, a pseudo first-order kinetic model was proposed.

$$-\frac{dC}{dt} = k_{ap}C \quad (3)$$

$$\ln\left(\frac{C}{C_0}\right) = k_{ap}t \quad (4)$$

where C_0: initial concentration, C: final concentration, k_{ap}: apparent rate constant and t: time in minutes.

4. Conclusions

In the present study, neat TiO_2 nanoparticles and its binary $CdTiO_2$ and ternary $NiCdTiO_2$ nanocomposites were synthesized by precipitation method. The synthesis of the TiO_2 and their binary and ternary nanocomposites were confirmed through various instrumental techniques. The synthesized materials have shown significant photocatalytic degradation of MB and MG dyes. The photodegradation results demonstrate that the dyes degradation increases with increasing irradiation time, catalyst dosage and pH of the medium. The increase in the degradation of the dyes with increasing photocatalyst dosage is due to more available active sites for dye adsorption followed by photodegradation. Similarly, the rapid photodegradation of the dyes at higher pH values could be a result of more reactive hydroxyl radicals formation. The coupled $CdTiO_2$ and $NiCdTiO_2$ nanocomposites result in the synergistic effect and have shown better photocatalytic efficiency compared to the neat TiO_2 nanoparticles. The $NiCdTiO_2$ nanocomposite is the most efficient photocatalyst for the degradation of MG (97.5% degraded in 100 min) as compared to MB (86% degraded in 100 min).

Author Contributions: Conceptualization and methodology, S.K., I.K.; Initial manuscript preparation, M.M.; Review and editing of the manuscript, M.S., A.N.; Investigation and supervision, N.M. All authors have read and agreed to the published version of the manuscript.

Funding: This research is supported by the Deanship of Scientific Research, Vice Presidency for Graduate Studies and Scientific Research, King Faisal University, Saudi Arabia (grant no. Grant2058).

Data Availability Statement: Not applicable.

Acknowledgments: The authors are grateful to Abdul Wali Khan University Mardan, for supporting the work.

Conflicts of Interest: The authors of this work have no conflict of interest to declare.

References

1. Saeed, M.; Muneer, M.; ul Haq, A.; Akram, N. Photocatalysis: An effective tool for photodegradation of dyes—A review. *Environ. Sci. Pollut. Res.* **2022**, *29*, 293–311. [CrossRef] [PubMed]
2. Köktürk, M.; Altindağ, F.; Ozhan, G.; Çalimli, M.H.; Nas, M.S. Textile dyes Maxilon blue 5G and Reactive blue 203 induce acute toxicity and DNA damage during embryonic development of *Danio rerio*. *Comp. Biochem. Physiol. Part C Toxicol. Pharmacol.* **2021**, *242*, 108947. [CrossRef] [PubMed]
3. Slama, H.B.; Bouket, A.C.; Pourhassan, Z.; Alenezi, F.N.; Silini, A.; Cherif-Silini, H.; Oszako, T.; Luptakova, L.; Golińska, P.; Belbahri, L. Diversity of Synthetic Dyes from Textile Industries, Discharge Impacts and Treatment Methods. *Appl. Sci.* **2021**, *11*, 6255. [CrossRef]
4. Khan, I.; Sadiq, M.; Khan, I.; Saeed, K. Manganese dioxide nanoparticles/activated carbon composite as efficient UV and visible-light photocatalyst. *Environ. Sci. Pollut. Res.* **2019**, *26*, 5140–5154. [CrossRef] [PubMed]

5. Gita, S.; Shukla, S.P.; Deshmukhe, G.; Choudhury, T.G.; Saharan, N.; Singh, A.K. Toxicity Evaluation of Six Textile Dyes on Growth, Metabolism and Elemental Composition (C, H, N, S) of Microalgae *Spirulina platensis*: The Environmental Consequences. *Bull. Environ. Contam. Toxicol.* **2021**, *106*, 302–309. [CrossRef]
6. Sharma, S.; Sharma, G.; Kumar, A.; AlGarni, T.S.; Naushad, M.; ALOthman, Z.A.; Stadler, F.J. Adsorption of cationic dyes onto carrageenan and itaconic acid-based superabsorbent hydrogel: Synthesis, characterization and isotherm analysis. *J. Hazard. Mater.* **2022**, *421*, 126729. [CrossRef]
7. Abdelhamid, A.E.; El-Sayed, A.A.; Khalil, A.M. Polysulfone nanofiltration membranes enriched with functionalized graphene oxide for dye removal from wastewater. *J. Polym. Eng.* **2020**, *40*, 833–841. [CrossRef]
8. Li, M.; He, Z.; Xu, J. A comparative study of ozonation on aqueous reactive dyes and reactive-dyed cotton. *Color. Technol.* **2021**, *137*, 376–388. [CrossRef]
9. Mcyotto, F.; Wei, Q.; Macharia, D.K.; Huang, M.; Shen, C.; Chow, C.W.K. Effect of dye structure on color removal efficiency by coagulation. *Chem. Eng. J.* **2021**, *405*, 126674. [CrossRef]
10. Srinivasan, S.; Sadasivam, S.K. Biodegradation of textile azo dyes by textile effluent non-adapted and adapted *Aeromonas hydrophila*. *Environ. Res.* **2021**, *194*, 110643. [CrossRef]
11. Ahila, K.G.; Ravindran, B.; Muthunarayanan, V.; Nguyen, D.D.; Nguyen, X.C.; Chang, S.W.; Nguyen, V.K.; Thamaraiselvi, C. Phytoremediation Potential of Freshwater Macrophytes for Treating Dye-Containing Wastewater. *Sustainability* **2020**, *13*, 329. [CrossRef]
12. Khan, I.; Zada, N.; Khan, I.; Sadiq, M.; Saeed, K. Enhancement of photocatalytic potential and recoverability of Fe_3O_4 nanoparticles by decorating over monoclinic zirconia. *J. Environ. Health Sci. Eng.* **2020**, *18*, 1473–1489. [CrossRef] [PubMed]
13. Kurian, M. Advanced oxidation processes and nanomaterials—A review. *Clean. Eng. Technol.* **2021**, *2*, 100090. [CrossRef]
14. Izghri, Z.; Enaime, G.; Eouarrat, M.; Chahid, L.; El Gaini, L.; Baçaoui, A.; Yaacoubi, A. Hydroxide Sludge/Hydrochar-Fe Composite Catalysts for Photo-Fenton Degradation of Dyes. *J. Chem.* **2021**, *2021*, 5588176. [CrossRef]
15. Khan, I.; Khan, A.A.; Khan, I.; Usman, M.; Sadiq, M.; Ali, F.; Saeed, K. Investigation of the photocatalytic potential enhancement of silica monolith decorated tin oxide nanoparticles through experimental and theoretical studies. *New J. Chem.* **2020**, *44*, 13330–13343. [CrossRef]
16. Khan, I.; Saeed, K.; Ali, N.; Khan, I.; Zhang, B.; Sadiq, M. Heterogeneous photodegradation of industrial dyes: An insight to different mechanisms and rate affecting parameters. *J. Environ. Chem. Eng.* **2020**, *8*, 104364. [CrossRef]
17. Dodoo-Arhin, D.; Asiedu, T.; Agyei-Tuffour, B.; Nyankson, E.; Obada, D.; Mwabora, J.M. Photocatalytic degradation of Rhodamine dyes using zinc oxide nanoparticles. *Mater. Today Proc.* **2021**, *38*, 809–815. [CrossRef]
18. Najjar, M.; Hosseini, H.A.; Masoudi, A.; Sabouri, Z.; Mostafapour, A.; Khatami, M.; Darroudi, M. Green chemical approach for the synthesis of SnO_2 nanoparticles and its application in photocatalytic degradation of Eriochrome Black T dye. *Optik* **2021**, *242*, 167152. [CrossRef]
19. Chairungsri, W.; Subkomkaew, A.; Kijjanapanich, P.; Chimupala, Y. Direct dye wastewater photocatalysis using immobilized titanium dioxide on fixed substrate. *Chemosphere* **2022**, *286*, 131762. [CrossRef]
20. Lee, S.Y.; Kang, D.; Jeong, S.; Do, H.T.; Kim, J.H. Photocatalytic Degradation of Rhodamine B Dye by TiO_2 and Gold Nanoparticles Supported on a Floating Porous Polydimethylsiloxane Sponge under Ultraviolet and Visible Light Irradiation. *ACS Omega* **2020**, *5*, 4233–4241. [CrossRef]
21. Pang, S.; Pang, S.; Lu, Y.; Lu, Y.; Cheng, L.; Cheng, L.; Liu, J.; Liu, J.; Ma, H.; et al. Facile synthesis of oxygen-deficient nano-TiO_2 coordinated by acetate ligands for enhanced visible-light photocatalytic performance. *Catal. Sci. Technol.* **2020**, *10*, 3875–3889. [CrossRef]
22. Padmanabhan, N.T.; Thomas, N.; Louis, J.; Mathew, D.T.; Ganguly, P.; John, H.; Pillai, S.C. Graphene coupled TiO_2 photocatalysts for environmental applications: A review. *Chemosphere* **2021**, *271*, 129506. [CrossRef] [PubMed]
23. Barzegar, M.H.; Sabzehmeidani, M.M.; Ghaedi, M.; Avargani, V.M.; Moradi, Z.; Roy, V.A.L.; Heidari, H. S-scheme heterojunction g-C_3N_4/TiO_2 with enhanced photocatalytic activity for degradation of a binary mixture of cationic dyes using solar parabolic trough reactor. *Chem. Eng. Res. Des.* **2021**, *174*, 307–318. [CrossRef]
24. Tian, X.; Cui, X.; Lai, T.; Ren, J.; Yang, Z.; Xiao, M.; Wang, B.; Xiao, X.; Wang, Y. Gas sensors based on TiO_2 nanostructured materials for the detection of hazardous gases: A review. *Nano Mater. Sci.* **2021**, *3*, 390–403. [CrossRef]
25. Heo, B.; Ha, J.; Kim, Y.T.; Choi, J. 10 μm-thick MoO_3-coated TiO_2 nanotubes as a volume expansion regulated binder-free anode for lithium ion batteries. *J. Ind. Eng. Chem.* **2021**, *96*, 364–370. [CrossRef]
26. Kumar, R.; Kumar, R.; Singh, B.K.; Soam, A.; Parida, S.; Sahajwalla, V.; Bhargava, P. In situ carbon-supported titanium dioxide (ICS-TiO_2) as an electrode material for high performance supercapacitors. *Nanoscale Adv.* **2020**, *2*, 2376–2386. [CrossRef]
27. Ziaeifar, F.; Alizadeh, A.; Shariatinia, Z. Dye sensitized solar cells fabricated based on nanocomposite photoanodes of TiO_2 and $AlMo_{0.5}O_3$ perovskite nanoparticles. *Sol. Energy* **2021**, *218*, 435–444. [CrossRef]
28. Kwon, S.H.; Kim, T.H.; Kim, S.M.; Oh, S.; Kim, K.K. Ultraviolet light-emitting diode-assisted highly sensitive room temperature NO_2 gas sensors based on low-temperature solution-processed ZnO/TiO_2 nanorods decorated with plasmonic Au nanoparticles. *Nanoscale* **2021**, *13*, 12177–12184. [CrossRef]
29. Badvi, K.; Javanbakht, V. Enhanced photocatalytic degradation of dye contaminants with TiO_2 immobilized on ZSM-5 zeolite modified with nickel nanoparticles. *J. Clean. Prod.* **2021**, *280*, 124518. [CrossRef]

30. Veziroglu, S.; Obermann, A.L.; Ullrich, M.; Hussain, M.; Kamp, M.; Kienle, L.; Leißner, T.; Rubahn, H.G.; Polonskyi, O.; Strunskus, T.; et al. Photodeposition of Au Nanoclusters for Enhanced Photocatalytic Dye Degradation over TiO_2 Thin Film. *ACS Appl. Mater. Interfaces* **2020**, *12*, 14983–14992. [CrossRef]
31. Bellè, U.; Pelizzari, F.; Lucotti, A.; Castiglioni, C.; Ormellese, M.; Pedeferri, M.; Diamanti, M.V. Immobilized Nano-TiO_2 Photocatalysts for the Degradation of Three Organic Dyes in Single and Multi-Dye Solutions. *Coatings* **2020**, *10*, 919. [CrossRef]
32. Zada, N.; Saeed, K.; Khan, I. Decolorization of Rhodamine B dye by using multiwalled carbon nanotubes/Co–Ti oxides nanocomposite and Co–Ti oxides as photocatalysts. *Appl. Water Sci.* **2020**, *10*, 40. [CrossRef]
33. Khlyustova, A.; Sirotkin, N.; Kusova, T.; Kraev, A.; Titov, V.; Agafonov, A. Doped TiO_2: The effect of doping elements on photocatalytic activity. *Mater. Adv.* **2020**, *1*, 1193–1201. [CrossRef]
34. Samuel, J.J.; Yam, F.K. Photocatalytic degradation of methylene blue under visible light by dye sensitized titania. *Mater. Res. Express* **2020**, *7*, 015051. [CrossRef]
35. Pino, E.; Calderón, C.; Herrera, F.; Cifuentes, G.; Arteaga, G. Photocatalytic Degradation of Aqueous Rhodamine 6G Using Supported TiO_2 Catalysts. A Model for the Removal of Organic Contaminants From Aqueous Samples. *Front. Chem.* **2020**, *8*, 365. [CrossRef]
36. Habibi-Yangjeh, A.; Feizpoor, S.; Seifzadeh, D.; Ghosh, S. Improving visible-light-induced photocatalytic ability of TiO_2 through coupling with Bi_3O_4Cl and carbon dot nanoparticles. *Sep. Purif. Technol.* **2020**, *238*, 116404. [CrossRef]
37. Rani, M.; Keshu; Shanker, U. Efficient degradation of organic pollutants by novel titanium dioxide coupled bismuth oxide nanocomposite: Green synthesis, kinetics and photoactivity. *J. Environ. Manag.* **2021**, *300*, 113777. [CrossRef]
38. Nyankson, E.; Efavi, J.K.; Agyei-Tuffour, B.; Manu, G. Synthesis of TiO_2-Ag_3PO_4 photocatalyst material with high adsorption capacity and photocatalytic activity: Application in the removal of dyes and pesticides. *RSC Adv.* **2021**, *11*, 17032–17045. [CrossRef]
39. Chakhtouna, H.; Benzeid, H.; Zari, N.; Qaiss, A.e.k.; Bouhfid, R. Recent progress on Ag/TiO_2 photocatalysts: Photocatalytic and bactericidal behaviors. *Environ. Sci. Pollut. Res.* **2021**, *28*, 44638–44666. [CrossRef]
40. Radoor, S.; Karayil, J.; Jayakumar, A.; Parameswaranpillai, J.; Siengchin, S. Release of toxic methylene blue from water by mesoporous silicalite-1: Characterization, kinetics and isotherm studies. *Appl. Water Sci.* **2021**, *11*, 110. [CrossRef]
41. Sun, Z.; Feng, T.; Zhou, Z.; Wu, H. Removal of methylene blue in water by electrospun PAN/β-CD nanofibre membrane. *E-Polymers* **2021**, *21*, 398–410. [CrossRef]
42. Khan, I.; Saeed, K.; Zekker, I.; Zhang, B.; Hendi, A.H.; Ahmad, A.; Ahmad, S.; Zada, N.; Ahmad, H.; Shah, L.A.; et al. Review on Methylene Blue: Its Properties, Uses, Toxicity and Photodegradation. *Water* **2022**, *14*, 242. [CrossRef]
43. Sultana, S.; Rafiuddin; Khan, M.Z.; Umar, K.; Ahmed, A.S.; Shahadat, M. SnO_2–SrO based nanocomposites and their photocatalytic activity for the treatment of organic pollutants. *J. Mol. Struct.* **2015**, *1098*, 393–399. [CrossRef]
44. Haque, M.M.; Khan, A.; Umar, K.; Mir, N.A.; Muneer, M.; Harada, T.; Matsumura, M. Synthesis, Characterization and Photocatalytic Activity of Visible Light Induced Ni-Doped TiO_2. *Energy Environ. Focus* **2013**, *2*, 73–78. [CrossRef]
45. Shaban, M.; Ahmed, A.M.; Shehata, N.; Betiha, M.A.; Rabie, A.M. Ni-doped and Ni/Cr co-doped TiO_2 nanotubes for enhancement of photocatalytic degradation of methylene blue. *J. Colloid Interface Sci.* **2019**, *555*, 31–41. [CrossRef]
46. Zhang, W.; Li, X.; Jia, G.; Gao, Y.; Wang, H.; Cao, Z.; Liu, J. Preparation, characterization, and photocatalytic activity of boron and lanthanum co-doped TiO_2. *Catal. Commun.* **2014**, *45*, 144–147. [CrossRef]
47. Kader, S.; Al-Mamun, M.R.; Suhan, M.B.K.; Shuchi, S.B.; Islam, M.S. Enhanced photodegradation of methyl orange dye under UV irradiation using MoO_3 and Ag doped TiO_2 photocatalysts. *Environ. Technol. Innov.* **2022**, *27*, 102476. [CrossRef]
48. Zhu, X.; Pei, L.; Zhu, R.; Jiao, Y.; Tang, R.; Feng, W. Preparation and characterization of Sn/La co-doped TiO_2 nanomaterials and their phase transformation and photocatalytic activity. *Sci. Rep.* **2018**, *8*, 12387. [CrossRef]
49. Stoyanova, A.; Ivanova, N.; Bachvarova-Nedelcheva, A.; Christov, C. Synthesis and photocatalytic activity of cerium-doped and cerium-boron co-doped TiO_2 nanoparticles. *J. Chem. Technol. Metall.* **2021**, *56*, 1294–1302.
50. Qutub, N.; Singh, P.; Sabir, S.; Sagadevan, S.; Oh, W.C. Enhanced photocatalytic degradation of Acid Blue dye using CdS/TiO_2 nanocomposite. *Sci. Rep.* **2022**, *12*, 5759. [CrossRef]
51. Ganesh, I.; Gupta, A.K.; Kumar, P.P.; Sekhar, P.S.C.; Radha, K.; Padmanabham, G.; Sundararajan, G. Preparation and Characterization of Ni-Doped TiO_2 Materials for photocurrent and photocatalytic applications. *Sci. World J.* **2012**, *2012*, 127326. [CrossRef] [PubMed]

Disclaimer/Publisher's Note: The statements, opinions and data contained in all publications are solely those of the individual author(s) and contributor(s) and not of MDPI and/or the editor(s). MDPI and/or the editor(s) disclaim responsibility for any injury to people or property resulting from any ideas, methods, instructions or products referred to in the content.

Article

Auramine O UV Photocatalytic Degradation on TiO$_2$ Nanoparticles in a Heterogeneous Aqueous Solution

Cristina Pei Ying Kong [1], Nurul Amanina A. Suhaimi [1], Nurulizzatul Ningsheh M. Shahri [1], Jun-Wei Lim [2,3], Muhammad Nur [4], Jonathan Hobley [5] and Anwar Usman [1,*]

[1] Department of Chemistry, Faculty of Science, Universiti Brunei Darussalam, Jalan Tungku Link, Gadong BE1410, Brunei
[2] HICoE-Centre for Biofuel and Biochemical Research, Institute of Self-Sustainable Building, Department of Fundamental and Applied Sciences, Universiti Teknologi PETRONAS, Seri Iskandar 32610, Perak Darul Ridzuan, Malaysia
[3] Department of Biotechnology, Saveetha School of Engineering, Saveetha Institute of Medical and Technical Sciences, Chennai 602105, India
[4] Center for Plasma Research, Integrated Laboratory, Universitas Diponegoro, Tembalang Campus, Semarang 50275, Indonesia
[5] Department of Biomedical Engineering, National Cheng Kung University, No. 1 University Road, Tainan City 701, Taiwan
* Correspondence: anwar.usman@ubd.edu.bn

Abstract: Amongst the environmental issues throughout the world, organic synthetic dyes continue to be one of the most important subjects in wastewater remediation. In this paper, the photocatalytic degradation of the dimethylmethane fluorescent dye, Auramine O (AO), was investigated in a heterogeneous aqueous solution with 100 nm anatase TiO$_2$ nanoparticles (NPs) under 365 nm light irradiation. The effect of irradiation time was systematically studied, and photolysis and adsorption of AO on TiO$_2$ NPs were also evaluated using the same experimental conditions. The kinetics of AO photocatalytic degradation were pseudo-first order, according to the Langmuir–Hinshelwood model, with a rate constant of 0.048 ± 0.002 min^{-1}. A maximum photocatalytic efficiency, as high as 96.2 ± 0.9%, was achieved from a colloidal mixture of 20 mL (17.78 µmol L^{-3}) AO solution in the presence of 5 mg of TiO$_2$ NPs. The efficiency of AO photocatalysis decreased nonlinearly with the initial concentration and catalyst dosage. Based on the effect of temperature, the activation energy of AO photocatalytic degradation was estimated to be 4.63 kJ mol^{-1}. The effect of pH, additional scavengers, and H$_2$O$_2$ on the photocatalytic degradation of AO was assessed. No photocatalytic degradation products of AO were observed using UV–visible and Fourier transform infrared spectroscopy, confirming that the final products are volatile small molecules.

Keywords: Auramine O; basic dye; titania nanoparticle; photocatalytic degradation; mechanism

1. Introduction

Water is indispensable for sustaining the environment, keeping entire ecosystems regulated. Water is also an important natural resource and a vital asset for daily human life, as it is used for drinking, hygiene, and cooking, as well as in agriculture and fisheries. Although accessible clean water is crucially important over all regions, about one-sixth of the global population have difficulties in accessing clean water [1]. Additionally, it has also been claimed that four billion people are now facing a severe scarcity of clean water due to extinction, depletion, and pollution in major rivers of the world [2]. The presence of sediments, soil, and aquatic organisms, which are naturally produced by erosion of rock and soil, and the breakdown and rotting of organic matters, is related to water quality [3]. However, the presence of organic pollutants in water systems is even more dangerous, as they contaminate entire ecosystems and endanger human health. With this in mind, water pollution continues to be one of the biggest issues humanity faces, especially with rapid population growth and increasing burdens from economic growth leading to increased industrial activity.

Although industrial pollutants are often handled in treatment or storage systems, they may eventually leach into the surrounding urban areas by rainfall, entering sewage systems and water reservoirs. This contamination degrades the quality of water, and has therefore been singled out as one of the reasons for economic slowdown in many developing countries [4]. One of the most serious classes of environmental pollutants found in water systems is synthetic dyes, which are used intensively in the dying process in the textile, paper, leather, plastic, and rubber industries, due to their vibrant colors and low cost [5,6]. Notably, over 50% of the global dye usage, and the resulting contamination, occurs in developing regions of Asia [7]. Even the presence of low concentrations of dyes can greatly affect aquatic life and ecosystems, in terms of eutrophication and perturbations. This is because of the color intensity of dyes, which has the ability to prevent penetration of sunlight through water, resulting in a clear decline in the rate of photosynthesis, lowering dissolved oxygen levels. This increases the biochemical oxygen demand [8]. Even though strict control of water quality has been implemented in many countries, in order to fulfil the regulation of minimum allowed concentrations of pollutants, better treatments to eliminate these persistent organic compounds from industrial wastewater must also be curated.

Typically, industrial dyes are highly water-soluble so that they are cheaper and easier to use in manufacturing dying processes [9–11]. These dyes are categorized as chromophoric or auxochromic dyes, containing different moieties and functional groups which are responsible for their color intensity [12]. Among them, Auramine Orange (AO) and its derivatives, which are cationic diarylmethane dyes having yellow fluorescence and vibrant color, are widely mass produced for use in food, textile, paint, ink, plastic, and cosmetic industries [13,14]. Due to AOs' carcinogenic nature, there have been studies that examined its biotransformation to reactive species in target organs of rats and humans when administered orally [15]. Considering that AO and its derivatives cause long-term impacts on aquatic environments, as well as causing other health risks [16,17], their removal from wastewater before discharge is imperative. The treatment of such effluents would not only protect the water systems and the entire ecosystem, but also encourage manufacturers to reuse the spent water from their dyeing processes.

A variety of methods, including electro-coagulation [18], chemical precipitation, coagulation and filtration [19], reverse osmosis membrane [20], ozonation [21], aerobic and anaerobic processes [22], adsorption on activated carbon [23], and photocatalytic degradation [24], have been devised for the treatment of industrial wastewater effluents. However, amongst these methods, adsorption and photocatalysis have attracted great interest due to their cost-efficiency, sustainability, and selectivity [25,26]. Heterogenous photocatalysis has been reported to be more desirable as this method shows several advantages in the decolorization of wastewater due to the high efficiency of photocatalytic degradation in the removal of dyes from complicated organic effluents [27,28], easy waste disposal, low cost, and complete mineralization [24]. Additionally, this process can be applied in ambient or mild pressure conditions, using solar energy for power, or pre-existing natural UV light in water purification systems, in order to degrade synthetic dyes completely into less harmful byproducts [29].

The key mode of action of heterogenous photocatalysis is the degradation of dyes during a chemical reaction with photochemically generated hydroxyl ($OH^\bullet$) and oxygen ($O_2^{-\bullet}$) radicals on the surface of the photocatalyst [30–32]. Therefore, the photocatalytic degradation of synthetic dyes is strongly governed by the dynamics of adsorption onto the catalyst surface and their reactivity with the radicals once there. This can be further controlled using several parameters, such as irradiation time, the catalyst dosage, dye concentration, and H_2O_2-mediated processes. Additionally, using nanoparticles (NPs) instead of micro-powders should greatly increase reaction rates and efficiencies.

The efficiency of dye photocatalytic degradation depends on their relative redox potentials with respect to those of the catalyst, allowing electron and hole transfer to generate $O_2^{-\bullet}$ and $OH^{\bullet}$ radicals [31]. In this sense, photocatalytic degradation of AO on TiO_2 NPs and its kinetics have been reported by Montazerozohori [33]. Photocatalysis of AO on semiconductor oxides has been intensively investigated [26,34], but several aspects of the photocatalytic degradation of the dye are still deficient. In particular, photochemical systems are complicated and it takes time to elucidate systems as the literature about them builds up. In general, a lot of works are required to generate a consensus as to what is actually going on.

Therefore, in this study, photocatalytic degradation of AO on anatase TiO_2 NPs under 365 nm light irradiation was investigated. The objective was to systematically evaluate the effect of irradiation time, the initial concentration of AO, and catalyst dosage on the photocatalytic degradation of the dye. The efficiency and rate constant of the photodegradation were estimated based on absorption spectra of AO before and after irradiation. The photocatalytic degradation data were analyzed with standard empirical models. The thermodynamics of the AO photodegradation process were assessed by monitoring the effect of temperature. The photocatalytic degradation mechanism was further assessed by observing the effect of pH and additional scavengers as well as H_2O_2.

This work should provide a baseline for future works, which may include using doped and sensitized TiO_2 in order to shift the absorbance further to the visible to improve catalytic efficiency [35–39]. It is important to highlight that there are several differences between this study and those in the literature, including the use of NPs, which should give better photocatalytic degradation rates. There are also several similarities and some agreements between this current study and the reported works, which give affirmative verification of many of the conclusions from the earlier work, which is a general duty of the traditional scientific approach.

2. Results

2.1. Photolysis and Adsorption of AO

The photolysis of 35.06 μmol L^{-1} of AO in aqueous solution in the absence of TiO_2 NPs under 365 nm light irradiation is shown in Figure S1A. The absorption spectra of AO have two main peaks at 432 nm and 370 nm. The spectrum shows that the absorbance of AO solution gradually and slightly decreased over time when the AO solution was exposed to the UV irradiation, confirming that the dye was slowly decomposing. From these absorption spectra, the concentration of AO was extracted and plotted against the irradiation time, as shown in Figure S1B. It is clearly seen that the concentration of AO after 180 min of irradiation is only slightly lower than it was before irradiation, and hence, the efficiency of the noncatalytic photolysis was determined to be less than 8% even after such a prolonged irradiation time. This is conclusive evidence that the direct photolysis of AO by the 365 nm light irradiation is not efficient.

In comparison, adsorption of AO onto TiO_2 NPs was revealed by a gradual decrease in the absorbance of AO with contact time, as shown in Figure S2. Based on this decrease in absorbance up to 180 min of contact time, the adsorption efficiency of AO was found to be less than 3%, indicating that the adsorption of AO by TiO_2 NPs is also inefficient.

2.2. Photocatalytic Degradation of AO

Figure S3 shows a colloidal mixture of 20 mL AO solution with 2.5 mg TiO_2 NPs before and after UV light irradiation for 100 min. It can clearly be seen that the colloidal mixture was completely decolored, suggesting that 100% efficient photocatalytic degradation of AO had occurred within this time.

The degradation of the dye in the heterogeneous aqueous solution was then monitored at different irradiation times from 0 to 150 min, and the reduction in the color with increasing irradiation time is shown in Figure S4. The absorption spectra of the heterogeneous aqueous solutions of AO and TiO_2 NPs, after being exposed to the UV light, were measured after centrifugation. These spectra are shown in Figure 1. The concentration of AO in the heterogeneous aqueous solutions was determined based on its absorbance at 432 nm, at which value the molar decadic extinction coefficient of the dye is 25,300 L mol^{-1} cm^{-1}. The photodegradation efficiency (η), also known as color removal rate, was then calculated as

$$\eta(\%) = \frac{(C_0 - C_t)}{C_0} \times 100\% \tag{1}$$

where C_0 and C_t are the initial and remaining concentration of AO in the mixtures at irradiation time, t.

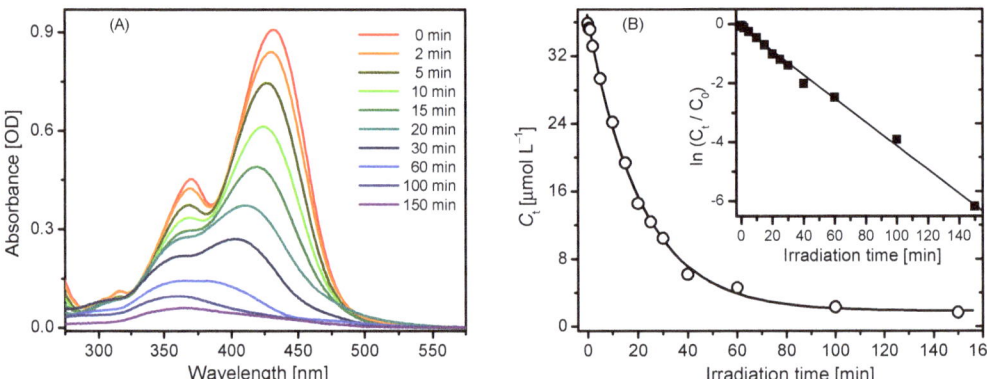

Figure 1. (**A**) Absorption spectra of AO (35.06 μmol L^{-1}) in an aqueous colloidal solution, in the presence of 5 mg TiO$_2$ NPs, after different irradiation times, as indicated; and (**B**) a plot of C_t against AO as a function of irradiation time, simulated using the Langmuir–Hinshelwood kinetic model. Inset: the linear plot of $ln(C_t/C_0)$ against irradiation time.

It was found that the dye was almost completely degraded within 150 min of irradiation, with the η value being 96.2 ± 0.9%. This is slightly higher than reported for methylene blue (MB) (93.1%) and rhodamine B (RhB) (96.1%) [28]. Considering that, alone, the non-photocatalytic photolysis and dark-adsorption of AO onto TiO$_2$ NPs were inefficient, the enhanced degradation of the dye in Figure 1 can be assigned to a photocatalytic process on the catalyst surface.

The photocatalytic degradation kinetics of AO were evaluated by simulating the experimental data using a single exponential decay function. Here, the degradation rate constant is considered to be linearly related to concentration of the dye, according to the Langmuir–Hinshelwood (L–H) model [40,41]. The L–H equation is expressed as [40–42]

$$C_t = C_0 \exp(-k_{obs}t) \tag{2}$$

where k_{obs} is the observed degradation rate constant, and is determined from the single exponential decay of C_t as a function of irradiation time, t.

As shown in Figure 1B, the data fit well with the L–H kinetic model, suggesting that the heterogeneous photocatalytic degradation of AO is a pseudo-first-order reaction. This is unambiguously supported by the linear correlation between lnC_t/C_0 as a function of irradiation time. From this best fit, the degradation rate constant, k_{obs}, of AO was estimated to be 0.048 ± 0.002 min^{-1}. In comparison, under the same experimental conditions, the k_{obs} value of AO is much slower than those of RHB (0.115 ± 0.005 min^{-1}) and MB (0.173 ± 0.019 min^{-1}) [28].

The photocatalytic degradation of AO must be proportional to the external mass transfer of the dye onto the catalyst surface. In this sense, the mass transfer behavior of AO was analyzed using the intraparticle diffusion model, given by [43]

$$C_0 - C_t = k_i t^{1/2} + C \tag{3}$$

Here, k_i is the diffusion rate and C is the boundary layer thickness on the catalyst surface.

The simulation plot shown in Figure 2A demonstrates that mass transfer occurred in three diffusion steps. There was slow diffusion with a k_i of 0.89 µmol L^{-1} min$^{-1/2}$ which occurred within 1 min of irradiation, and is associated with early diffusion of AO onto the catalyst surface. This was followed by a fast and effective diffusion with a k_i of 5.69 mmol L^{-1} min$^{-1/2}$. Finally, another slow diffusion step occurs, with a k_i of 0.89 µmol L^{-1} min$^{-1/2}$ at irradiation times longer than 40 min until the complete degradation of AO in the solution is achieved. It is noteworthy that extrapolation of the simulation plot at early irradiation times passed the origin, implying that boundary layer thickness can be assumed to be negligible. In other words, the diffusion rate of the dye in the solution was comparable to that on the catalyst surface. This also provides an interpretation that photodegradation byproducts of AO did not disturb the diffusion of the dye onto the catalyst surface.

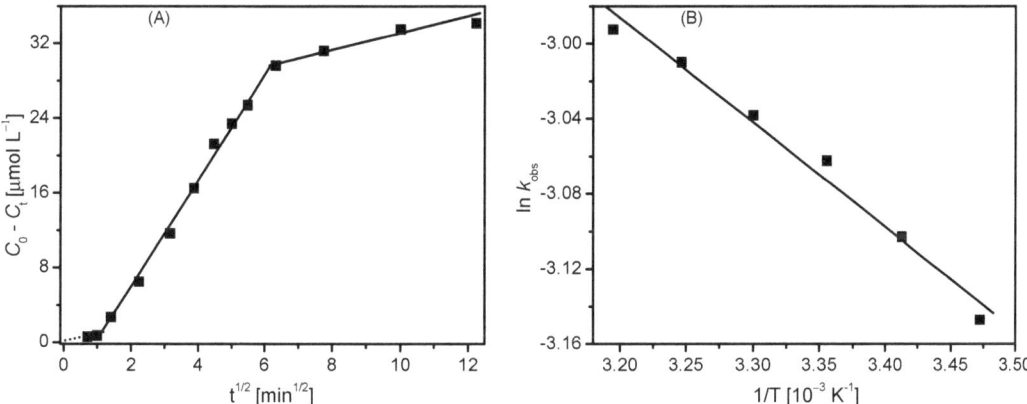

Figure 2. (**A**) Simulation of kinetic data with the intraparticle diffusion model, and (**B**) an Arrhenius plot of lnk_{obs} against 1/T for the photocatalytic degradation of AO in the heterogenous aqueous solution. The lines are the best fits. The activation energy of the photocatalytic degradation was deduced based on the Arrhenius equation.

2.3. Effect of Temperature

The photocatalytic degradation of AO depends on diffusion and immobilization of the dye onto the TiO$_2$ NPs; hence, it should be affected by temperature. Figure S5 shows absorption spectra of AO solutions before and after UV light irradiation for 30 min at different temperatures. The effect of temperature was then further analyzed based on the k_{obs} of AO photocatalytic decomposition. These results demonstrated that k_{obs} increased with the temperature, suggesting that diffusion and immobilization of the dye on the catalyst surface were accelerated at higher temperature. Additionally, electron–hole recombination is also believed to accelerate with increased temperature [27,44–46].

The activation energy (E_a) of the photocatalytic degradation of the dye was then evaluated based on the effect of temperature (15–40 °C) on k_{obs} by using the Arrhenius equation;

$$k_{obs} = A \exp(-E_a/RT) \tag{4}$$

where A is the pre-exponential factor, R is the gas constant, and T is the temperature.

Based on the Arrhenius plot of lnk_{obs} as a function of 1/T shown in Figure 2B, the E_a of photocatalytic degradation of AO on the TiO$_2$ NPs was estimated to be 4.63 kJ mol^{-1}. For comparison, under the same experimental conditions, the E_a value of the photocatalytic degradation of MB was 37.3 kJ mol^{-1} [27]. Thus, the potential barrier of the photocatalytic degradation of AO on the catalyst surface is much lower than that of MB. Therefore, it can be concluded that the oxidation reaction between AO and the generated $O_2^{-\bullet}$ and OH$^\bullet$ radicals on the catalyst surface is much more energetically favorable than it is for MB.

2.4. Effect of Various Parameters on the Photocatalytic Degradation of AO

As this photocatalytic degradation is an oxidation reaction of the dye on the catalyst surface, at a certain temperature, the reaction should depend on various parameters, including the dye concentration and catalyst dosage. Figure S6 shows the spectra of AO with different initial AO concentrations (C_0) before and after photocatalytic degradation. Based on these spectra, the concentration of AO that was degraded during the photocatalysis, and the η value, increased and reached an optimum condition with respect to concentration (C_0). This was followed by a nonlinear decrease, as seen in Figure 3. This finding highlighted that the photocatalytic activity of the dye is related to the number of dye molecules in the heterogeneous colloidal mixture, and the low η value at high initial concentration is attributed to the well-known screening effect [47–49].

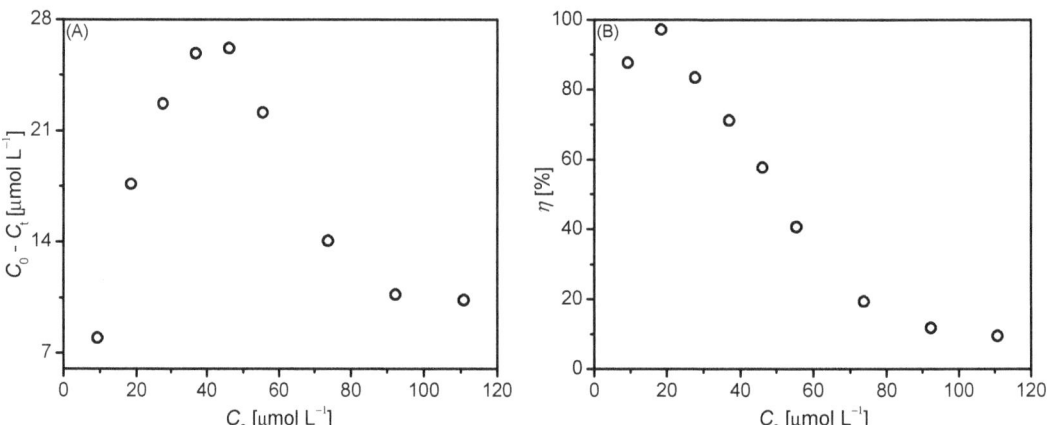

Figure 3. (**A**) Plot of $C_0 - C_t$; and (**B**) the η values of photocatalytic degradation of AO as a function of the initial AO concentration (C_0) in the colloidal mixture with 5 mg TiO$_2$ NPs after irradiation for 30 min.

As shown in Figure S7, the catalyst dosage also affects the photocatalytic degradation of AO. Based on the absorption spectra of AO before and after irradiation in the presence of different dosages of TiO$_2$ NPs, the degradation efficiency was found to decrease nonlinearly with the catalyst dosage. This phenomenon is assigned to the inefficient photocatalytic degradation of the dye at high catalyst dosages.

The photocatalytic degradation of AO was also followed in the presence of a small amount (1–5%) of benzoquinone (BQ) and tert-butanol (t-BuOH) which scavenge $O_2^{-\bullet}$ and OH$^\bullet$ radicals, respectively. It was found that the η value of AO decreases abruptly with the addition of BQ and t-BuOH, as shown in Figure 4A. This result confirms that the degradation mechanism by UV/TiO$_2$ NPs depends on the oxidation reaction of the dye with both $O_2^{-\bullet}$ and OH$^\bullet$ radicals, as has been described in several studies [50,51]. The formation of $O_2^{-\bullet}$, by reduction of solvated oxygen in the aqueous solution, is an important step to prevent the recombination of the photogenerated electrons and holes [52]. High concentrations of oxygen in the solution should reduce the recombination process and hence assist the formation of both $O_2^{-\bullet}$ and OH$^\bullet$ radicals. To explore this possibility, the effect of adding a small amount of H_2O_2 on the k_{obs} of the photocatalytic degradation of AO was evaluated, as presented in Figure 4B. The dissociation of H_2O_2 enhances the concentration of oxidants. This accelerates the generation of $O_2^{-\bullet}$ and $OH^\bullet$ radicals [53], leading to a higher photodegradation rate, although the efficiency of AO after prolonged irradiation time was almost unchanged (96.2–97.3%). Thus, the results further suggest that the generation of $O_2^{-\bullet}$ and $OH^\bullet$ radicals is the rate-determining step of the photocatalytic degradation of the dye.

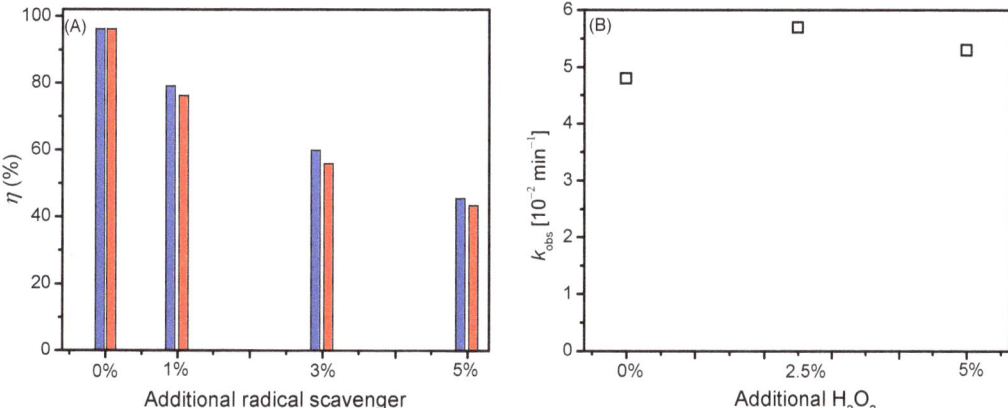

Figure 4. Plots of η values of photodegradation of AO (35.06 µmol L^{-1}) in aqueous colloidal solution in the presence of 5 mg TiO$_2$ NPs after irradiation for 30 min with the addition of (**A**) BQ (blue bars) and t-BuOH (red bars), and (**B**) H$_2$O$_2$.

The effect of the pH of the medium on the η value of photocatalytic degradation of AO is shown in Figure S8. At pH lower than 9, the η value increased with pH. The η value reached a maximum value at pH 8–9, and then abruptly decreased at pHs above 10.

2.5. FTIR Analysis

Steady-state FTIR spectroscopy was used to search for large molecular fragmentation of the products from the photocatalytic degradation of AO. For this analysis, the AO solution after irradiation (see Figure S3) was collected and dried. The vibrational spectrum was then measured in the spectral range of 4000 to 450 cm^{-1}, as shown in Figure 5. For comparison, the spectrum of AO before irradiation is also presented. The main vibrational bands of AO before irradiation were observed at 3407, 3004, 1691, 1602, 1374, 1156, 941, and 821 cm^{-1}, which are assigned to NH stretching of dimethyl amine, C=N stretching, CH of aromatic rings, C=C stretching of aromatic rings, CH bending of aromatic rings, C–N stretching, C–C stretching, and CH out-of-plane bending vibrations of the dye, respectively. Similar spectral features of AO were reported by Mallakpour et al. [54].

It is important to note that the FTIR spectrum of AO after irradiation is similar to that before irradiation. No new additional bands are clearly observed, except a broad band at 600–900 cm^{-1} which could be assigned to the symmetric stretching vibrations of O–Ti–O of anatase TiO$_2$ NPs [55] remaining after the photocatalysis. This confirms that the steady-state FTIR spectroscopy did not detect any photoproducts of AO; instead, it detected the remaining AO and TiO$_2$ NPs. This provides an interpretation that either the photocatalytic products have low infrared cross sections or they are volatile and evaporated during the photoirradiation or drying process, so that none of them were detected by the steady-state measurement. In support of this latter argument, UV–visible spectra of AO (and many other reported dyes) only show a reduction in absorption of AO, with no new peaks being identifiable as coming from any large fragment degradation products.

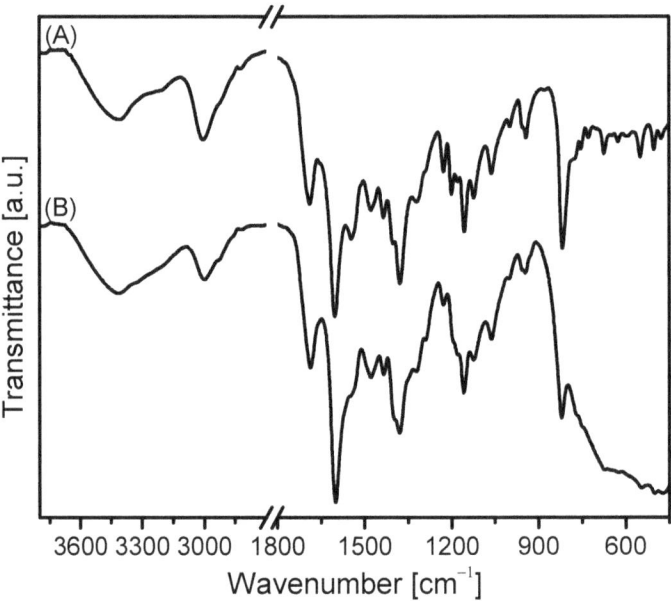

Figure 5. FTIR spectra of AO (**A**) before and (**B**) after photocatalytic degradation.

3. Discussion

In this study, the TiO$_2$ NPs used as photocatalyst were pure anatase crystals, which was confirmed based on FTIR spectra (not shown) and XRD patterns (see Figure S9). The particle size of TiO$_2$ NPs is approximately 100 nm with a BET surface area and pore volume of 12.791 m^2 g^{-1} and 0.05733 cm^3 g^{-1}, respectively [27]. With the bandgap energy being 3.20 eV, the 365 nm light irradiation easily excites the TiO$_2$ NPs, generating electron–hole pairs [56]. This is an advantage in the degradation of AO, because AO absorbs mainly in the visible region. In this sense, the UV irradiation mostly excites the catalyst and, in any case, the photolysis of the dye is inefficient.

As has been discussed in many studies, separation and migration of the photo-generated charge carriers onto the catalyst surface is essential for the photocatalysts to generate the OH$^\bullet$ and O$_2^-{}^\bullet$ radicals. With this in mind, the anatase phase of TiO$_2$ has been theoretically and experimentally revealed to possess high charge-carrier mobility and low charge resistance [55,57], and hence it has high potential as photocatalyst. The photocatalytic degradation of organic dyes is not only governed by the formation rate of OH$^\bullet$ and O$_2^-{}^\bullet$ radicals on the catalyst surface. It should also be governed by the diffusion and immobilization of dyes onto the catalyst surface as well as by the potential energy barrier of the oxidation reaction of the dyes.

The diffusion of a dye in the colloidal solution depends on its hydrodynamic size (related to its molecular structure) as given by the Einstein–Stokes relation. Although there is no report of this for AO in the literature, the structure and size of MB and RhB are approximately comparable to AO. Therefore, the diffusion constant (D) value of AO in aqueous solution can be expected to be close to those of MB (6.74 × 10^{-6} cm^2 s^{-2}) [58] or RhB (4.50 × 10^{-6} cm^2 s^{-2}) [59]. The D value is positively related to the diffusion-limited rate constant (k_D) by the generalized Smoluchowski equation, as given by

$$k_D = 4\pi\sigma D \quad (5)$$

where σ is the encounter distance. As k_{obs} is a proportionally related to k_D, the k_{obs} value of AO can be considered to be comparable to those of MB and RhB. In fact, it was at least two or three times lower in this case compared to those of MB and RhB. A plausible reason for the lower than expected k_{obs} value measured for AO is its nonplanar structure, which is due to the few degrees of rotation of both *N,N*-dimethylaniline moieties with respect to their C–C bond. In order to examine the planarity of AO, ab initio structural optimization was performed using Gaussian basis sets in Chem3D. The structure was optimized for 500 iterations, and the minimum RMS gradient was 0.1. As shown in Figure 6, the MM2 force field suggests that the structure of AO in the gas phase is nonplanar with a torsion angle between the two aromatic ring systems being ~40°. A similar result was also observed using an MMF94 force field, but the torsion angle of the optimized structure was larger (~65°). Although the polar environment in aqueous solution might suppress the rotation of aromatic rings and alter the charge redistribution on AO, the calculation suggests a relatively nonpolar structure of AO. The torsional dynamics could cause intramolecular charge redistribution on AO, leading to strong to medium intermolecular friction, thus slowing the dynamics of the photocatalytic reaction.

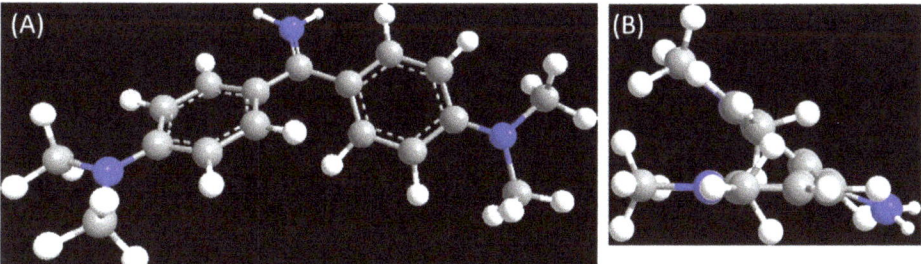

Figure 6. (**A**) Nonplanar structure of AO optimized using Chem3D (the force field MM2), and (**B**) the torsion angle of 40° between the two aromatic systems.

The photocatalytic degradation behavior of AO should also be considered based on the driving force to immobilize the dye on the catalyst surface. Considering that AO is nonplanar and that the dimethylamino groups attached to the aromatic rings are not favorable for hydrogen bonding interactions, AO could only possibly approach the surface of TiO_2 NPs through the methaniminium (=NH_2^+) group through hydrogen bonding or electrostatic interactions.

This description is supported by the observed pH dependence. It is well known that the solution pH is effective in modifying the net charge on the surface of TiO_2 NPs which is known to be amphoteric [26]. The net surface charge turns from positive to negative charge at pH 6.1 [60,61]. At a solution pH lower than 6.1, the positive charge on the catalyst surface is not effective to support immobilization of AO towards the catalyst surface. On the other hand, for pH higher than 6.1, there is electrostatic attraction of the surface to AO, enhancing the photocatalysis of AO [62].

A similar observation was reported for the photocatalytic degradation of MB and RhB in the presence of TiO_2 NPs [27,28] or ZrO_2 NPs [63]. The trend of photocatalytic degradation efficiency was, therefore, that it increased from pH 7 to pH 10, as the ionic state of AO was unchanged, until the solution pH reached a pKa value (pKa 9.8–10.7) [64]. The change of ionic state of AO above pKa is also inferred by the decrease of the photocatalytic degradation efficiency at pH higher than 10.

To obtain an accurate description of the photocatalytic degradation of dyes, the reaction should be followed by liquid chromatography–mass spectroscopy or ultrafast spectroscopy [65,66], but such a detailed study of AO has not been reported. In fact, from this work, it is not clear that significant degradation products remain in the solution, as they may gasify. It is important to recall that AO has a methaniminium and two *N*-dimethylamino groups attached to its aromatic rings. Based on steady-state vibrational spectroscopy (FTIR Figure 5) and the photodegradation mechanism of related organic compounds, such as MB and RhB, and crystal violet [65], it is proposed that oxidation of AO should form Michler's ketone, which further undergoes N-demethylation by successive oxidation reactions to form various intermediates, followed by destruction of the conjugated structure into small compounds, such as CO_2 and NH_4^+, as shown in Figure 7. The fact that small volatile molecular products are the final form of the photodecomposition products is supported by FTIR and UV–visible spectroscopy, because no spectroscopic evidence for larger degradation products is seen. This is a null result and yet is has significance. The assignment of the final product to gaseous small molecules is also backed up by reference [26,67].

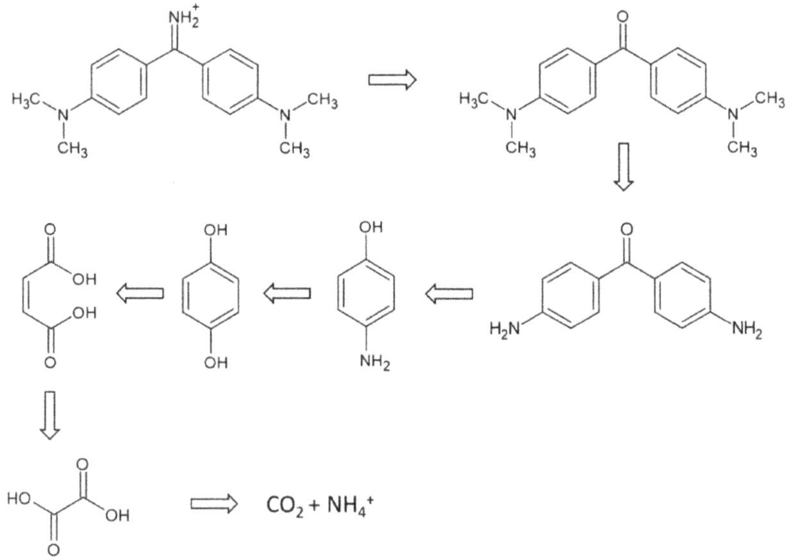

Figure 7. The proposed photodegradation of Auramine O using TiO_2 NPs catalyst.

All of these oxidation steps, which are mediated by $OH^\bullet$ and $O_2^{-\bullet}$ radicals, would occur on or close to the immobilized dyes, where direct interactions are possible between the organic molecules and the photochemically generated radicals on the catalyst surface. A similar degradation mechanism was proposed in the electrochemical degradation of AO by Hmani et al. [67], where the final product of the oxidation was CO_2 gas.

4. Materials and Methods

4.1. Materials and Reagents

The chemicals used in the present experiment were analytical reagent grade of TiO_2 NPs and AO chloride salt ($C_{17}H_{21}N_3 \cdot HCl$; 303.83 g mol^{-1}; CAS: 2465-27-2) which were purchased from Sigma-Aldrich Co. (St.Louis, MO, USA) and were used without any further purification. A stock dye solution was prepared by dissolving 100 mg of powdered AO chloride in distilled water to obtain a concentration of 100 mg L^{-1}. Experimental solutions of a desired concentration were obtained by suitable dilutions.

4.2. Characterization of TiO_2 Catalyst

In this study, the commercial TiO_2 NPs were characterized in a previous study by Suhaimi et al. [28]. The crystalline phase of the TiO_2 NPs was determined based on their X-ray diffraction (XRD) pattern which was measured using an XRD-7000 (Shimadzu, Kyoto, Japan) with collimated Cu Kα radiation (λ = 0.15418 nm). As seen in Figure S9A, the XRD pattern indicated a typical pure anatase phase with the main peak being observed at $2\theta = 25°$. This is in good agreement with the standard XRD pattern of anatase TiO_2 (#JCPDS 84-1286). An SEM image scanned with an SEM-JSM-7600D (JEOL, Tokyo, Japan) indicated that the TiO$_2$ NPs have regular spherical shapes with little agglomeration. Their size was approximately 100 nm (see Figure S9B), which is similar to the report of Amini and Ashrafi [68]. With this loose agglomeration, the TiO$_2$ NPs in a colloidal solution can be considered to have a high surface area to interact with AO molecules, thereby improving its photocatalytic activity [66].

4.3. Photocatalysis Setup

The experimental setup for photocatalysis was reported previously by Suhaimi et al. [28]. Briefly, AO solution (20 mL) was mixed with a few milligrams of TiO$_2$ NPs in a Petri dish with a diameter of 7.5 cm and covered with a UV-transparent glass. They were gently stirred on a temperature-controlled stage. The colloidal mixtures were irradiated from above 10 cm distance using a UV fluorescent lamp (Vilber Lourmat, 6 W, 211 mm; Marne-la-Vallée cedex 3, France). The light power was reduced using an ND filter (6.25%), so that the light power on the solution was 0.28 mW/cm^2. After selected irradiation times, the mixtures were centrifuged at 3000 rpm for 15 min. The filtrates were collected and analyzed by a UV–visible absorption measurement in a 1 cm cuvette cell. All absorption measurements were performed using a UV-1900 spectrophotometer (Shimadzu, Kyoto, Japan).

Prior to the photocatalysis experiments, photolysis (in the absence of TiO$_2$ NPs) and adsorption (in the dark) of AO on TiO$_2$ NPs were evaluated otherwise using the same experimental conditions. The direct photolysis of AO was monitored by irradiating the dye solution, in the absence of catalyst, with 365 nm UV light using the same irradiation geometry and power described above. After a desired irradiation time, the solution was analyzed using a UV–Vis absorption spectroscopic measurement. "Dark" adsorption of AO onto the surface of TiO$_2$ NPs was evaluated by keeping the colloidal mixture in the dark to equilibrate. After a selected contact time, the mixture was centrifuged using an Eppendorf 8504 Centrifuge (Hamburg, Germany) at 3000 rpm for 15 min. The filtrate was collected and analyzed using a UV–visible absorption measurement.

The effect of contact time was evaluated for the photocatalysis of AO in a colloidal mixture of 20 mL (9.5 ppm or equivalent to 35.06 µmol L^{-1}) of the dye with 5 mg TiO$_2$ NPs. The effect of the initial AO concentration was studied by adjusting the concentration to be within 9.3 µmol L^{-1} and 110.8 µmol L^{-1} at a constant mass of TiO$_2$ NPs (5 mg). On the other hand, the effect of the catalyst dosage was evaluated by adjusting the mass of TiO$_2$ NPs (0.5–20 mg) in the mixture with a constant initial concentration of AO (35.06 µmol L^{-1}). Finally, the effect of temperature was investigated from the photocatalytic degradation of AO in a mixture of 20 mL AO (35.06 µmol L^{-1}) and TiO$_2$ NPs (5 mg) at different temperatures from 15 °C to 40 °C.

Fourier transform infrared (FTIR) spectroscopy was used to search for large-fragment photoproducts potentially formed during the photocatalytic reaction. Here, after the photoirradiation, the colloidal mixture was centrifuged, and the precipitated solid was collected and dried in an oven at 40 °C. The vibrational spectrum of the dried solid then was recorded on an FTIR (IRPrestige-21, Shimadzu, Kyoto, Japan) in a KBr disc.

5. Conclusions

In this study, the photocatalytic degradation of toxic cationic Auramine O (AO) in aqueous solution on TiO$_2$ nanoparticles (NPs) under 365 nm light irradiation was investigated. Prior to the photocatalysis experiments, photolysis-alone and dark-adsorption of AO on TiO$_2$ NPs were evaluated using the same experimental conditions. From this it was found that both photolysis-alone and dark-adsorption were inefficient at reducing the aqueous burden of AO. The effects of irradiation time, initial AO concentration, and catalyst dosage on the photocatalytic degradation were evaluated in detail. The results revealed that the photodegradation kinetics of AO can be described using the Langmuir–Hinshelwood model, emphasizing that the oxidation reaction is pseudo-first order. The photodegradation rate constant is 0.048 ± 0.002 min^{-1}, which is slower than that of MB (0.173 ± 0.019 min^{-1}), due to the nonplanar structure of AO. At ambient pH, the photocatalytic efficiency depends on the initial concentration of AO, and a maximum efficiency as high as 96.2 ± 0.9% was achieved from a colloidal mixture of 20 mL (17.78 µmol L^{-1}) AO solution in the presence of 5 mg of TiO$_2$ NPs. The photocatalytic efficiency of the dye decreases nonlinearly with increasing the initial concentration and catalyst dosage. The activation energy of the photocatalytic degradation of AO on the TiO$_2$ NPs was estimated to be 4.63 kJ mol^{-1}. The photocatalytic degradation of AO was assessed by observing the effect of pH, additional scavengers and H$_2$O$_2$, from which it was confirmed that the degradation is due to an oxidation reaction of the immobilized dyes on the catalyst surface, where they have direct interactions with photochemically generated OH$^\bullet$ and O$_2^{-\bullet}$. The nature of the degradation products of photocatalytic removal of AO was evaluated using Fourier transform infrared (FTIR) spectroscopy. The steady-state FTIR spectroscopy did not show any detectable byproducts of photocatalytic degradation of AO. This implies that, although the catalytic reaction may involve many organic intermediates, through N-demethylation and successive oxidation reactions, the final photocatalytic products were volatile compounds, such as CO$_2$ and NH$_4^+$, which escaped from the solution. The overall results provide detailed description of photocatalytic degradation of toxic cationic AO, a dimethylmethane fluorescent dye, in an aqueous heterogeneous solution on TiO$_2$ NPs irradiated using 365 nm light. Finally, we confirmed that using TiO$_2$ in the form of NPs greatly enhanced the rate of AO removal from solution compared to the use of micro-sized powders.

Supplementary Materials: The following supporting information can be downloaded online at: https://www.mdpi.com/article/10.3390/catal12090975/s1, Figure S1: UV–Vis absorption spectra following photolytic decomposition of AO as a function of irradiation time; Figure S2: Absorption spectra of AO in aqueous colloidal solution in the presence of 5 mg TiO2 NPs in the dark at different contact times; Figure S3: Images of a colloidal mixture of 20 mL of AO solution before irradiation and after irradiation; Figure S4: AO solution before irradiation and after irradiation before and after centrifugation; Figure S5: Absorption spectra of AO in aqueous colloidal mixture with 5 mg TiO2 NPs after irradiation for 30 min at different temperatures; Figure S6: Absorption spectra of different concentrations AO in an aqueous colloidal mixture with 5 mg TiO2 NPs before and after irradiation; Figure S7: Absorption spectra of different concentrations AO in aqueous colloidal mixture with different masses of TiO$_2$ NPs, and a plot of the remaining AO concentration C_t as a function of the catalyst dosage; Figure S8: The plot of η value of photocatalytic degradation of AO in aqueous colloidal mixture with 5 mg TiO2 NPs at different pHs; Figure S9: XRD patterns of anatase TiO$_2$ NPs, with comparison to standard data (#JCPDS 84-1286) and SEM image of TiO$_2$ NPs at ×50,000 magnification.

Author Contributions: Conceptualization, A.U.; methodology, A.U., C.P.Y.K. and N.A.A.S.; validation, A.U., J.H., J.-W.L. and M.N.; investigation, C.P.Y.K., N.A.A.S. and N.N.M.S.; writing—original draft preparation, A.U.; writing—review and editing, A.U. and J.H.; supervision, A.U., J.H., J.-W.L. and M.N. All authors have read and agreed to the published version of the manuscript.

Funding: This research received no external funding.

Acknowledgments: Jonathan Hobley is grateful to National Cheng Kung University's NCKU90 Distinguished Visiting Scholar Program for hosting his research and to MOST for providing research funding under project number 111-2222-E-006-007.

Conflicts of Interest: The authors declare no conflict of interest.

References

1. Mekonnen, M.M.; Hoekstra, A.Y. Sustainability: Four billion people facing severe water scarcity. *Sci. Adv.* **2016**, *2*, e1500323. [CrossRef] [PubMed]
2. Schwarzenbach, R.P.; Egli, T.; Hofstetter, T.B.; von Gunten, U.; Wehrli, B. Global water pollution and human health. *Annu. Rev. Environ. Resour.* **2010**, *35*, 109–136. [CrossRef]
3. Akhtar, N.; Ishak, M.I.S.; Bhawani, S.A.; Umar, K. Various natural and anthropogenic factors responsible for water quality degradation: A review. *Water* **2021**, *13*, 2660. [CrossRef]
4. Boretti, A.; Rosa, L. Reassessing the projections of the World Water Development Report. *NPJ Clean Water* **2019**, *2*, 15. [CrossRef]

5. Ghaly, A.E.; Ananthashankar, R.; Alhattab, M.; Ramakrishnan, V.V. Production, characterization and treatment of textile effluents: A critical review. *J. Chem. Eng. Process Technol.* **2014**, *5*, 1000182.
6. Kumar, A.; Dixit, U.; Singh, K.; Gupta, S.P.; Beg, M.S.J. Structure and properties of dyes and pigments. In *Dyes and Pigments-Novel Applications and Waste Treatment*; Papadakis, R., Ed.; IntechOpen Limited: London, UK, 2021; pp. 1–19.
7. Routoula, E.; Patwardhan, S.V. Degradation of anthraquinone dyes from effluents: A review focusing on enzymatic dye degradation with industrial potential. *Environ. Sci. Technol.* **2020**, *54*, 647–664. [CrossRef]
8. Lellis, B.; Fávaro-Polonio, C.Z.; Pamphile, J.A.; Polonio, J.C. Effects of textile dyes on health and the environment and bioremediation potential of living organisms. *Biotechnol. Res. Innov.* **2019**, *3*, 275–290. [CrossRef]
9. Yaseen, D.A.; Scholz, M. Textile dye wastewater characteristics and constituents of synthetic effluents: A critical review. *Int. J. Environ. Sci. Technol.* **2019**, *16*, 1193–1226. [CrossRef]
10. Ledakowicz, S.; Paździor, K. Recent achievements in dyes removal focused on advanced oxidation processes integrated with biological methods. *Molecules* **2021**, *26*, 870. [CrossRef]
11. Slama, H.B.; Bouket, A.C.; Pourhassan, Z.; Alenezi, F.N.; Silini, A.; Cherif-Silini, H.; Oszako, T.; Luptakova, L.; Golińska, P.; Belbahri, L. Diversity of synthetic dyes from textile industries, discharge impacts and treatment methods. *Appl. Sci.* **2021**, *11*, 6255. [CrossRef]
12. Berradi, M.; Hsissou, R.; Khudhair, M.; Assouag, M.; Cherkaoui, O.; El Bachiri, A.; El Harfi, A. Textile finishing dyes and their impact on aquatic environs. *Heliyon* **2019**, *5*, e02711. [CrossRef] [PubMed]
13. Vatchalan, L.; Kesavan, B.; Selvam, P. Adsorption and photocatalytic degradation of auramine–O dye using carbon nanoparticles and Carbon–CaO nanocomposites. *Nanotechnol. Environ. Eng.* **2022**. [CrossRef]
14. Khatri, N.; Tyagi, S.; Rawtani, D. Removal of basic dyes auramine yellow and auramine O by halloysite nanotubes. *Int. J. Waste Manag.* **2016**, *17*, 44. [CrossRef]
15. Martelli, A.; Campart, G.B.; Canonero, R.; Carrozzino, R.; Mattioli, F.; Robbiano, L.; Cavanna, M. Evaluation of auramine genotoxicity in primary rat and human hepatocytes and in the intact rat. *Mutat. Res.–Genet. Toxicol. Environ. Mutagen.* **1998**, *414*, 37–47. [CrossRef]
16. Carmen, Z.; Daniel, S. Textile organic dyes—Characteristics, polluting effects and separation/elimination procedures from industrial effluents—A critical overview. *Org. Pollut. Ten Years Stock. Conv.-Environ. Anal. Updat.* **2012**, *1*, 56–86.
17. Ardila-Leal, L.D.; Poutou-Piñales, R.A.; Pedroza-Rodríguez, A.M.; Quevedo-Hidalgo, B.E. A brief history of colour, the environmental impact of synthetic dyes and removal by using laccases. *Molecules* **2021**, *26*, 3813. [CrossRef]
18. Ali, I.; Khan, T.A.; Asim, M. Removal of arsenate from groundwater by electrocoagulation method. *Environ. Sci. Pollut. Res.* **2012**, *19*, 1668–1676. [CrossRef]
19. Ma, B.; Xue, W.; Hu, C.; Liu, H.; Qu, J.; Li, L. Characteristics of microplastic removal via coagulation and ultrafiltration during drinking water treatment. *Chem. Eng. J.* **2019**, *359*, 159–167. [CrossRef]
20. Al-Bastaki, N. Removal of methyl orange dye and Na_2SO_4 salt from synthetic waste water using reverse osmosis. *Chem. Eng. Process. Process Intensif.* **2004**, *43*, 1561–1567. [CrossRef]
21. Bilińska, L.; Blus, K.; Foszpańczyk, M.; Gmurek, M.; Ledakowicz, S. Catalytic ozonation of textile wastewater as a polishing step after industrial scale electrocoagulation. *J. Environ. Manag.* **2020**, *265*, 110502. [CrossRef]
22. Mohan, S.V.; Bhaskar, Y.V.; Karthikeyan, J. Biological decolourisation of simulated azo dye in aqueous phase by algae Spirogyra species. *Int. J. Environ. Pollut.* **2004**, *21*, 211–222. [CrossRef]
23. Benjelloun, M.; Miyah, Y.; Evrendilek, G.A.; Zerrouq, F.; Lairini, S. Recent advances in adsorption kinetic models: Their application to dye types. *Arab. J. Chem.* **2021**, *14*, 103031. [CrossRef]
24. Mills, A.; Davies, R.H.; Worsley, D. Water purification by semiconductor photocatalysis. *Chem. Soc. Rev.* **1993**, *22*, 417–425. [CrossRef]
25. Ali, R.; Mahmood, T.; Naeem, A.; Ullah, A.; Aslam, M.; Khan, S. Process optimization of Auramine O adsorption by surfactant-modified activated carbon using Box–Behnken design of response surface methodology. *Desalin. Water Treat.* **2021**, *217*, 367–390. [CrossRef]
26. Poulios, I.; Avranas, A.; Rekliti, E.; Zouboulis, A. Photocatalytic oxidation of Auramine O in the presence of semiconducting oxides. *J. Chem. Technol. Biotechnol.* **2000**, *75*, 205–212. [CrossRef]
27. Zulmajdi, S.L.N.; Zamri, N.I.I.; Yasin, H.M.; Kusrini, E.; Hobley, J.; Usman, A. Comparative study on the adsorption, kinetics, and thermodynamics of the photocatalytic degradation of six different synthetic dyes on TiO_2 nanoparticles. *React. Kinet. Mech. Catal.* **2020**, *129*, 519–534. [CrossRef]
28. Suhaimi, N.A.A.; Shahri, N.N.M.; Samat, J.H.; Kusrini, E.; Lim, J.-W.; Hobley, J.; Usman, A. Domination of methylene blue over rhodamine B during simultaneous photocatalytic degradation by TiO_2 nanoparticles in an aqueous binary solution under UV irradiation. *React. Kinet. Mech. Catal.* **2022**, *135*, 511–527. [CrossRef]
29. Lee, Y.; von Gunten, U. Oxidative transformation of micropollutants during municipal wastewater treatment: Comparison of kinetic aspects of selective (chlorine, chlorine dioxide, ferrate VI, and ozone) and non-selective oxidants (hydroxyl radical). *Water Res.* **2010**, *44*, 555–566. [CrossRef]
30. Herrmann, J.-M. Heterogeneous photocatalysis: Fundamentals and applications to the removal of various types of aqueous pollutants. *Catal. Today* **1999**, *53*, 115–129. [CrossRef]

31. Janczarek, M.; Kowalska, E. On the origin of enhanced photocatalytic activity of copper-modified titania in the oxidative reaction systems. *Catalysts* **2017**, *7*, 317. [CrossRef]
32. Chiu, Y.-H.; Chang, T.-F.M.; Chen, C.-Y.; Sone, M.; Yung-Jung Hsu, Y.-J. Mechanistic insights into photodegradation of organic dyes using heterostructure photocatalysts. *Catalysts* **2019**, *9*, 430. [CrossRef]
33. Montazerozohori, M.; Nasr-Esfahani, M.; Moradi-Shammi, Z.; Malekhoseini, A. Photocatalytic decolorization of auramine and its kinetics study in the presence of two different sizes titanium dioxide nanoparticles at various buffer and non-buffer media. *J. Ind. Eng. Chem.* **2015**, *21*, 1044–1050. [CrossRef]
34. Pandurungan, A.; Kamala, P.; Uma, S.; Palanichamy, M.; Murugesan, V. Degradation of basic yellow Auramine O—A textile dye by semiconductor photocatalysis. *Indian J. Chem. Technol.* **2001**, *8*, 496–499.
35. Ako, R.T.; Ekanayake, P.; Young, D.J.; Hobley, J.; Chellappan, V.; Tan, A.L.; Gorelik, S.; Subramanian, G.S.; Lim, C.M. Evaluation of surface energy state distribution and bulk defect concentration in DSSC photoanodes based on Sn, Fe, and Cu doped TiO_2. *Appl. Surf. Sci.* **2015**, *351*, 950–961. [CrossRef]
36. O'Regan, B.; Grätzel, M. A low-cost, high-efficiency solar cell based on dye-sensitized colloidal TiO_2 films. *Nature* **1991**, *353*, 737–740. [CrossRef]
37. Furube, A.; Du, L.; Hara, K.; Katoh, R.; Tachiya, M. Ultrafast plasmon-induced electron transfer from gold nanodots into TiO_2 nanoparticles. *J. Am. Chem. Soc.* **2007**, *129*, 14852–14853. [CrossRef] [PubMed]
38. Nair, R.V.; Gummaluri, V.S.; Matham, M.V.; Vijayan, C. A review on optical bandgap engineering in TiO_2 nanostructures via doping and intrinsic vacancy modulation towards visible light applications. *J. Phys. D Appl. Phys.* **2022**, *55*, 313003. [CrossRef]
39. Khan, M.S.; Shah, J.A.; Riaz, N.; Butt, T.A.; Khan, A.J.; Khalifa, W.; Gasmi, H.H.; Latifee, E.R.; Arshad, M.; Al-Naghi, A.A.A.; et al. Synthesis and characterization of Fe-TiO_2 nanomaterial: Performance evaluation for RB5 decolorization and in vitro antibacterial studies. *Nanomaterials* **2021**, *11*, 436. [CrossRef]
40. Kumar, K.V.; Porkodi, K.; Rocha, F. Langmuir-Hinshelwood kinetics—A theoretical study. *Catal. Commun.* **2008**, *9*, 82–84. [CrossRef]
41. Armenise, S.; García-Bordejé, E.; Valverde, J.L.; Romeo, E.; Monzón, A. A Langmuir-Hinshelwood approach to the kinetic modelling of catalytic ammonia decomposition in an integral reactor. *Phys. Chem. Chem. Phys.* **2013**, *15*, 12104–12117. [CrossRef] [PubMed]
42. Shaban, Y.A. Solar light-induced photodegradation of chrysene in seawater in the presence of carbon-modified n-TiO_2 nanoparticles. *Arab. J. Chem.* **2019**, *12*, 652–663. [CrossRef]
43. Cruz, G.J.F.; Gómez, M.M.; Solis, J.L.; Rimaycuna, J.; Solis, R.L.; Cruz, J.F.; Rathnayake, B.; Keiski, R.L. Composites of ZnO nanoparticles and biomass based activated carbon: Adsorption, photocatalytic and antibacterial capacities. *Water Sci. Technol.* **2018**, *2017*, 492–508. [CrossRef] [PubMed]
44. Setarehshenas, N.; Hosseini, S.H.; Ahmadi, G. Optimization and kinetic model development for photocatalytic dye degradation. *Arab. J. Sci. Eng.* **2018**, *43*, 5785–5797. [CrossRef]
45. Bloh, J.Z. A holistic approach to model the kinetics of photocatalytic reactions. *Front. Chem.* **2019**, *7*, 1–13. [CrossRef]
46. Riaz, N.; Hassan, M.; Siddique, M.; Mahmood, Q.; Farooq, U.; Sarwar, R.; Khan, M.S. Photocatalytic degradation and kinetic modeling of azo dye using bimetallic photocatalysts: Effect of synthesis and operational parameters. *Environ. Sci. Pollut. Res.* **2020**, *27*, 2992–3006. [CrossRef] [PubMed]
47. Zulmajdi, S.L.N.; Ajak, S.N.F.H.; Hobley, J.; Duraman, N.; Harunsani, M.H.; Yasin, H.M.; Nur, M.; Usman, A. Kinetics of photocatalytic degradation of methylene blue in aqueous dispersions of TiO_2 nanoparticles under UV-LED irradiation. *Am. J. Nanomater.* **2017**, *5*, 1–6.
48. Paul, D.R.; Sharma, R.; Nehra, S.P.; Sharma, A. Effect of calcination temperature, pH and catalyst loading on photodegradation efficiency of urea derived graphitic carbon nitride towards methylene blue dye solution. *RSC Adv.* **2019**, *9*, 15381–15391. [CrossRef]
49. Blažeka, D.; Car, J.; Klobučar, N.; Jurov, A.; Zavašnik, J.; Jagodar, A.; Kovačević, E.; Krstulović, N. Photodegradation of methylene blue and rhodamine B using laser-synthesized ZnO nanoparticles. *Materials* **2020**, *13*, 4357. [CrossRef]
50. Divya, N.; Bansal, A.; Jana, A.K. Photocatalytic degradation of azo dye Orange II in aqueous solutions using copper-impregnated titania. *Int. J. Environ. Sci. Technol.* **2013**, *10*, 1265–1274. [CrossRef]
51. Mencigar, D.P.; Strlič, M.; Štangar, U.L.; Korošec, R.C. Hydroxyl radical scavenging-based method for evaluation of TiO_2 photocatalytic activity. *Acta Chim. Slov.* **2013**, *60*, 908–912.
52. Liu, Y.; Li, Y.-H.; Li, X.; Zhang, Q.; Yu, H.; Peng, X.; Peng, F. Regulating electron–hole separation to promote photocatalytic H_2 evolution activity of nanoconfined Ru/MXene/TiO_2 Catalysts. *ACS Nano* **2020**, *14*, 14181–14189. [CrossRef] [PubMed]
53. Hirakawa, T.; Yawata, K.; Nosaka, Y. Photocatalytic reactivity for O_2^- and OH radical formation in anatase and rutile TiO_2 suspension as the effect of H_2O_2 addition. *Appl. Catal. A Gen.* **2007**, *325*, 105–111. [CrossRef]
54. Mallakpour, S.; Dinari, M.; Hadadzadeh, H. Insertion of fluorophore dyes between cloisite Na^+ layered for preparation of novel organoclays. *J. Incl. Phenom. Macrocycl. Chem.* **2013**, *77*, 463–470. [CrossRef]
55. Zulmajdi, S.L.N.; Zamri, N.I.I.; Mahadi, A.H.; Rosli, M.Y.H.; Ja'afar, F.; Yasin, H.M.; Kusrini, E.; Hobley, J.; Usman, A. Sol-gel preparation of different crystalline phases of TiO_2 nanoparticles for photocatalytic degradation of methylene blue in aqueous solution. *Am. J. Nanomater.* **2019**, *7*, 39–45. [CrossRef]

56. Saalinraj, S.; Ajithprasad, K.C. Effect of calcination temperature on non-linear absorption co-efficient of nano sized titanium dioxide (TiO_2) synthesised by sol-gel method. *Mater. Today Proc.* **2017**, *4*, 4372–4379. [CrossRef]
57. Luttrell, T.; Halpegamage, S.; Tao, J.; Kramer, A.; Sutter, E.; Batzill, M. Why is anatase a better photocatalyst than rutile?—Model studies on epitaxial TiO_2 films. *Sci. Rep.* **2015**, *4*, 4043. [CrossRef]
58. Selifonov, A.A.; Shapoval, O.G.; Mikerov, A.N.; Tuchin, V.V. Determination of the diffusion coefficient of methylene blue solutions in dentin of a human tooth using reflectance spectroscopy and their antibacterial activity during laser exposure. *Opt. Spectrosc.* **2019**, *126*, 758–768. [CrossRef]
59. Gendron, P.O.; Avaltroni, F.; Wilkinson, K.J. Diffusion coefficients of several rhodamine derivatives as determined by pulsed field gradient–nuclear magnetic resonance and fluorescence correlation spectroscopy. *J. Fluoresc.* **2008**, *18*, 1093. [CrossRef]
60. Kamaluddin, M.R.; Zamri, N.I.I.; Kusrini, E.; Prihandini, W.W.; Mahadi, A.H.; Usman, A. Photocatalytic activity of kaolin-titania composites to degrade methylene blue under UV light irradiation; Kinetics, mechanism and thermodynamics. *React. Kinet. Mech. Catal.* **2021**, *113*, 517–529. [CrossRef]
61. Nguyen, L.T.; Nguyen, H.T.; Pham, T.D.; Tran, T.D.; Chu, H.T.; Dang, H.T.; Nguyen, V.H.; Nguyen, K.M.; Pham, T.T.; van der Bruggen, B. UV–visible light driven photocatalytic degradation of ciprofloxacin by N,S co-doped TiO_2: The effect of operational parameters. *Top. Catal.* **2020**, *63*, 985–995. [CrossRef]
62. Reza, K.M.; Kurny, A.S.W.; Gulshan, F. Parameters affecting the photocatalytic degradation of dyes using TiO_2: A review. *Appl. Water Sci.* **2017**, *7*, 1569–1578. [CrossRef]
63. Rani, S.; Aggarwal, M.; Kumar, M.; Sharma, S.; Kumar, D. Removal of methylene blue and rhodamine B from water by zirconium oxide/graphene. *Water Sci.* **2016**, *30*, 51–60. [CrossRef]
64. Sabnis, R.W. Synthesis and industrial applications. In *Handbook of Biological Dyes and Stains*; John Wiley & Sons Inc. Publication: Hoboken, NJ, USA, 2010; pp. 27–29.
65. Hisaindee, S.; Meetani, M.A.; Rauf, M.A. Application of LC-MS to the analysis of advanced oxidation process (AOP) degradation of dye products and reaction mechanisms. *TrAC-Trends Anal. Chem.* **2013**, *49*, 31–44. [CrossRef]
66. Wang, X.Q.; Han, S.F.; Zhang, Q.W.; Zhang, N.; Zhao, D.D. Photocatalytic oxidation degradation mechanism study of methylene blue dye waste water with GR/I/TiO_2. *MATEC Web Conf.* **2018**, *238*, 03006. [CrossRef]
67. Hmani, E.; Samet, Y.; Abdelhédi, R. Electrochemical degradation of Auramine-O dye at boron-doped diamond and lead dioxide electrodes. *Diam. Relat. Mater.* **2012**, *30*, 1–8. [CrossRef]
68. Amini, M.; Ashrafi, M. Photocatalytic degradation of some organic dyes under solar light irradiation using TiO_2 and ZnO nanoparticles. *Nano. Chem. Res.* **2016**, *1*, 79–86.

Review

Dynamics of Diffusion- and Immobilization-Limited Photocatalytic Degradation of Dyes by Metal Oxide Nanoparticles in Binary or Ternary Solutions

Nurul Amanina A. Suhaimi [1], Cristina Pei Ying Kong [1], Nurulizzatul Ningsheh M. Shahri [1], Muhammad Nur [2], Jonathan Hobley [3] and Anwar Usman [1,*]

[1] Department of Chemistry, Faculty of Science, Universiti Brunei Darussalam, Jalan Tungku Link, Gadong BE1410, Brunei
[2] Center for Plasma Research, Integrated Laboratory, Universitas Diponegoro, Tembalang Campus, Semarang 50275, Indonesia
[3] Department of Biomedical Engineering, National Cheng Kung University, No. 1 University Road, Tainan City 701, Taiwan
* Correspondence: anwar.usman@ubd.edu.bn

Abstract: Photocatalytic degradation employing metal oxides, such as TiO_2 nanoparticles, as catalysts is an important technique for the removal of synthetic dyes from wastewater under light irradiation. The basic principles of photocatalysis of dyes, the effects of the intrinsic photoactivity of a catalyst, and the conventional non-fundamental factors are well established. Recently reported photocatalysis studies of dyes in single, binary, and ternary solute solutions opened up a new perspective on competitive photocatalytic degradation of the dyes. There has not been a review on the photocatalytic behavior of binary or ternary solutions of dyes. In this regard, this current review article summarizes the photocatalytic behavior of methylene, rhodamine B, and methyl orange in their binary or ternary solutions. This brief overview introduces the importance of the dynamics of immobilization and reactivity of the dyes, the vital roles of molecular conformation and functional groups on their diffusion onto the catalyst surface, and photocatalytic degradation, and provides an understanding of the simultaneous photocatalytic processes of multiple dyes in aqueous systems.

Keywords: synthetic dyes; titania nanoparticle; photocatalytic degradation; mechanism

Citation: Suhaimi, N.A.A.; Kong, C.P.Y.; Shahri, N.N.M.; Nur, M.; Hobley, J.; Usman, A. Dynamics of Diffusion- and Immobilization-Limited Photocatalytic Degradation of Dyes by Metal Oxide Nanoparticles in Binary or Ternary Solutions. *Catalysts* 2022, 12, 1254. https://doi.org/10.3390/catal12101254

Academic Editor: John Vakros

Received: 20 September 2022
Accepted: 13 October 2022
Published: 17 October 2022

Publisher's Note: MDPI stays neutral with regard to jurisdictional claims in published maps and institutional affiliations.

Copyright: © 2022 by the authors. Licensee MDPI, Basel, Switzerland. This article is an open access article distributed under the terms and conditions of the Creative Commons Attribution (CC BY) license (https://creativecommons.org/licenses/by/4.0/).

1. Introduction

The charming colors of the world originate from the light absorption and/or emission of natural and synthetic dyes. Due to their attractiveness, these coloring agents are applied in all products used in daily life, and throughout history, they have been utilized especially in textile clothes, fabrics, paints, and foods. In ancient times, coloring agents were based on natural dyes from plants and animals. Synthetic dyes have been intensively used since their discovery in the nineteenth century [1]. The production and use of synthetic dyes are continuously increasing, and in dying industries, these man-made dyes have taken over natural dyes due to their brighter colors and relatively low production costs. Another important reason is their stable colors, which can last for long periods of time. In addition to having bright colors, synthetic dyes dissolve easily in water, resulting in easier and cheaper manufacturing processes for the coloration of papers, plastics, leather, and foods and the formation of strong chemical bonds with cottons, fabrics, polymers, and food ingredients. There are about 10^5 varieties of dyes, and around 10% of them are used for coloring cottons, fabrics, polymers, and food and/or printing [2,3], with their consumption rate being about 36 kilotons per year [4,5]. Notably, over 50% of the global consumption of dyes caused an increase in dye contamination in water systems, particularly in developing regions of Asia [6].

Although textile, paper, leather, plastic, and rubber industries, in principle, handle their respective effluents using treatment or storage systems, it was believed that up to 20% of the effluent is released into wastewater and leached into the surrounding urban areas by rainfall. The dye effluents are resistant to microbial attack, persist in the environment, prevent light penetration, and lower the rate of photosynthesis, thus decreasing dissolved oxygen levels, increasing the biochemical oxygen demand, and resulting in negative effects on living organisms in the water systems [7]. The large-scale production and large industries utilizing toxic dyes on a massive scale eventually cause the contamination of water systems and soil [8]. This extensive environmental pollution is highly poisonous to the ecosystem and decreases agricultural productivity.

As dyes are mutagenic, carcinogenic, and toxic, they have chronic effects. Many dye effluents have been associated with several acute health risks, such as hepatocarcinoma, splenic sarcomas, occupational asthma, and bladder cancer, causing chromosomal aberrations and nuclear anomalies in animal cells [9]. Therefore, the removal of them from wastewater before discharging them into the water systems is imperative [8–11], and removing the dye effluents from wastewater could protect the water systems from harmful pollution, and could recycling the spent water in the dyeing processes [11]. Although synthetic dyes are resistant to many extreme conditions and cannot be removed using any conventional wastewater treatments, research efforts have been devoted to the development of biological, physical, and chemical methods [12]. Among the successful methods, adsorption, photocatalysis, and ozonation have attracted great attention for eliminating organic pollutants from wastewater due to their high removal efficiency, sustainability, and selectivity [13–15].

Adsorption has been predominantly considered as the most promising and favorable method for removing synthetic dyes from wastewater because of its simple technological operation, as this method provides a simple route, operates at room temperature, and achieves relatively high removal efficiency [16]. In this sense, activated carbon, silica gel, polymeric ion exchange resins, and their nanocomposites and aerogels have been widely used as adsorbents to remove a wide variety of dyes [17–21]. Although these materials are considered excellent adsorbents due to their surface hydrophobicity, high surface area, well-developed porous structure, and special surface reactivity [17,18], the process to produce them is often rather expensive and forms secondary contaminated products. The selection of adsorbent material becomes the most essential consideration in any adsorption process. As a result, low-cost, sustainable, and renewable adsorbents have been intensively explored [22].

Although agricultural wastes and clays account for the most explored materials for cost-effective adsorbents, their adsorption ability to remove dyes is much lower than other materials, such as composites and biopolymers [23,24]. The main reason is that, unlike the adsorption of gasses, which is strongly governed by the surface area and porous structure of the absorbents, the adsorption of dyes relies mainly on the intermolecular hydrogen bonding and electrostatic interactions between the functional groups on the adsorbent surface and dyes, as well as the conformational flexibility of the adsorbent when dispersed in water [25]. Therefore, the search for new sustainable adsorbents for the removal of dyes and their ability to be regenerated is still a research challenge [26,27]. Most importantly, the adsorption process from wastewater should be optimized toward the simultaneous removal of dyes from industrial effluents, which often contain multiple contaminants [11,27–31]. Similarly, biological methods are cost-effective, excellent at reducing odor and color, and environmental-friendly [32,33]. However, these methods still require strict and inflexible control of temperature and pH; otherwise, the synthetic dyes will take a long time to degrade and a large amount of sludge will need to be disposed of in the process [33]. Another important point that remains an issue is the application of biological methods in the removal of high concentrations of organic wastes.

On the other hand, heterogeneous photocatalysis has been demonstrated to be more desirable in the removal of dyes from wastewater [34,35]. The photocatalysis can be operated using UV or white light in water treatment systems at ambient temperature and pressure conditions. This method shows several advantages in the decolorization of wastewater due to the low cost, easy waste disposal, environmental friendliness [36], and highly efficient photocatalytic degradation [15]. Most importantly, the removal of dyes using photocatalysis has been claimed to produce nontoxic or less harmful by-products [37–40].

Although most of the studies of photocatalysis of dyes have been reported since the discovery of titanium oxide (TiO_2) in the early 1970s, important contributions to the understanding of this particular field were developed since the earliest establishment of photocatalysis in water treatment systems [41]. The research on the phenomenon of photocatalysis is continuously established to explore either new photocatalysts, new types of dyes, new concept of photocatalysis, or the photodegradation mechanism of the synthetic dyes. In general, all the contributions have developed knowledge of the photocatalysis of dyes. It is well-known that the removal of dyes is due to their consecutive chemical reactions with photochemically generated hydroxyl ($OH^\bullet$) and oxygen ($O_2^-{}^\bullet$) radicals, resulting in the degradation of dyes on the surface of the catalyst [42,43]. It is conceivable that the rate and efficiency of photocatalytic degradation of dyes depend on many factors; most of them are the intrinsic photoactivity of the catalyst [44], the dynamics of immobilization and the reactivity of the dyes with the photogenerated radicals on the surface of the catalyst [45]. These aspects of the photocatalytic degradation of the dyes are still deficient, while one could always consider the conventional non-fundamental factors, such as irradiation time, the ratio between dye concentration and the catalyst dosage, and the mediation of oxidative agents. This current review article summarizes the vital roles of the crystal structure of catalysts in their photocatalytic activity. It also highlights the molecular conformation and functional groups of synthetic dyes that affect the rate and efficiency of photocatalytic degradation of the dyes based on recently reported photocatalysis of dyes in single, binary, and ternary solute solutions. Therefore, the objective of this review article is to provide an insight into the importance of the dynamics of immobilization dyes (which are attributed to their molecular conformation and functional groups) on their diffusion onto the catalyst surface and their photocatalytic degradation.

2. Metal Oxide Semiconductors as Photocatalysts

The photophysical processes in photocatalysts are depicted in Figure 1. In general, photocatalysts are regarded as harvesting visible and/or near UV light energy to promote the separation of charge carriers and to utilize them to degrade dyes. Most of the studies have, therefore, been focused on metal oxide semiconductors due to their good photoactivity, relatively low bandgap, biologically and chemically inert nature, strong catalytic activity, high charge mobility and long charge carrier diffusion length, low cost, and non-toxicity [46,47]. In this sense, when the sizes of catalysts are in nanometer scales, the photocatalytic activity of the metal oxide semiconductors should be greatly increased.

With a bandgap energy of around 3.0–3.5 eV, the photocatalytic activity of TiO_2, ZnO, Mn_3O_4, CeO_2, $BiVO_4$, and $BiOI/Bi_2WO_6$ nanoparticles (NPs) has been intensively investigated. These catalysts absorb photon energy in the UV region less than 420 nm [48–51], and hence, their photocatalytic activity relies on UV light irradiation. This is advantageous for the degradation of the dyes that absorb photon light in the visible region, as the UV light irradiation can solely excite the catalyst and the spontaneous photolysis of the dyes can be avoided. As the bandgap of the metal oxides can be modified by incorporating a small number of metallic dopants, many researchers have successfully engineered the bandgap energy so that the metal oxide NPs could have absorption in the visible region, allowing the utilization of visible light or sunlight. By using this band engineering strategy, a large group of visible-light-responsive metal-oxide-based catalysts have been synthesized, as have been summarized in a number of review articles [52,53]. This means that these types of metal oxide NPs can be excited by visible light, in contrast to TiO_2, ZnO, Mn_3O_4, CeO_2, $BiVO_4$, and $BiOI/Bi_2WO_6$ NPs, which are limited by UV light excitation. As dyes also absorb photon energy in the visible region, their photocatalytic degradation on visible-light-responsive metal-oxide-based catalysts should compete with their spontaneous photolysis [54,55].

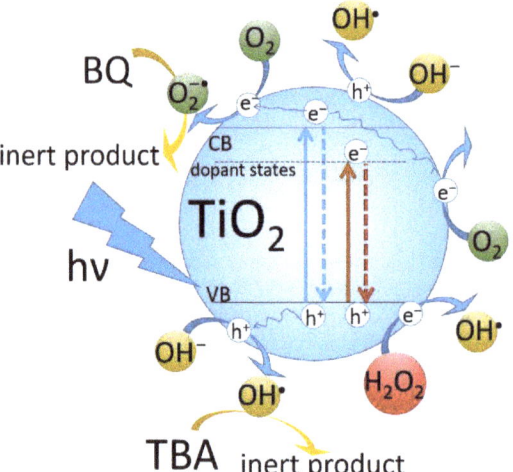

Figure 1. A schematic illustration of the photophysical processes in photocatalysts. BQ and TBA stand for benzoquinone and tert-butyl alcohol.

Many studies have been focused on finding new metal oxides or incorporating metallic dopants in the metal oxides as photocatalysts. It should be noted that the photocatalytic activity of metal oxide NPs, which depends on the charge carrier separation, allowing electron and hole transfers to generate OH• and O_2^-• radicals [42,56], must be explained by the redox reaction of the charge carriers with water molecules and solvated oxygen. Therefore, the potentials of oxidation–reduction of the valence band and the conduction band with respect to those of H_2O and O_2 should be evaluated. In this sense, the metal oxide materials are expected to have a redox potential of the photogenerated conduction band to be more negative in order to reduce adsorbed O_2 to superoxide. Simultaneously, the redox potential of the photogenerated valence band hole has to be positive to produce OH• radicals in order to subsequently oxidize the organic pollutant [57,58]. For instance, the high photocatalytic activity of TiO_2 NPs has a reduction potential (-0.5 V vs. NHE) lower than O_2 (-0.33 vs. NHE), allowing the electron transfer to generate a superoxide radical (O_2^-•), and an oxidation potential (+2.7 V ns. NHE) higher than H_2O (+2.53 V ns. NHE), allowing a hole transfer to generate a hydroxyl radical (OH•) [42]. It is obvious that the redox potentials of TiO_2 make it a promising photocatalyst, though the O_2^-• and OH• radicals are not spontaneously generated upon photoexcitation. The most important factor for the formation of the radicals is the mobility of the charge carrier inside the catalyst [54,56]. The crystal structure of the catalyst plays an important role in accelerating the charge carrier mobility and in suppressing the recombination of an electron–hole. With this in mind, Luttrell et al. pointed out that, among the different crystal phases of TiO_2 crystals, the anatase phase has the best performance in suppressing the recombination of an electron–hole [56], as has been experimentally observed [59].

The O_2^-• and OH• radicals are generated by an exciton and hole after their migration onto the catalyst surface [56], which are then involved in the oxidative reaction of synthetic dyes [60]. The photocatalytic activity of TiO_2 NPs could be improved by enhancing the rate of immobilization of dyes onto the catalyst surface and by suppressing the electron–hole recombination by designing them in the form of composites with polymers [61], clay minerals [62], and ZnS shell structures [63]. In the presence of TiO_2 NPs as catalysts, regardless of their crystalline phase, the photocatalytic degradation rate of synthetic dyes has been found to depend on diffusion, immobilization, and oxidation reactions with $O_2^{-\bullet}$ and $OH^\bullet$ radicals on the catalyst surface [64]. In this sense, there are many metal oxides that exhibit low photocatalytic activity because of the exciton-hole recombination. Such recombination results in low quantum efficiency and, hence, a low probability of generating $O_2^{-\bullet}$ and $OH^\bullet$ radicals on the catalyst surface. The other important parameter is the surface area of the catalyst. It has been revealed that the photocatalysis is an ultrafast process [65], which requires the dyes to be in chemical contact with the radicals on the catalyst surface. This means that if the diffusion of the synthetic dyes onto the catalyst surface is much slower than the radical formation, the generated radicals are dissolved in the solution and such solvated radicals will quickly decay into non-reactive species; thus, no photocatalytic degradation reaction occurs. This results in the breakdown of the conjugated system of the dyes to produce final products such as CO_2, H_2O and other small organic molecules [66–68].

It is very interesting to highlight the effect of a dopant on the electron–hole recombination. In several photocatalytic studies, the bandgap engineering has been successfully performed by the incorporation of metallic dopants, thus lowering the bandgap energy, which has been claimed to optimize the photoexcitation utilizing visible light or sunlight and enhance the photocatalysis of dyes [69]. However, many reports stated otherwise [70,71]. The latter is rationalized by considering that defects in the crystalline structure due to the incorporation of metallic dopants create trapping states (or dopant states), enhancing the rate of the electron–hole recombination. This has also been clearly demonstrated in the photocatalytic conversion of acetic acid to carbon dioxide on metal-doped graphitic carbon [72].

Photocatalysis can be assisted by the presence of a small amount of hydrogen peroxide (H_2O_2) due to its good capability of oxidizing the photogenerated electron on the catalyst surface to form an $OH^\bullet$ radical [73,74]. The peroxide dissociation is also accelerated under UV light irradiation, and eventually leads to the formation of $OH^\bullet$ radicals in the solution that readily oxidizes and degrades the dyes [74]. Therefore, in the presence of H_2O_2, in addition to the photocatalysis process on the catalyst surface when irradiated with light energy, direct oxidation by the formation of an $OH^\bullet$ radical in the solution also contributes to the degradation of synthetic dyes, although the direct degradation produces different mechanisms and products from those resulted from the photocatalysis process. The degradation of dyes due to the spontaneous dissociation of H_2O_2 is relatively slow and inefficient. Nevertheless, the additional H_2O_2 in the photocatalysis is used to examine if the formation of $O_2^{-\bullet}$ and $OH^\bullet$ radicals are the rate-determining step of the photocatalytic degradation of dyes.

It is conceivable that, due to their photoactivity, the catalysts should always generate electron–hole pairs. After their migration onto the catalyst surface, depending on the redox potential of the catalysts, the exciton and hole could form the $O_2^{-\bullet}$ and $OH^\bullet$ radicals [56]. Radical scavengers intentionally added into the mixture of dyes with photocatalysts in an aqueous system, such as tert-butanol and benzoquinone, which scavenge $OH^\bullet$ and $O_2^{-\bullet}$ radicals, respectively, inhibit the oxidation reaction of the dyes. The mechanism of photocatalytic degradation of dyes is therefore evaluated based on the observation of whether the degradation efficiency of the dyes is affected by the additional radical scavengers or not. The bottom line is to find out which kinds of radicals are formed by the catalysts and which kinds of radicals play a role in the oxidative decomposition of the dyes.

Another important successful approach is the combination of the photocatalysis with sonication [75,76]. The additional treatment by applying 20–1000 kHz ultrasonic waves in the photocatalysis has been demonstrated to induce cavitation bubbles, which could generate $H^\bullet$ and $OH^\bullet$ radicals, enhancing the degradation of organic dyes. It should be understood that, in this so-called sonophotocatalysis, the intrinsic photocatalytic activity of catalysts and photocatalytic degradation of hydrophilic dyes are unchanged. However, the additional sonication is very useful in degrading hydrophobic compounds in wastewater containing multiple synthetic dyes via sonolysis [75,76].

As mentioned above, owing to their good photoactivity and suitable bandgap with the incident light excitation, metal oxides such as TiO_2, ZnO, WO_3, Mn_3O_4, CeO_2, and $BiVO_4$ NPs are excellent photocatalysts. These photoactive metal oxides have some advantages of being easy to manufacture into a variety of morphologies and nanostructures (such as nanospheres, nanorods, nanowires, and nanofibers), abundant, and inexpensive [77]. In addition to their suitable redox potential to reduce or oxidize oxygen and water, the reusability and stability of photocatalytic materials have also been of interest [78]. In this sense, some of the metal oxides suffer from several drawbacks. For instance, TiO_2 and ZnO NPs are amphoteric and have poor chemical stability at high pH of a medium [79], $BiVO_4$ and Fe_2O_3 NPs have poor charge carrier mobility in the crystal lattice and slow oxygen kinetics on the particle surface, whereas WO_3 NPs undergo peroxidation and photocorrosion during the photocatalytic process [80]. Nevertheless, as the crystal structure and the surface of the mentioned metal oxides remain unchanged after the photoexcitation and washing with water, the photoactive catalysts can be recovered, regenerated, and reused [81]. Recycling of the nanostructured photocatalysts from electronic waste, such as printed circuit boards and batteries, has also opened up a new pathway of photocatalysis of recalcitrant dyes to be more efficient in terms of synthesis cost, environmental risks, electronic waste valorization, and waste disposal management [82].

3. Synthetic Dyes

Most of the synthetic dyes utilized by the textile clothes, fabrics, paints, and foods available in the market are categorized as chromophoric or auxochromic dyes. These dyes are classified into direct dyes, reactive dyes, indigo dyes, mordant dyes, nitro dyes, azo dyes, sulfonate dyes, basic dyes, and acidic dyes, depending on their chemical structure and their functional groups [9,83,84]. Common chromophoric dyes include azo (-N=N-), carbonyl (-C=O), ethenyl (-C=C-), nitro (-NO_2 or =NO-OH), imino (-C=N-), thiocarbonyl (-CH=S), nitroso (-N=O) moieties. Auxochromic dyes are related to their functional groups, such as amino (-NH_2), carboxylic (-COOH), suphonyl (-SO_3H), alkoxyl (-OR), hydroxyl (-OH), and quinoid rings [9,85]. The number of functional groups present in dyes is responsible for the intensity of color. Amongst these chemically synthesized dyes, azo dyes containing auxochromes are widely used by textile, paper, leather, plastic, and rubber industries in large quantities, e.g., 60–70% of the total synthetic dyes, due to their push–pull effect on intramolecular electron delocalization and, hence, intense light absorption, bright emission, and excellent speed properties [9,83,84,86]. The azo dyes are well-known for their good solubility in water and their harmful impacts on aquatic organisms [87] and human health [88] when discharged directly into water systems.

In the literature, there are still no reports of the effluents of synthetic dyes that are mixed in industrial wastewater, but it is believed that textile industries may use multiple coloring agents at the same time in the dyeing process and produce multiple dyes in their wastewater [11]. Although the impacts of water containing multiple dyes have also not been reported, one could consider that possible synergistic effects of the dye effluents might cause more harmful health effects than wastewater of single solute solutions of dyes. The multiple dyes also have different photocatalytic degradation reactions due to their chemical structures and functional groups. In many studies, auxochromic dyes are demonstrated to undergo photocatalytic degradation reaction easily [37–40]. As it has been discussed in the literature for methylene blue, rhodamine B, crystal violet, and auramine O, the auxochromic dyes become unstable upon reaction with $O_2^-\bullet$ and $OH^\bullet$ radicals and undergo N-demethylation or N-deethylation through successive oxidation reactions [37–40]. The conjugated structure of the dyes is destructed via various intermediates into small volatile compounds, such as CO_2, NH_2, and NH_4^+, which may gasify and escape from the solution [37–40,89].

Comparing the rate and efficiency of the photocatalytic degradation of different auxochromic dyes is not an easy task because of the different photocatalytic setups and experimental conditions in each reported paper. By comparing the photocatalytic degradation of six different auxochromic dyes, i.e., methylene blue, rhodamine B, crystal violet, methyl violet, brilliant green, and malachite green, Zulmajdi et al. reported that although the dyes have slightly different degradation rates, they have almost similar efficiencies, and all of them can be completely depleted [15]. However, this finding still does not resolve the factors that control the photocatalytic degradation of auxochromic dyes, which could be related to their chemical structures. The important findings are given by the photocatalytic degradation behavior of methylene blue and rhodamine B in the binary solute solution [45] and methylene blue, methyl orange, and rhodamine B in the ternary solute solution [90], ensuring all the experimental conditions are the same. It was revealed that the dyes with planar structure and smaller size have a faster rate and a higher efficiency of photocatalytic degradation [45,90]. This provides an interpretation that the rate and efficiency of photocatalytic degradation of dyes are controlled by their diffusion in the solution and their immobilization dynamics onto the catalyst surface, in addition to the potential energy barrier of the oxidation reaction of the dyes.

4. Parameters Affecting the Rate of Photodegradation of Dyes

In heterogeneous photocatalysis, a typical experimental setup should consist of a suitable light source (UV lamp, visible lamp, or natural light source such as sunlight), magnetic stirrer, and a reactor or petri dish containing a dye solution, photocatalyst, and a magnetic bead. Although sunlight is naturally available in abundance, which can be regarded as an alternative cost-effective light source, artificial UV irradiation provides better reproducible results than sunlight with a higher degradation efficiency of dyes [91]. As mentioned above, the use of UV irradiation in photocatalysis could minimize the spontaneous photolysis of dyes.

As mentioned above, the intrinsic properties of photocatalysts, including the crystallinity and surface chemistry, play a role in the photocatalytic activity of metal oxide catalysts. In addition, it is well-known that catalysts with smaller particle sizes, which have a higher surface-to-volume ratio, have higher photocatalytic activity. In this sense, the energy band structure remains intact, suggesting that the particle size does not change the electronic structure of the catalyst; however, one could consider that the surface phonon is enhanced, and the excited energy is highly confined. Therefore, it is also widely accepted that the particle size determines the charge-carrier dynamics. In this sense, the electron–hole recombination tends to reduce in the catalysts with smaller sizes, and hence, their photocatalytic activity is enhanced [92,93].

The performance of catalysts and photocatalytic degradation of dyes in single, binary, or ternary solute solutions are also determined by several conventional non-fundamental parameters, such as the irradiation time, initial concentration, catalyst dosage, and temperature. The effect of irradiation time is used to decipher the photocatalytic degradation kinetics. Based on the assumption that dyes degrade according to the first-order kinetic reaction, as described by the Langmuir–Hinshelwood (L–H) model [94–96], the kinetics of the heterogeneous photocatalytic degradation of each dye is usually evaluated by a single exponential decay function, as given by

$$C_t = C_0 \ \exp(-k_{obs}t) \tag{1}$$

where k_{obs} is the observed photocatalytic degradation rate constant and is determined from the single exponential decay of C_t as a function of irradiation time, t. It should be noted that this L–H model is, however, a simplified process of photocatalysis, which involves complex multistep processes. The most important consideration of this model is that there is no interaction between dyes in solution, and all of them would be completely degraded, regardless of the k_{obs} value. In other words, the dyes will reach 100% efficiency of photocatalytic degradation. In fact, in many cases, the dyes are not completely degraded, and thus have lower photodegradation efficiency (η), much lower than 100%, even under elongated continuous light irradiation. In this sense, it is proposed that the degradation reaction of dyes involves many transition states and various intermediates, or the reaction undergoes different pathways so that there is a backward reaction, and the photocatalytic degradation eventually reaches an equilibrium. By comparing the effects of the conventional non-fundamental parameters on the photocatalytic degradation of dyes, such as k_{obs} and η, in their single solutions and in binary or tertiary solutions, as summarized in Table 1, one could anticipate simultaneous photocatalysis of the multiple dyes with respect to their individual photocatalytic degradation.

Table 1. Photocatalytic experimental conditions and observed photocatalytic degradation efficiency and rate of methylene blue, rhodamine B, and methyl orange in their single solute, binary, or ternary solutions.

Experimental Conditions	Single Solute Solution			Binary or Ternary Solutions	Ref.
pH 6.7 Concentration: Methylene blue = 2.0 × 10⁻⁵ M Rhodamine B = 2.0 × 10⁻⁵ M Catalyst: TiO₂ 0.25 g/L Irradiation time: 180 min	Methylene blue η = 93.1% k_{obs} = 0.173 min⁻¹ D = 6.74 × 10⁻⁶ cm² s⁻¹	Rhodamine B η = 96.1% k_{obs} = 0.115 min⁻¹ D = 4.50 × 10⁻⁶ cm² s⁻¹		Methylene blue η = 92.3% k_{obs} = 0.151 min⁻¹ Rhodamine B η = 20.5% k_{obs} = 0.025 min⁻¹	[45]
pH 6.5 Concentration: Methylene blue = 1.56 × 10⁻⁵ M Rhodamine B = 1.04 × 10⁻⁵ M Methyl orange = 1.53 × 10⁻⁵ M Catalyst:BaTiO₃/Bi₂WO₆ 0.03 g/L Irradiation time: 120 min	Methylene blue η = 99.9% k_{obs} = 0.017 min⁻¹ D = 6.74 × 10⁻⁶ cm² s⁻¹	Rhodamine B η = 94.1% k_{obs} = 0.023 min⁻¹ D = 4.50 × 10⁻⁶ cm² s⁻¹	Methyl orange η = 69.8% k_{obs} = 0.010 min⁻¹ D = 7.53 × 10⁻⁶ cm² s⁻¹	Methylene blue η = 34.9% Rhodamine B η = 5.5% Methyl orange η = 83.5%	[90]

As has been reported in many photocatalysis of dyes in single or binary solute solutions, the initial concentration of the dyes affects their photocatalytic degradation. It is rational to consider that higher initial concentration of dyes generates a competitive effect of the dye molecules to diffuse in close vicinity to the catalyst surface. As a result, their photodegradation rate decreases. It is also noteworthy that high initial concentration of dyes can hinder light penetration into the solution. This so-called screen effect may reduce the number of incident photons to excite the catalysts. Therefore, the removal rate tends to decrease exponentially with the initial dye concentration at the same irradiation time [97].

It was noted that an increase in catalyst dosage results in an increase in photodegradation rate up to a certain weight and saturates at high catalyst dosages [34]. An increase in catalyst dosage also indicates a larger surface area, which provides more exposure for $O_2^{-\bullet}$ and $OH^{\bullet}$ radicals. Because the dye degradation rate is directly proportional to the mass of photocatalyst, above the established saturation point, the reaction rates may be decelerated. This can be suggested due to the prevention of light from penetrating into the solution due to an agglomeration of the photocatalyst, and as a consequence of overdosing, the number of radicals is decreased, thus resulting in a decrease in photocatalytic performance [15,98].

The effect of temperature on photocatalytic degradation can be considered as an important factor in term of diffusion. In many photocatalysis studies of dyes in single and binary solute solutions, increasing the reaction temperature in the range of 10–50 °C enhances k_{obs} [45,89], suggesting that, within this temperature range, the photocatalytic degradation rate was increased due to higher diffusion of the dyes onto the catalyst surface [99,100]. Based on the temperature-dependent k_{obs}, one could always determine the activation energy (E_a) of the photocatalytic degradation using the Arrhenius equation;

$$k_{obs} = A \exp(-E_a/RT) \quad (2)$$

where A is the pre-exponential factor, R is the gas constant, and T is the temperature. In this sense, the E_a value just reflects an average energy barrier and thermodynamic favorability of the photocatalytic degradation reaction of individual dyes, but it is not necessarily that of the oxidation reaction between the dyes and the generated $O_2^{-\bullet}$ and $OH^{\bullet}$ radicals.

The effect of pH on photocatalytic degradation has been explored, but it is not straightforward, as the pH of the medium affects the chemical properties of both catalysts and dyes. In particular, charge on dyes and the characteristics of the catalyst surface can be modified by pH. Consequently, the diffusion and photocatalytic degradation of the dyes, as well as the reaction mechanism of the photoinduced dye degradation, are affected, as summarized by Reza et al. [101]. In this sense, the interaction between the dyes and the catalyst surface, and hence, the dynamics of the immobilization and reactivity of the dyes with the photogeneration on the surface of the catalyst, are modified. The pH also shows a pronounced effect on the electrostatic interaction between dyes and the catalyst surface. Abdellah et al. [102], Suhaimi et al. [45], and Kong et al. [89] reported that the photocatalytic degradation of methylene blue, rhodamine B, and auramine O on TiO_2 NPs generally increased at a pH value of 3 to 7. In this sense, these results are explained by considering the pKa of the dyes and the point of zero charge of anatase TiO_2 NPs (pH 6.1) [103]. Thus, TiO_2 NPs have a positive charge at low pH and gain a negative charge at pH > 6.1 towards alkalinity, modifying the electrostatic interactions and dynamics of immobilization of dyes on the catalyst surface. It is interesting to note that Suhaimi et al. showed a similar effect of pH on the photocatalytic degradation of methylene blue and rhodamine B in a binary solution, suggesting that each dye maintains its respective diffusional mass transfer onto the catalyst surface over a wide range of pH [45].

In other studies, the dynamics of immobilization of dyes has been anticipated, so that the effect of the adsorption of the dyes on their photocatalytic degradation was evaluated after keeping the mixture of dyes with TiO_2 NPs in an aqueous system in the dark [104]. In this sense, the photocatalytic degradation of dyes effectively occurred after the mixture was kept for 48 h in the dark, suggesting that deposition of the dyes onto the catalyst surface due to the adsorption process played an important role. In other words, photocatalysis of the preexisting dyes on the catalyst surface due to Le Chatelier's equilibrium principle improves the degradation efficiency. However, during the photocatalytic degradation, it would be more reasonable to consider the dynamics of immobilization rather than the adsorption of the dyes onto the catalyst surfaces, as the adsorption–desorption equilibrium should be disrupted under light irradiation.

5. The Effect of Magnetism of the Photocatalyst in the Photocatalytic Activity

Recently, the effect of the magnetic field on the photocatalytic degradation of dyes on TiO_2 NPs has been reported by Bian et al. [104]. For instance, an applied magnetic field of as low as 0.28 T enhances the degradation rate of methyl orange by 24% as compared with that under zero magnetic field [104]. It was revealed that the macroscopic and microscopic structures of the TiO_2 catalysts are unchanged in the presence or in the absence of a magnetic field, suggesting that the enhanced degradation of the dyes is due to an increase in the photocatalytic process rather than to the nature of the catalyst. Firstly, under an applied magnetic field, the recombination of the photogenerated electron–hole could be suppressed, and thus the charge carriers migrate onto the catalyst surfaces more efficiently [105]. Moreover, the applied magnetic field could polarize the chemical bonds on the catalyst surfaces and induce the Lorentz force on the generated charge carriers [106]. As a result, the applied magnetic field accelerated the movement of charge carriers and the adsorption or immobilization of dyes onto the catalyst surfaces, leading to higher degradation rate of the photocatalysis of the dyes [104].

It is noteworthy that the magnitude of the Lorentz force is linearly related to the magnetic field intensity. Thus, the electron–hole recombination, the migration of charge carriers, the adsorption of dyes onto the catalyst surfaces, and hence, the photocatalytic degradation of dyes are influenced by the intensity of the applied magnetic field [107,108]. A relatively high magnetic field intensity has been revealed to suppress the degradation of dyes. One possible explanation is that the formations of $O_2^{-\bullet}$ and $OH^{\bullet}$ radicals on the catalyst surface are interrupted by the strong Lorentz force exerted on the charge carriers so that the oxidation reaction of the immobilized dyes does not occur. In addition to the magnetic field intensity, the diffusion length of the charge carriers in the crystal structure of the catalyst should also be considered. In this sense, Bian et al. estimated that the diffusion length of the generated electron and hole in a TiO_2 lattice is approximately 10 nm [104]. Therefore, the effect of the Lorentz force exerted on the charge carriers should be more obviously observable when the particle size of TiO_2 is around 10 nm or less, which is comparable to the diffusion length of the generated electron and hole. This conclusion has been demonstrated in the study reported by Bian et al. [104].

6. Diffusion of Dyes onto the Catalyst Surface

With the assumption that the degradation of dyes is proportional to the external mass transfer, the diffusion of dyes onto the catalyst surface can be analyzed based on the time-dependent photocatalytic degradation of the dyes in single, binary, or ternary solute solutions. Among the established diffusion models, the Weber–Morris intraparticle model provides the rate and profile of the external mass transfer of dyes onto the catalyst surface [109];

$$C_0 - C_t = k_i t^{1/2} + C \quad (3)$$

Here, k_i is the diffusion rate, and C is the boundary layer thickness on the catalyst surface.

The profile of external mass transfer of dyes was deduced based on the plot of the intraparticle model. Typically, a fast diffusion and effective degradation of the dyes was observed at an early irradiation time, followed by a slower diffusion at longer irradiation times until the dyes in the solution reached an equilibrium. Extrapolation of the Weber–Morris plot at time zero can provide an interpretation of the boundary layer thickness as well as the different diffusion rate of dyes in the solution and on the catalyst surface.

The diffusion of dyes in solution should be related to their respective hydrodynamic size, as reflected by their diffusion constant (D) value (see Table 1). As the photocatalytic degradation depends on the diffusion and migration of dyes onto the catalyst surface, the diffusion-limited rate constant (k_D) of the dyes should be proportional to their respective D value, as given by the generalized Smoluchowski equation;

$$k_D = 4\pi\sigma D \quad (4)$$

Here, σ is the encounter distance. Consequently, the k_{obs} or k_i value of the dyes is positively related to the respective D value.

By comparing the k_{obs} value of each dye in a single solute solution and those in binary or ternary solute solutions, one could interpret the role of the molecular and conformational structures of the dyes with the rate and efficiency of their photocatalytic degradation. With this in mind, Suhaimi et al. suggested that the planar conformation and relatively smaller size of methylene blue, as reflected by its higher D value, are the reasons for its domination in photocatalytic degradation over rhodamine B in the binary solution of these dyes [45]. Similar findings have been reported by Wang et al. in the photocatalytic degradation of the ternary solution of methylene blue, rhodamine B, and methyl orange using a novel $BaTiO_3/Bi_2WO_6$ heterojunction photocatalyst, where the photodegradation performance was dominated by methyl orange [90]. Recently, Kong et al. pointed out that the non-planar structure of auramine O, with a torsion angle between its two aromatic ring systems being ~40°, and the non-favorable dimethylamino groups attached to the aromatic rings for hydrogen bonding interactions to approach the surface of TiO$_2$ NPs are the reasons for the low k_{obs} value of this dye [89].

Based on the above argument, one could anticipate that the catalytic degradation of organic pesticides on photocatalysts should also be governed by the diffusion of the pesticides onto the catalyst surfaces [110]. However, the photocatalytic degradation of the organic pesticides was suppressed by the presence of activated carbon support [111]. The organic compounds are well adsorbed and fill up the pore structures on the surfaces of the activated carbon/TiO$_2$ composites NPs. The adsorption then hinders the diffusion of the organic compounds onto the catalyst surfaces. This has also been supported by the lower degradation of organic pesticides at their higher concentrations, which acts as a driving force in the adsorption process, most probably due to the faster saturation of the composite surfaces [112].

7. Conclusions

The heterogeneous photocatalysis, which is initiated by light radiation in the presence of metal oxide nanoparticles, has been proven as one of the promising methods for the removal of a large variety of dyes. This brief review focuses on the photocatalytic degradation of dyes for evaluating their overall rate and efficiency based on recently reported photocatalysis of the dyes in single, binary, and ternary solute solutions. While the basic principles of photocatalysis of dyes are explained, the effects of the intrinsic photoactivity of catalyst and the conventional non-fundamental factors are elucidated in detail. The importance of the dynamics of immobilization and reactivity of the dyes with the photogenerated radicals on the surface of the catalyst has been highlighted. The diffusion of dyes onto the catalyst surface is reflected by the average degradation rate of the dyes and can be analyzed based on the degradation kinetics. By comparing the degradation kinetics and diffusion of the dyes, it is revealed that those with planar conformation and smaller size dominate in photocatalytic degradation in their binary or ternary solutions. Overall, this current review article emphasizes the vital roles of the molecular conformation and functional groups of synthetic dyes in their photocatalytic degradation. Nevertheless, further studies of the molecular structure, steric hindrance, diffusion, immobilization onto the catalyst surface, and synergetic-photosensitization effect of the dyes are some plausible factors to be explained and still remain a research challenge in photocatalysis. Another challenge in the future is the recovery, regeneration, and reuse of the photoactive catalysts, as their crystal structures and surfaces should remain unchanged after the photoexcitation. Moreover, deriving nanostructured metal oxides from electronic waste, such as printed circuit boards and batteries, is a challenging research area and opens up a new pathway of photocatalysis of recalcitrant dyes to be more efficient in terms of synthesis cost, environmental risks, electronic waste valorization, and waste disposal management, so that the photocatalysis keeps its record as the cleanest and most eco-friendly process to remove dyes from wastewater.

Author Contributions: Conceptualization, A.U.; resources, A.U., N.A.A.S. and C.P.Y.K.; data curation, N.N.M.S.; writing—original draft preparation, N.A.A.S., C.P.Y.K. and A.U.; validation, A.U., J.H. and M.N.; writing—review and editing, A.U. and J.H.; supervision, A.U. and M.N. All authors have read and agreed to the published version of the manuscript.

Funding: This research received no external funding.

Institutional Review Board Statement: Not applicable.

Informed Consent Statement: Not applicable.

Data Availability Statement: Not applicable.

Acknowledgments: Jonathan Hobley is grateful to National Cheng Kung University's NCKU90 Distinguished Visiting Scholar Program for hosting his research and to MOST for providing research funding under project number 111-2222-E-006-007.

Conflicts of Interest: The authors declare no conflict of interest.

References

1. Hagan, E.; Poulin, J. Statistics of the early synthetic dye industry. *Herit. Sci.* **2021**, *9*, 33. [CrossRef]
2. Madhav, S.; Ahamad, A.; Singh, P.; Mishra, P.K. A review of textile industry: Wet processing, environmental impacts, and effluent treatment methods. *Environ. Qual. Manag.* **2018**, *27*, 31–41. [CrossRef]
3. Chequer, F.M.D.; de Oliveira, G.A.R.; Ferraz, E.R.A.; Cardoso, J.C.; Zanoni, M.V.B.; de Oliveira, D.P. Textile dyes: Dyeing process and environmental impact. In *Eco-Friendly Textile Dyeing and Finishing*; Melih, G.M., Ed.; IntechOpen: London, UK, 2013; pp. 151–176.
4. Verma, A.K.; Dash, R.R.; Bhunia, P. A Review on chemical coagulation/flocculation technologies for removal of colour from textile wastewaters. *J. Environ. Manag.* **2012**, *93*, 154–168. [CrossRef]
5. Berradi, M.; Hsissou, R.; Khudhair, M.; Assouag, M.; Cherkaoui, O.; El Bachiri, A.; El Harfi, A. Textile finishing dyes and their impact on aquatic environs. *Heliyon* **2019**, *5*, e02711. [CrossRef] [PubMed]
6. Routoula, E.; Patwardhan, S.V. Degradation of anthraquinone dyes from effluents: A review focusing on enzymatic dye degradation with industrial potential. *Environ. Sci. Technol.* **2020**, *54*, 647–664. [CrossRef] [PubMed]
7. Lellis, B.; Fávaro-Polonio, C.Z.; Pamphile, J.A.; Polonio, J.C. Effects of textile dyes on health and the environment and bioremediation potential of living organisms. *Biotechnol. Res. Innov.* **2019**, *3*, 275–290. [CrossRef]
8. Carmen, Z.; Daniel, S. Textile organic dyes—Characteristics, polluting effects and separation/elimination procedures from industrial effluents—A critical overview. *Org. Pollut. Ten Years After Stock. Conv. Environ. Anal. Updat.* **2012**, *1*, 56–86.
9. Slama, H.B.; Bouket, A.C.; Pourhassan, Z.; Alenezi, F.N.; Silini, A.; Cherif-Silini, H.; Oszako, T.; Luptakova, L.; Golińska, P.; Belbahri, L. Diversity of synthetic dyes from textile industries, discharge impacts and treatment methods. *Appl. Sci.* **2021**, *11*, 6255. [CrossRef]
10. Carneiro, P.A.; Umbuzeiro, G.A.; Oliveira, D.P.; Zanoni, M.V.B. Assessment of water contamination caused by a mutagenic textile effluent/dyehouse effluent bearing disperse dyes. *J. Hazard. Mater.* **2010**, *174*, 694–699. [CrossRef]
11. Yaseen, D.A.; Scholz, M. Textile dye wastewater characteristics and constituents of synthetic effluents: A critical review. *Int. J. Environ. Sci. Technol.* **2019**, *16*, 1193–1226. [CrossRef]
12. Palani, G.; Arputhalatha, A.; Kannan, K.; Lakkaboyana, S.K.; Hanafiah, M.M.; Kumar, V.; Marella, R.K. Current trends in the application of nanomaterials for the removal of pollutants from industrial wastewater treatment-A review. *Molecules* **2021**, *26*, 2799. [CrossRef] [PubMed]
13. Shahrin, E.W.E.S.; Narudin, N.A.H.; Shahri, N.N.M.; Verinda, S.B.; Nur, M.; Hobley, J.; Usman, A. Adsorption behavior and dynamic interactions of anionic Acid Blue 25 on agricultural waste. *Molecules* **2022**, *27*, 1718. [CrossRef] [PubMed]
14. Verinda, S.B.; Muniroh, M.; Yulianto, E.; Maharani, N.; Gunawan, G.; Amalia, N.F.; Hobley, J.; Usman, A.; Nur, M. Degradation of ciprofloxacin in aqueous solution using ozone microbubbles: Spectroscopic, kinetics, and antibacterial analysis. *Heliyon* **2022**, *8*, e10137. [CrossRef]
15. Zulmajdi, S.L.N.; Zamri, N.I.I.; Yasin, H.M.; Kusrini, E.; Hobley, J.; Usman, A. Comparative study on the adsorption, kinetics, and thermodynamics of the photocatalytic degradation of six different synthetic dyes on TiO_2 nanoparticles. *React. Kinet. Mech. Catal.* **2020**, *129*, 519–534. [CrossRef]
16. Wawrzkiewicz, M.; Hubicki, Z. Anion exchange resins as efective sorbents for removal of acid, reactive, and direct dyes from textile wastewaters. In *Ion Exchange-Studies and Applications*; Kilislioglu, A., Ed.; IntechOpen: Rijeka, Croatia, 2015.
17. Bläker, C.; Muthmann, J.; Pasel, C.; Bathen, D. Characterization of activated carbon adsorbents—State of the art and novel approaches. *Chem. Bio. Eng. Rev.* **2019**, *6*, 119–138. [CrossRef]
18. Okoniewska, E. Removal of selected dyes on activated carbons. *Sustainability* **2021**, *13*, 4300. [CrossRef]
19. Yadav, S.; Asthana, A.; Chakraborty, R.; Jain, B.; Singh, A.K.; Carabineiro, S.A.C.; Susan, M.A.B.H. Cationic dye removal using novel magnetic/activated charcoal/β-cyclodextrin/alginate polymer nanocomposite. *Nanomaterials* **2020**, *10*, 170. [CrossRef]
20. Yadav, S.; Asthana, A.; Singh, A.K.; Chakraborty, R.; Vidya, S.S.; Susan, M.A.B.H.; Carabineiro, S.A.C. Adsorption of cationic dyes, drugs and metal from aqueous solutions using a polymer composite of magnetic/β-cyclodextrin/activated charcoal/Na alginate: Isotherm, kinetics and regeneration studies. *J. Hazard. Mater.* **2021**, *409*, 124840. [CrossRef]
21. Yadav, S.; Asthana, A.; Singh, A.K.; Patel, J.; Vidya, S.S.; Carabineiro, S.A.C. Facile preparation of methionine-functionalized graphene oxide/chitosan polymer nanocomposite aerogel for the efficient removal of dyes and metal ions from aqueous solutions. *Environ. Nanotechnol. Monit. Manag.* **2022**, *18*, 100743. [CrossRef]
22. Crini, G.; Lichtfouse, E.; Wilson, L.D.; Morin-Crini, N. Conventional and non-conventional adsorbents for wastewater treatment. *Environ. Chem. Lett.* **2019**, *17*, 195–213. [CrossRef]
23. Samat, J.H.; Shahri, N.N.M.; Abdullah, M.A.; Suhaimi, N.A.A.; Padmosoedarso, K.M.; Kusrini, E.; Mahadi, A.H.; Hobley, J.; Usman, A. Adsorption of Acid Blue 25 on agricultural wastes: Efficiency, kinetics, mechanism, and regeneration. *Air Soil Water Res.* **2021**, *14*, 11786221211057496. [CrossRef]
24. Zamri, N.I.I.; Zulmajdi, S.L.N.; Daud, N.Z.A.; Mahadi, A.H.; Kusrini, E.; Usman, A. Insight into the adsorption kinetics, mechanism, and thermodynamics of methylene blue from aqueous solution onto pectin-alginate-titania composite microparticles. *SN Appl. Sci.* **2021**, *3*, 222. [CrossRef]

25. Shahrin, E.W.E.S.; Narudin, N.A.H.; Padmosoedarso, K.M.; Kusrini, E.; Mahadi, A.H.; Shahri, N.N.M.; Usman, A. Pectin derived from pomelo pith as a superior adsorbent to remove toxic Acid Blue 25 from aqueous solution. *Carbohydr. Polym. Technol. Appl.* **2021**, *2*, 100116. [CrossRef]
26. Hamad, H.N.; Idrus, S. Recent developments in the application of bio-waste-derived adsorbents for the removal of methylene blue from wastewater: A review. *Polymers* **2022**, *14*, 783. [CrossRef] [PubMed]
27. Hynes, N.R.J.; Kumar, J.S.; Kamyab, H.; Sujana, J.A.J.; Al-Khashman, O.A.; Kuslu, Y.; Ene, A.; Kumar, B.S. Modern enabling techniques and adsorbents based dye removal with sustainability concerns in textile industrial sector—A comprehensive review. *J. Clean. Prod.* **2020**, *272*, 122636. [CrossRef]
28. Asbollah, M.A.; Mahadi, A.H.; Kusrini, E.; Usman, A. Synergistic effect in concurrent removal of toxic methylene blue and acid red-1 dyes from aqueous solution by durian rind: Kinetics, isotherm, thermodynamics, and mechanism. *Int. J. Phytorem.* **2021**, *23*, 1432–1443. [CrossRef] [PubMed]
29. Asbollah, M.A.; Sahid, M.S.M.; Padmosoedarso, K.M.; Mahadi, A.H.; Kusrini, E.; Hobley, J.; Usman, A. Individual and competitive adsorption of negatively charged acid blue and acid red 1 onto raw Indonesian kaolin clay. *Arab. J. Sci. Eng.* **2022**, *47*, 6617–6630. [CrossRef]
30. Asbollah, M.A.; Sahid, M.S.M.; Shahrin, E.W.E.S.; Narudin, N.A.H.; Kusrini, E.; Shahri, N.N.M.; Hobley, J.; Usman, A. Dynamics and thermodynamics for competitive adsorptive removal of methylene blue and rhodamine B from binary aqueous solution onto durian rind. *Environ. Monit. Assess.* **2022**, *194*, 645. [CrossRef]
31. Rehman, M.Z.U.; Aslam, Z.; Shawabkeh, R.A.; Hussein, I.A.; Mahmood, N. Concurrent adsorption of cationic and anionic dyes from environmental water on amine functionalized carbon. *Water Sci. Technol.* **2020**, *81*, 466–478. [CrossRef]
32. Shirazi, E.K.; Metzger, J.W.; Fischer, K.; Hessam, A. Removal of textile dyes from single and binary component systems by Persian bentonite and a mixed adsorbent of bentonite/charred dolomite. *Colloids Surf. A Physicochem. Eng. Asp.* **2020**, *598*, 124807. [CrossRef]
33. Mohan, S.V.; Bhaskar, Y.V.; Karthikeyan, J. Biological decolourisation of simulated azo dye in aqueous phase by algae Spirogyra species. *Int. J. Environ. Pollut.* **2004**, *21*, 211–222. [CrossRef]
34. Herrmann, J.-M. Heterogeneous photocatalysis: Fundamentals and applications to the removal of various types of aqueous pollutants. *Catal. Today* **1999**, *53*, 115–129. [CrossRef]
35. Chen, D.; Cheng, Y.; Zhou, N.; Chen, P.; Wang, Y.; Li, K.; Huo, S.; Cheng, P.; Peng, P.; Zhang, R.; et al. Photocatalytic degradation of organic pollutants using TiO_2-based photocatalysts: A review. *J. Clean. Prod.* **2020**, *268*, 121725. [CrossRef]
36. Lau, G.E.; Abdullah, C.A.C.; Ahmad, W.A.N.W.; Assaw, S.; Zheng, A.L.T. Eco-friendly photocatalysts for degradation of dyes. *Catalysts* **2020**, *10*, 1129. [CrossRef]
37. Anucha, C.B.; Altin, I.; Bacaksiz, E.; Stathopoulos, V.N. Titanium dioxide (TiO_2)-based photocatalyst materials activity enhancement for contaminants of emerging concern (CECs) degradation: In the light of modification strategies. *Chem. Eng. J. Adv.* **2022**, *10*, 100262. [CrossRef]
38. Hisaindee, S.; Meetani, M.A.; Rauf, M.A. Application of LC-MS to the analysis of advanced oxidation process (AOP) degradation of dye products and reaction mechanisms. *TrAC—Trends Anal. Chem.* **2013**, *49*, 31–44. [CrossRef]
39. Wang, X.Q.; Han, S.F.; Zhang, Q.W.; Zhang, N.; Zhao, D.D. Photocatalytic oxidation degradation mechanism study of methylene blue dye waste water with GR/I/TiO_2. *MATEC Web Conf.* **2018**, *238*, 03006. [CrossRef]
40. Guo, S.; Zhu, X.; Zhang, H.; Gu, B.; Chen, W.; Liu, L.; Alvarez, P.J.J. Improving photocatalytic water treatment through nanocrystal engineering: Mesoporous nanosheet-assembled 3D BiOCl hierarchical nanostructures that induce unprecedented large vacancies. *Environ. Sci. Technol.* **2018**, *52*, 6872–6880. [CrossRef]
41. Yatmaz, H.C.; Dizge, N.; Kurt, M.S. Combination of photocatalytic and membrane distillation hybrid processes for reactive dyes treatment. *Environ. Technol.* **2017**, *38*, 2743–2751. [CrossRef]
42. Janczarek, M.; Kowalska, E. On the origin of enhanced photocatalytic activity of copper-modified titania in the oxidative reaction systems. *Catalysts* **2017**, *7*, 317. [CrossRef]
43. Chiu, Y.-H.; Chang, T.-F.M.; Chen, C.-Y.; Sone, M.; Hsu, Y.-J. Mechanistic insights into photodegradation of organic dyes using heterostructure photocatalysts. *Catalysts* **2019**, *9*, 430. [CrossRef]
44. Addamo, M.; Augugliaro, V.; Di Paola, A.; García-López, E.; Loddo, V.; Marcì, G.; Molinari, R.; Palmisano, L.; Schiavello, M. Preparation, characterization, and photoactivity of polycrystalline nanostructured TiO_2 catalysts. *J. Phys. Chem. B* **2004**, *108*, 3303–3310. [CrossRef]
45. Suhaimi, N.A.A.; Shahri, N.N.M.; Samat, J.H.; Kusrini, E.; Lim, J.-W.; Hobley, J.; Usman, A. Domination of methylene blue over rhodamine B during simultaneous photocatalytic degradation by TiO_2 nanoparticles in an aqueous binary solution under UV irradiation. *React. Kinet. Mech. Catal.* **2022**, *135*, 511–527. [CrossRef]
46. Bhatkhande, D.S.; Pangarkar, V.G.; Beenackers, A.A.C.M. Photocatalytic degradation for environmental applications—A review. *J. Chem. Technol. Biotechnol.* **2002**, *77*, 102–116. [CrossRef]
47. Li, J.; Wu, N. Semiconductor-based photocatalysts and photoelectrochemical cells for solar fuel generation: A review. *Catal. Sci. Technol.* **2015**, *5*, 1360–1384. [CrossRef]
48. Wang, J.L.; Xu, L.J. Advanced oxidation processes for wastewater treatment: Formation of hydroxyl radical and application. *Crit. Rev. Environ. Sci. Technol.* **2012**, *42*, 251–325. [CrossRef]

49. Li, H.; Cui, Y.; Hong, W. High photocatalytic performance of BiOI/Bi$_2$WO$_6$ toward toluene and reactive Brilliant Red. *Appl. Surf. Sci.* **2013**, *264*, 581–588. [CrossRef]
50. Dong, P.; Hou, G.; Xi, X.; Shao, R.; Dong, F. WO$_3$-based photocatalysts: Morphology control, activity enhancement and multifunctional applications. *Environ. Sci. Nano* **2017**, *4*, 539–557. [CrossRef]
51. Malathi, A.; Madhavan, J.; Ashokkumar, M.; Arunachalam, P. A review on BiVO$_4$ photocatalyst: Activity enhancement methods for solar photocatalytic applications. *Appl. Catal. A Gen.* **2018**, *555*, 47–74.
52. Mehtab, A.; Ahmed, J.; Alshehri, S.M.; Mao, Y.; Ahmad, T. Rare earth doped metal oxide nanoparticles for photocatalysis: A perspective. *Nanotechnology* **2022**, *33*, 142001. [CrossRef]
53. Etacheri, V.; Di Valentin, C.; Schneider, J.; Bahnemann, D.; Pillai, S.C. Visible-light activation of TiO$_2$ photocatalysts: Advances in theory and experiments. *J. Photochem. Photobiol. C Photochem. Rev.* **2015**, *25*, 1–29. [CrossRef]
54. Xu, C.; Rangaiah, G.P.; Zhao, X.S. Photocatalytic degradation of methylene blue by titanium dioxide: Experimental and modeling study. *Ind. Eng. Chem. Res.* **2014**, *53*, 14641–14649. [CrossRef]
55. Bae, S.; Kim, S.; Lee, S.; Choi, W. Dye decolorization test for the activity assessment of visible light photocatalysts: Realities and limitations. *Catal. Today* **2014**, *224*, 21–28. [CrossRef]
56. Luttrell, T.; Halpegamage, S.; Tao, J.; Kramer, A.; Sutter, E.; Batzill, M. Why is anatase a better photocatalyst than rutile?—Model studies on epitaxial TiO$_2$ films. *Sci. Rep.* **2015**, *4*, 4043. [CrossRef] [PubMed]
57. Mills, A.; Davies, R.H.; Worsley, D. Water purification by semiconductor photocatalysis. *Chem. Soc. Rev.* **1993**, *22*, 417–425. [CrossRef]
58. Qin, Y.L.; Zhao, W.W.; Sun, Z.; Liu, X.Y.; Shi, G.L.; Liu, Z.Y.; Ni, D.R.; Ma, Z.Y. Photocatalytic and adsorption property of ZnS–TiO2/RGO ternary composites for methylene blue degradation. *Adsorpt. Sci. Technol.* **2019**, *37*, 764–776. [CrossRef]
59. Zulmajdi, S.L.N.; Zamri, N., II; Mahadi, A.H.; Rosli, M.Y.H.; Ja'afar, F.; Yasin, H.M.; Kusrini, E.; Hobley, J.; Usman, A. Sol-gel preparation of different crystalline phases of TiO$_2$ nanoparticles for photocatalytic degradation of methylene blue in aqueous solution. *Am. J. Nanomater.* **2019**, *7*, 39–45. [CrossRef]
60. Baudys, M.; Zlámal, M.; Krýsa, J.; Jirkovský, J.; Kluson, P. Notes on heterogeneous photocatalysis with the model azo dye acid orange 7 on TiO$_2$. *Reac. Kinet. Mech. Cat.* **2012**, *106*, 297–311. [CrossRef]
61. Kubacka, A.; Fernández-García, M.; Cerrada, M.L.; Fernández-García, M. Titanium dioxide–polymer nanocomposites with advanced properties. In *Nano-Antimicrobials*; Cioffi, N., Rai, M., Eds.; Springer: Heidelberg, Germany, 2012.
62. Kamaluddin, M.R.; Zamri, N.I.I.; Kusrini, E.; Prihandini, W.W.; Mahadi, A.H.; Usman, A. Photocatalytic activity of kaolin-titania composites to degrade methylene blue under UV light irradiation; Kinetics, mechanism and thermodynamics. *Reac. Kinet. Mech. Cat.* **2021**, *113*, 517–529. [CrossRef]
63. Li, X.; Shi, Z.; Liu, J.; Wang, J. Shell thickness dependent photocatalytic activity of TiO$_2$/ZnS core-shell nanorod arrays. *Mater. Res. Express* **2019**, *6*, 1250b3. [CrossRef]
64. Rochkind, M.; Pasternak, S.; Paz, Y. Using dyes for evaluating photocatalytic properties: A critical review. *Molecules* **2015**, *20*, 88–110. [CrossRef] [PubMed]
65. Andreozzi, R.; Caprio, V.; Insola, A.; Marotta, R. Advanced oxidation processes (AOP) for water purification and recovery. *Catal. Today* **1999**, *53*, 51–59. [CrossRef]
66. Ren, G.; Han, H.; Wang, Y.; Liu, S.; Zhao, J.; Meng, X.; Li, Z. Recent advances of photocatalytic application in water treatment: A review. *Nanomaterials* **2021**, *11*, 1804. [CrossRef] [PubMed]
67. Javaid, R.; Qazi, U.Y. Catalytic oxidation process for the degradation of synthetic dyes: An overview. *Int. J. Environ. Res. Public Health* **2019**, *16*, 2066. [CrossRef] [PubMed]
68. Hmani, E.; Samet, Y.; Abdelhédi, R. Electrochemical degradation of auramine-O dye at boron-doped diamond and lead dioxide electrodes. *Diam. Relat. Mater.* **2012**, *30*, 2006. [CrossRef]
69. Huang, F.; Yan, A.; Zhao, H. Influences of doping on photocatalytic properties of TiO2 photocatalyst. In *Semiconductor Photocatalysis-Materials, Mechanisms and Applications*; Cao, W., Ed.; IntechOpen Limited: London, UK, 2016; pp. 31–80.
70. Smyth, D.M. The effects of dopants on the properties of metal oxides. *Solid State Ion.* **2000**, *129*, 5–12. [CrossRef]
71. Gatti, T.; Lamberti, F.; Mazzaro, R.; Kriegel, I.; Schlettwein, D.; Enrichi, F.; Lago, N.; Di Maria, E.; Meneghesso, G.; Vomiero, A.; et al. Opportunities from doping of non-critical metal oxides in last generation light-conversion devices. *Adv. Energy Mater.* **2021**, *11*, 2101041. [CrossRef]
72. Sakuna, P.; Ketwong, P.; Ohtani, B.; Trakulmututa, J.; Kobkeatthawin, T.; Luengnaruemitchai, A.; Smit, S.M. The influence of metal-doped graphitic carbon nitride on photocatalytic conversion of acetic acid to carbon dioxide. *Front Chem.* **2022**, *10*, 825786. [CrossRef]
73. Habib, I.Y.; Burhan, J.; Jaladi, F.; Lim, C.H.; Usman, A.; Kumara, N.T.R.N.; Tsang, S.C.E.; Mahadi, A.H. Effect of Cr doping in CeO$_2$ nanostructures on photocatalysis and H$_2$O$_2$ assisted methylene blue. *Catal. Today* **2021**, *375*, 506–513. [CrossRef]
74. Hirakawa, T.; Yawata, K.; Nosaka, Y. Photocatalytic reactivity for O$_2^-$ and OH radical formation in anatase and rutile TiO$_2$ suspension as the effect of H$_2$O$_2$ addition. *Appl. Catal. A General* **2007**, *325*, 105–111. [CrossRef]
75. Patel, J.; Singh, A.K.; Jain, B.; Yadav, S.; Carabineiro, S.A.C.; Susan, M.A.B.H. Iochrome dark blue azo dye removal by sonophoto-catalysis using Mn^{2+} doped ZnS quantum dots. *Catalysts* **2021**, *11*, 1025. [CrossRef]
76. Panda, D.; Manickam, S. Recent advancements in the sonophotocatalysis (SPC) and doped-sonophotocatalysis (DSPC) for the treatment of recalcitrant hazardous organic water pollutants. *Ultrason. Sonochem.* **2017**, *36*, 481–496. [CrossRef] [PubMed]

77. Borjigin, T.; Schmitt, M.; Morlet-Savary, F.; Xiao, P.; Lalevée, J. Low-cost and recyclable photocatalysts: Metal oxide/polymer composites applied in the catalytic breakdown of dyes. *Photochem* **2022**, *2*, 733–751. [CrossRef]
78. Danish, M.S.S.; Bhattacharya, A.; Stepanova, D.; Mikhaylov, A.; Grilli, M.L.; Khosravy, M.; Senjyu, T. A systematic review of metal oxide applications for energy and environmental sustainability. *Metals* **2020**, *10*, 1604. [CrossRef]
79. Wang, J.; Van Ree, T.; Wu, Y.; Zhang, P.; Gao, L. Metal oxide semiconductors for solar water splitting. In *Metal Oxides Energy Technologies*; Wu, Y., Ed.; Elsevier: Amsterdam, The Netherlands, 2018; pp. 205–249.
80. Zheng, G.; Wang, J.; Liu, H.; Murugadoss, V.; Zu, G.; Che, H.; Lai, C.; Li, H.; Ding, T.; Gao, Q. Tungsten oxide nanostructures and nanocomposites for photoelectrochemical water splitting. *Nanoscale* **2019**, *11*, 18968–18994. [CrossRef]
81. Bockenstedt, J.; Vidwans, N.A.; Gentry, T.; Vaddiraju, S. Catalyst recovery, regeneration and reuse during large-scale disinfection of water using photocatalysis. *Water* **2021**, *13*, 2623. [CrossRef]
82. Bahadoran, A.; De Lile, J.R.; Masudy-Panah, S.; Sadeghi, B.; Li, J.; Sabzalian, M.H.; Ramakrishna, S.; Liu, Q.; Cavaliere, P.; Gopinathan, A. Photocatalytic materials obtained from e-waste recycling: Review, techniques, critique, and update. *J. Manuf. Mater. Process.* **2022**, *6*, 69. [CrossRef]
83. Pandey, G.P.; Singh, A.K.; Deshmukh, L.; Asthana, A.; Deo, S.R. Photocatalytic degradation of an azo dye with ZnO nanoparticles. *AIP Conf. Proc.* **2013**, *1536*, 243–244.
84. Ghaly, A.E.; Ananthashankar, R.; Alhattab, M.; Ramakrishnan, V.V. Production, characterization and treatment of textile effluents: A critical review. *J. Chem. Eng. Process Technol.* **2014**, *5*, 1000182.
85. Kress, K.C.; Fischer, T.; Stumpe, J.; Frey, W.; Raith, M.; Beiraghi, O.; Eichhorn, S.H.; Tussetschläger, S.; Laschat, S. Influence of chromophore length and acceptor groups on the optical properties of rigidified merocyanine dyes. *Chem. Plus. Chem.* **2014**, *79*, 223–232.
86. Zhang, L.; Shao, Q.; Xu, C. Enhanced azo dye removal from wastewater by coupling sulfidated zero-valent iron with a chelator. *J. Clean. Prod.* **2019**, *213*, 753–761. [CrossRef]
87. Almroth, B.C.; Cartine, J.; Jönander, C.; Karlsson, M.; Langlois, J.; Lindström, M.; Lundin, J.; Melander, N.; Pesqueda, A.; Rahmqvist, I.; et al. Assessing the effects of textile leachates in fish using multiple testing methods: From gene expression to behavior. *Ecotoxicol. Environ. Saf.* **2021**, *207*, 111523. [CrossRef] [PubMed]
88. Al-Tohamy, R.; Ali, S.S.; Li, F.; Okasha, K.M.; Mahmoud, Y.A.-G.; Elsamahy, T.; Jiao, H.; Fu, Y.; Sun, J. A critical review on the treatment of dye-containing wastewater: Ecotoxicological and health concerns of textile dyes and possible remediation approaches for environmental safety. *Ecotoxicol. Environ. Saf.* **2022**, *231*, 113160. [CrossRef] [PubMed]
89. Kong, C.P.Y.; Suhaimi, N.A.A.; Shahri, N.N.M.; Lim, J.-W.; Nur, M.; Hobley, J.; Usman, A. Auramine O UV photocatalytic degradation on TiO$_2$ nanoparticles in a heterogeneous aqueous solution. *Catalysts* **2022**, *12*, 975. [CrossRef]
90. Wang, Y.; Sun, X.; Xian, T.; Liu, G.; Yang, H. Photocatalytic purification of simulated dye wastewater in different pH environments by using BaTiO$_3$/Bi$_2$WO$_6$ heterojunction photocatalysts. *Opt. Mater.* **2021**, *113*, 110853. [CrossRef]
91. Wang, L.; Zhao, J.; Liu, H.; Huang, J. Design, Modification and application of semiconductor photocatalysts. *J. Taiwan Inst. Chem. Eng.* **2018**, *93*, 590–602. [CrossRef]
92. Fujishima, A.; Zhang, X.; Tryk, D.A. TiO$_2$ photocatalysis and related surface phenomena. *Surf. Sci. Rep.* **2008**, *63*, 515–582. [CrossRef]
93. Zhang, H.; Banfield, J.F. Understanding polymorphic phase transformation behavior during growth of nanocrystalline aggregates: Insights from TiO$_2$. *J. Phys. Chem. B* **2000**, *104*, 3481–3487. [CrossRef]
94. Kumar, K.V.; Porkodi, K.; Rocha, F. Langmuir-Hinshelwood kinetics—A theoretical study. *Catal. Commun.* **2008**, *9*, 82–84. [CrossRef]
95. Armenise, S.; García-Bordejé, E.; Valverde, J.L.; Romeo, E.; Monzón, A. A Langmuir-Hinshelwood approach to the kinetic modelling of catalytic ammonia decomposition in an integral reactor. *Phys. Chem. Chem. Phys.* **2013**, *15*, 12104–12117. [CrossRef]
96. Shaban, Y.A. Solar light-induced photodegradation of chrysene in seawater in the presence of carbon-modified n-TiO$_2$ nanoparticles. *Arab. J. Chem.* **2019**, *12*, 652–663. [CrossRef]
97. Chen, X.; Wu, Z.; Liu, D.; Gao, Z. Preparation of ZnO photocatalyst for the efficient and rapid photocatalytic degradation of azo dyes. *Nanoscale Res. Lett.* **2017**, *12*, 4–13. [CrossRef] [PubMed]
98. Nezamzadeh-Ejhieh, A.; Shirvani, K. CdS loaded an Iranian clinoptilolite as a heterogeneous catalyst in photodegradation of p-Aminophenol. *J. Chem.* **2013**, *2103*, 541736. [CrossRef]
99. Miura, M.; Cole, C.-A.; Monji, N.; Hoffman, A.S. Temperature-dependent adsorption/desorption behavior of lower critical solution temperature (LCST) polymers on various substrates. *J. Biomater. Sci. Polym. Ed.* **1994**, *5*, 555–568. [CrossRef] [PubMed]
100. Panić, V.V.; Šešlija, S.I.; Nešić, A.R.; Veličković, S.J. Adsorption of azo dyes on polymer materials. *Hem. Ind.* **2013**, *67*, 881–900. [CrossRef]
101. Reza, K.M.; Kurny, A.; Gulshan, F. Parameters affecting the photocatalytic degradation of dyes using TiO$_2$: A review. *Appl. Water Sci.* **2015**, *7*, 1569–1578. [CrossRef]
102. Abdellah, M.H.; Nosier, S.A.; El-Shazly, A.H.; Mubarak, A.A. Photocatalytic decolorization of methylene blue using TiO$_2$/UV system enhanced by air sparging. *Alexandria Eng. J.* **2018**, *57*, 3727–3735. [CrossRef]
103. Nguyen, L.T.; Nguyen, H.T.; Pham, T.D.; Tran, T.D.; Chu, H.T.; Dang, H.T.; Nguyen, V.H.; Nguyen, K.M.; Pham, T.T.; van der Bruggen, B. UV–visible light driven photocatalytic degradation of ciprofloxacin by N, S co-doped TiO$_2$: The effect of operational parameters. *Top. Catal.* **2020**, *63*, 985–995. [CrossRef]

104. Bian, Y.; Zheng, G.; Ding, W.; Hu, L.; Sheng, Z. Magnetic field effect on the photocatalytic degradation of methyl orange by commercial TiO_2 powder. *RSC Adv.* **2021**, *11*, 6284. [CrossRef]
105. Gao, W.; Lu, J.; Zhang, S.; Zhang, X.; Wang, Z.; Qin, W.; Wang, J.; Zhou, W.; Liu, H.; Sang, Y. Suppressing photoinduced charge recombination via the Lorentz force in a photocatalytic system. *Adv. Sci.* **2019**, *6*, 1901244. [CrossRef]
106. Hsieh, T.H.; Keh, H.J. Electrokinetic motion of a charged colloidal sphere in a spherical cavity with magnetic fields. *J. Chem. Phys.* **2011**, *134*, 044125. [CrossRef] [PubMed]
107. Okumura, H.; Endo, S.; Joonwichien, S.; Yamasue, E.; Ishihara, K.N. Magnetic field effect on heterogeneous photocatalysis. *Catal. Today* **2015**, *258*, 634–647. [CrossRef]
108. Li, J.; Pei, Q.; Wang, R.; Zhou, Y.; Zhang, Z.; Cao, Q.; Wang, D.; Mi, W.; Du, Y. Enhanced photocatalytic performance through magnetic field boosting carrier transport. *ACS Nano* **2018**, *12*, 3351–3359. [CrossRef] [PubMed]
109. Sağ, Y.; Aktay, Y. Mass transfer and equilibrium studies for the sorption of chromium ions onto chitin. *Process Biochem.* **2000**, *36*, 157–173. [CrossRef]
110. Shahawy, A.; Ragab, A.H.; Mubarak, M.F.; Ahmed, I.A.; Mousa, A.E.; Bader, D.M.D. Removing the oxamyl from aqueous solution by a green synthesized $HTiO_2$@AC/SiO_2 nanocomposite: Combined effects of adsorption and photocatalysis. *Catalysts* **2022**, *12*, 163. [CrossRef]
111. Pinna, M.; Binda, G.; Altomare, M.; Marelli, M.; Dossi, C.; Monticelli, D.; Spanu, D.; Recchia, S. Biochar nanoparticles over TiO_2 nanotube arrays: A green co-catalyst to boost the photocatalytic degradation of organic pollutants. *Catalysts* **2021**, *11*, 1048. [CrossRef]
112. Wang, Y.; Lin, C.; Liu, X.; Ren, W.; Huang, X.; He, M.; Ouyang, W. Efficient removal of acetochlor pesticide from water using magnetic activated carbon: Adsorption performance, mechanism, and regeneration exploration. *Sci. Total Environ.* **2021**, *778*, 146353. [CrossRef]

Article

Nitrogen and Sulfur Co-Doped Graphene Quantum Dots Anchored TiO₂ Nanocomposites for Enhanced Photocatalytic Activity

Jishu Rawal [1], Urooj Kamran [1,2], Mira Park [3,4], Bishweshwar Pant [3,4,*] and Soo-Jin Park [1,*]

[1] Department of Chemistry, Inha University, 100 Inharo, Incheon 22212, Korea; rawaljishu26@gmail.com (J.R.); malikurooj9@gmail.com (U.K.)
[2] Department of Mechanical Engineering, College of Engineering, Kyung Hee University, Yongin 17104, Korea
[3] Carbon Composite Energy Nanomaterials Research Center, Woosuk University, Wanju, Chonbuk 55338, Korea; wonderfulmira@woosuk.ac.kr
[4] Woosuk Institute of Smart Convergence Life Care (WSCLC), Woosuk University, Wanju, Chonbuk 55338, Korea
* Correspondence: bisup@woosuk.ac.kr (B.P.); sjpark@inha.ac.kr (S.-J.P.)

Abstract: Herein, nitrogen (N) and sulfur (S) co-doped graphene quantum dots (GQDs) using different one-dimensional (1-D) carbon nanomaterials as precursors were synthesized, followed by heterojunction formation with TiO₂. GQDs exhibit unlike physiochemical properties due to the disproportionate ratio of N and S heteroatoms and dissimilar reaction parameters. Tailored type-II band gap (E_g) alignment was formed with narrowed E_g value that improves photogenerated electron transfer due to π-conjugation. GQDs-TiO₂ nanocomposites exhibit remarkably high methylene blue (MB) degradation up to 99.78% with 2.3–3 times elevated rate constants as compared with TiO₂. CNF-GQDs-TiO₂ demonstrates the fastest MB degradation (60 min) due to the synergistic effect of nitrogen and sulfur doping, and is considered the most stable photocatalyst among prepared nanocomposites as tested up to three cyclic runs. Whereas, C–O–Ti bonds were not only responsible for nanocomposites strengthening but also provide a charge transfer pathway. Moreover, charge transport behavior, generation of active species, and reaction mechanism were scrutinized via free-radical scavenger analysis.

Keywords: photocatalytic degradation; graphene quantum dots; nanocomposites; free radical generation

1. Introduction

Environmental catastrophes and energy crises have become the main challenges in recent decades because of their severe social and economic impact on humankind [1,2]. Heterogeneous photocatalysis has recently attracted enormous attention because of its green attributes and efficient applicability. Due to their favorable properties such as suitable bandgaps, good chemical stability, and strong oxidizing ability, semiconductor photocatalysts [3–5] (e.g., ZnO, WO₃, CdS, CdSe, PbS, TiO₂, and SnO₂) and some biocomposites [6,7] have been extensively used to decompose organic pollutants [8]. Among the semiconductor photocatalysts, titanium dioxide (TiO₂) appears to be a highly efficient, stable, nontoxic, and inexpensive photocatalyst [9,10]. TiO₂ containing 80% anatase and 20% rutile, named Degussa P25 TiO₂, is widely known to exhibit high performance because the mixed-phase material exhibits a slower recombination rate, higher photo efficiency, and lower light activation energy as compared with pure-phase anatase or rutile [11,12]. Despite its wide range of applications, good performance, and high stability, the use of TiO₂ is restricted by its narrow ultraviolet light range response resulting from the intrinsic wide energy bandgap of the anatase phase [13] and its low quantum efficiency due to the high recombination rate of photogenerated charge carriers in the rutile phase [14].

Various methods have been used to further expand the optical UV-vis absorption region of TiO_2 and to enhance the efficient light-induced charge separation. Numerous approaches to modifying TiO_2 via doping [15,16], semiconductor coupling [17], composite formation with carbon nanomaterials [18–21], and chemical treatment [22] have been explored to overcome the aforementioned shortcomings [23]. The best option reported thus far for spanning the UV-vis spectrum is combining TiO_2 with typical narrow-bandgap quantum dots (QDs) of different semiconductors and carbon quantum dots (CQDs) [24–28]. To establish a proper channel among the QDs and TiO_2 for effective induced charge separation, heterojunction formation is critical. Despite the incorporation of QDs, photocatalytic activity has not been remarkably improved because of the high instability, rich surface traps, and inevitable photo-oxidation of QDs [29]. Moreover, the use of heavy-metal ions (e.g., Zn, Cu, Cd, and Pb) which are toxic, raises concerns about environmental and human safety [30]. Therefore, the development of an eco-friendly, highly efficient, and recyclable catalyst that can efficiently harvest solar energy and form heterojunctions under suitable conditions is a formidable challenge.

Recently, CQDs have emerged as materials that introduce new possibilities for the development of heterojunctions because of their novel electronic and optical properties with improved photocatalytic activity. As a unique class of CQDs, graphene quantum dots (GQDs) are a nontoxic metal-free zero-dimensional (0D) material usually composed of one or a few layers of graphene and with diameters in the nanometer range [31]. GQDs have a wide range of applications in energy conversion, bio-imaging, drug delivery, sensing, solar energy conversion [32], and light-emitting diodes [33]. Moreover, they exhibit strong fluorescence and semiconducting properties that arise from their pronounced edge effects, quantum confinement, and functionalities [34]. The substantial UV-vis absorption capability of GQDs is provided by the π-plasmon absorption effect [35]. GQDs possess a tunable wide energy bandgap that can be altered by changing the size and functionalities on edge sites [36]. In addition to these merits, pure GQDs have a large exciton binding energy (0.8 eV calculated for 2 nm GQDs), which enhances the recombination rate, resulting in poor catalytic behavior [37]. Doping of different atoms such as nitrogen (N) and sulfur (S) into the lattice of GQDs has been reported to be the most effective approach to modifying the electronic, optical, and transport properties of GQDs to achieve fast electron transfer [38,39]. Shen et al. have reported that coupling of N- and S-GQDs with other materials results in the formation of an advanced hybrid photocatalyst that promotes photocatalysis [40]. Previous reports have indicated that TiO_2 nanoribbons and carbon nanotube composites exhibit faster photocatalytic degradation of methylene blue (MB) than TiO_2 alone [41]. Graphene and titania interfaces as Schottky-like nanocomposites for efficient photocatalysis have been reported [42]. Zhu et al. suggested a self-assembly method for preparing WSe_2-graphene-TiO_2 for dye degradation and hydrogen production [43]. C–O–Ti bond formation has been reported in TiO_2-graphene composites as a result of free-electron interactions [44]. However, to the best of our knowledge, a pathway for different synthesis routes of doped GQDs and forming TiO_2 heterojunctions has not yet been addressed. Thus, we propose an essential mechanism, the simplest method of doping, and oxygen functionalities' role on edge states of GQDs via acid treatment and the formation of GQDs nanocomposite with TiO_2 as a promising stable photocatalyst.

Here, a novel top-down synthesis of N, S co-doped GQDs via simple cutting methods using different one-dimensional carbon nanomaterials (carbon nanotubes (CNT), reflux carbon nanotubes (RCNT) [45,46], and carbon nanofibers (CNF) [47,48]) as precursors is reported. In the present study, we elucidated the comparative behavior of N, S co-doped GQDs and their respective GQDs-TiO_2 nanocomposites, named as (CNT-GQDs-TiO_2, RCNT-GQDs-TiO_2, and CNF-GQDs-TiO_2) as a highly robust visible-light sensitizer with a tunable bandgap, optical, and capacitive parameters. As-prepared GQDs were used to sensitize a TiO_2 photocatalyst for degradation of MB as a model pollutant dye. The mechanism of heterojunction formation between the GQDs and TiO_2 for charge separation at the interface was proposed through MB degradation and cyclic runs of samples to express the photocatalytic properties of nanocomposites. To elucidate the role of free radicals in the photodegradation of MB, scavenger tests on a complete set of as-prepared nanocomposites were also performed. The degradation pathway, interfacial charge transfer mechanism, and significance of the metal-free carbon nanomaterial-based photocatalyst are discussed.

2. Results and Discussion

2.1. Physiochemical Properties

Figure 1 shows the morphology of GQDs and GQDs-TiO$_2$ nanocomposites via images captured by FE-TEM and FE-SEM instruments. FE-SEM and FE-TEM image of the CNT-GQDs shows ring formation due to strong sp^2 bonding of carbon atoms, indicated by a white hexagonal outline of the shape of GQDs exhibit an average diameter of 8 nm. Similarly, FE-TEM images for the RCNT-GQDs and CNF-GQDs shows average diameters of 15 nm and 100 nm, respectively. Particle size of CNT-GQDs varied from 3 to 8 nm and average particle size is approximately 5 nm. Whereas, RCNT-GQDs size range lies between 8 to 20 nm with 13 nm average size and CNF-GQDs size is between 70–120 nm with average size of 90 nm. Agglomeration and overlap phenomena are observed because the high surface-area-to-volume ratio favors large surface energy and instability in the GQDs, especially in the case of CNT-GQDs and RCNT-GQDs [49]. To prepare a nanocomposite in which these effects of GQDs are diminished, we strongly sonicated GQDs and stirred them with TiO$_2$ to balance the surface energy and stability of the nanocomposites, which also promotes the formation of heterojunctions between the GQDs and TiO$_2$. Well distributed and firmly decorated GQDs are observed on TiO$_2$ surface. CNT-GQDs-TiO$_2$, RCNT-GQDs-TiO$_2$, and CNF-GQDs-TiO$_2$ exhibit lattice spacings of 0.2873 nm, 0.3368 nm, and 0.4275 nm corresponding to the (004), (112) and (101) planes, respectively. The FE-TEM images of nanocomposites demonstrate that the GQDs are amorphous phases and the TiO$_2$ is crystalline phases; these phases are clearly seen, where the GQDs (amorphous) lie on the TiO$_2$ (crystalline) lattice to form the well-structured nanocomposites.

Pristine phases of the crystal and structural behavior of as-prepared GQDs and GQDs-TiO$_2$ nanocomposites were characterized by XRD on the basis of Bragg's law, $2d\sin\theta = n\lambda$ ($n = 1$, $\lambda = 0.154$ nm). Detailed XRD patterns of GQDs and GQD-TiO$_2$ nanocomposites are shown in Figure 2a. Intense peaks of CNFs, CNTs at $2\theta \approx 25°$ and $27.95°$ are predominant GQDs data (JCPDS-ICDD file No. 75-1621), these peaks represent the (002) and (101) planes of sp^2-hybridized hexagonal carbon atoms, respectively [50]. Whereas, crystalline phases of TiO$_2$ represented by $2\theta \approx 25°$ and $48°$ corresponding to (110) and (200) planes, respectively and indicating good agreement with the standard spectrum (JCPDS-ICDD file No. 21-1272. Traces of S in the CNF-GQDs with 40% crystallinity and the RSWCNT-GQDs with 22% crystallinity are confirmed by the (220) plane with an interlayer spacing of 0.39 nm. Planes (110) represent oxygen abundance in GQDs. Peaks at $2\theta \approx 24.9°$ and $27.2°$ correspond to the anatase phase of TiO$_2$ and the graphitic structure of the GQD-TiO$_2$ nanocomposite, with d-spacings of 0.35 nm and 0.32 nm, respectively [51]. An intense (101) peak represents a high weight percentage of TiO$_2$ in the prepared nanocomposite as compared with the GQDs. Prominent peaks for GQDs in the XRD of the GQDs-TiO$_2$ were absent because of the lesser proportion of GQDs and the high surface coverage by TiO$_2$. We concluded that the GQDs' peaks are obscured by the dominant TiO$_2$ peaks. Similarly, peaks related to the presence of S and N in the GQDs are also diminished in the patterns of the nanocomposite samples. Additional peaks observed at $2\theta \approx 33.5°$ and $37.4°$ denote strong C–O–Ti bonding [52]. Because of interface formation and proper dispersion of the GQDs over the TiO$_2$ surface, the crystallinity of samples appears to increase to approximately 37% for the CNT-GQDs-TiO$_2$, RCNT-GQDs-TiO$_2$, and 47% for the CNF-GQDs-TiO$_2$ but cannot be increased further because of the abundance of O and other impurities.

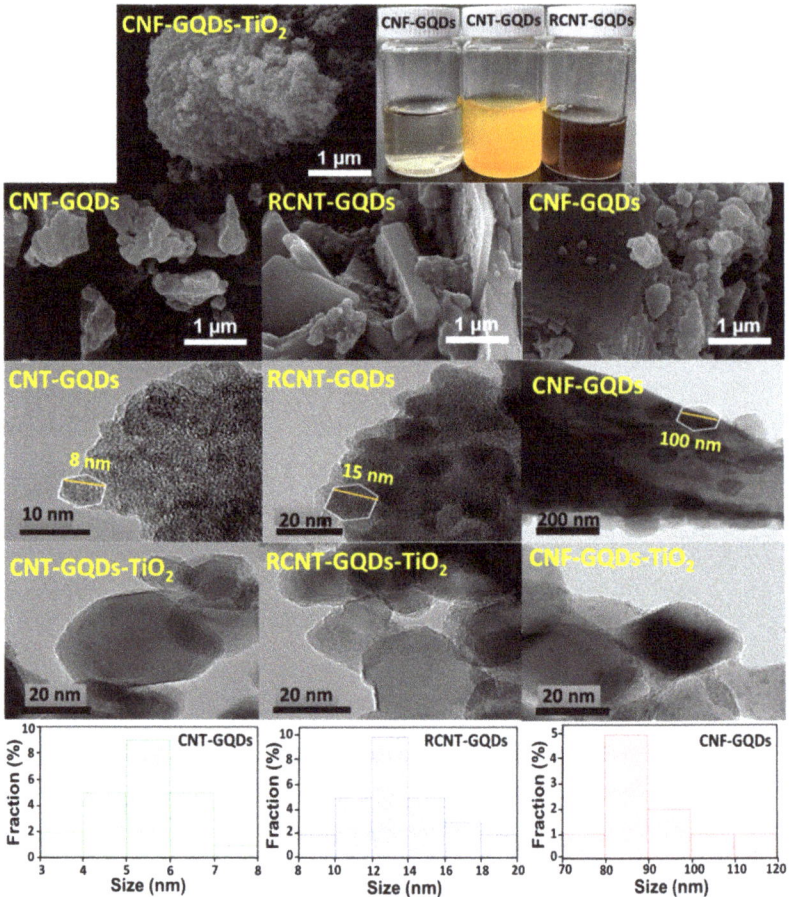

Figure 1. FE-SEM micrographs and FE-TEM images of as-synthesized GQDs with size distribution and their respective GQDs-TiO$_2$ nanocomposites.

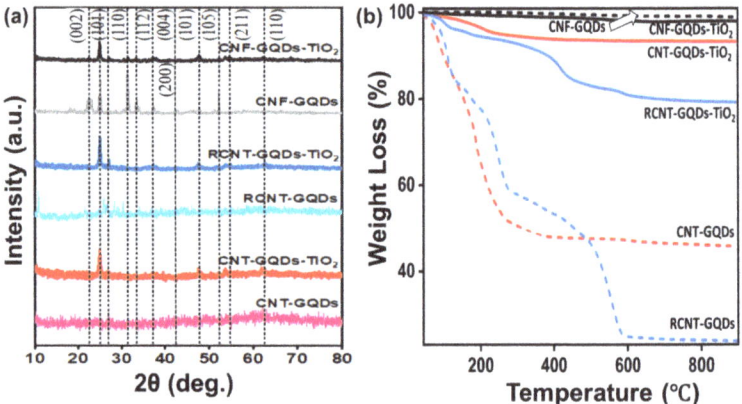

Figure 2. (**a**) XRD pattern of as-synthesized GQDs and their respective GQDs-TiO$_2$ nanocomposites. (**b**) TGA analysis of as-synthesized GQDs and their respective GQDs-TiO$_2$ nanocomposites.

TGA was carried out on GQDs-TiO$_2$ nanocomposites to clarify the structural thermal stability, bonding with GQDs, and the significance of the nanocomposite properties. Thermograph temperature in the range of 50–150 °C, 150–350 °C, 350–550 °C, and above 550 °C denote physio-adsorbed water, dehydration–decarboxylation of bonded water and functional groups, oxidation of amorphous groups, and oxidation of graphitic structures, respectively [53]. Figure 2b shows the TGA curve of the GQDs-TiO$_2$ nanocomposites. Interestingly, these thermographs show nanocomposite formations that denote low weight loss and indicate strong bonding of the GQDs onto the TiO$_2$ surface. The CNT-GQDs-TiO$_2$, RCNT-GQDs-TiO$_2$, and CNF-GQDs-TiO$_2$ demonstrate 6.81%, 20.81%, and 2.09% mass loss, respectively. These results undoubtedly predict the formation of thermally stable structures of nanocomposites. Minor weight loss in CNF-GQDs-TiO$_2$ indicates strongest covalent bonding between CNF-GQDs and TiO$_2$. Specifically, it indicates chemisorption in the investigated temperature range, rearrangement of lattices when in-plane (C–O) functional groups were transferred to oxygen-deficient vacancies in the TiO$_2$ plane, and the formation of stable C–O–Ti bonds. This explanation reflects the roles of functional groups in GQDs and oxygen vacancies in TiO$_2$ in determining the configuration of nanocomposites.

The heterostructure interfacial chemical bonding and interaction in GQDs-TiO$_2$ samples were investigated by XPS analysis. Peaks appeared at binding energies of 531, 284, 406, and 168 eV in the spectra of the GQDs, corresponding to O 1s, C 1s, N 1s, and S 2p, respectively. Identical peaks with a slight change in intensity were observed in the spectrum of the GQDs-TiO$_2$ nanocomposites, along with an additional peak of Ti 2p at 458 eV, as shown in Figure 3a–c. High-resolution C 1s spectra were fitted with two Gaussian peaks with maxima at 284.28 and 287.48 eV in the spectrum of CNT-GQDs-TiO$_2$ as shown in Figure 3d, signifying C=C bonds of a graphitic honeycomb structure and other show a Ti–O–C bond peak associated with titanium carbonate and strong C=C peak of GQD, respectively. Deconvoluted C 1s peaks for the spectrum of the RCNT-GQDs-TiO$_2$ are attributed to titanium carbide and C=C as mentioned in Figure 3e. These results strongly evidence the formation of C–Ti bonds and C–O–Ti bonds. Ti–C bonds exhibit greater stability and better interfacing for RCNT-GQDs-TiO$_2$ nanocomposite. However, the C 1s peak fitting in Figure 3f results for CNF-GQDs-TiO$_2$ that are identical to RCNT-GQDs-TiO$_2$, which indicates C–Ti and C–O–Ti bond formation. C 1s range of GQDs-TiO$_2$, C=C bond formation indicates that the synthesis of nanocomposites using high sonication of GQDs and extended stirring times at a certain temperature led to a well-decorated GQDs-TiO$_2$ interface with strong bonding between Ti–O and Ti–O–C.

Meanwhile, a core-level study of the high-resolution O 1s spectra of the GQDs-TiO$_2$ nanocomposites as shown in Figure 3g–i revealed relevant peaks centered at 529.18, 530.6, and 532.08 eV, which are attributed to Ti–O–Ti, C=O, and Ti–O–C interfacial bonds, respectively [54]. The Ti–O–Ti bond peaks are prominent because the TiO$_2$ content of nanocomposites was much greater than their GQD and O^{2-} contents. After analyzing the spectroscopic data, it can be concluded that reaction parameters such as temperature, chemical treatment, and time cause changes in the ratio of N, S- heteroatom doping extent, oxygen functionalities, and physiochemical properties of as-prepared GQDs that affects overall bonding strength of the nanocomposites. N and S heteroatom dopant traces can be clearly seen in CNF-GQDs-TiO$_2$ XPS data, whereas absence or negligence of S doping in CNT-GQDs-TiO$_2$ and N doping in RCNT-GQDs-TiO$_2$ were observed. Due to which it can be stated that CNF-GQDs-TiO$_2$ can experience the synergistic effect that leads to the better performance of nanocomposite because of strong absorption capability and photosensitizer ability, as discussed in the next sections. These XPS results are fully consistent with the XRD analyses.

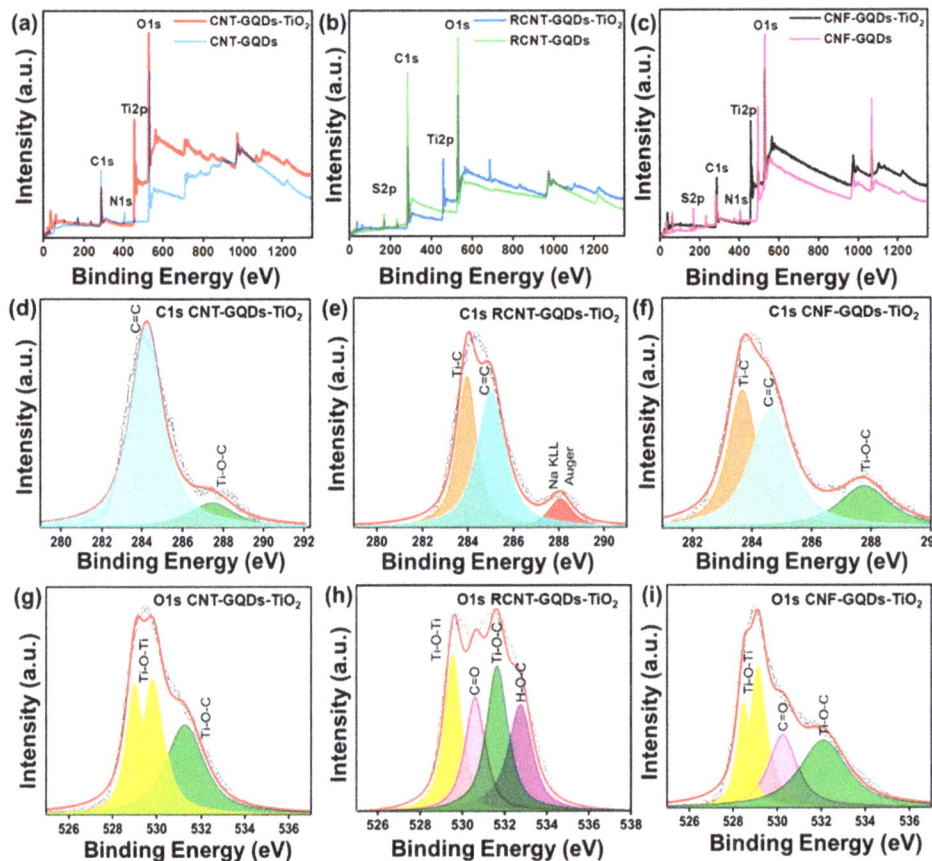

Figure 3. Synoptic XPS survey scan of (**a**) CNT-GQDs and CNT-GQDs-TiO$_2$ nanocomposites, (**b**) RCNT-GQDs and RCNT-GQDs-TiO$_2$ nanocomposites, (**c**) CNF-GQDs and CNF-GQDs-TiO$_2$ nanocomposites. High resolution XPS spectra of C1s peak (**d**) CNT-GQDs-TiO$_2$, (**e**) RCNT-GQDs-TiO$_2$, (**f**) CNF-GQDs-TiO$_2$. High resolution XPS spectra of O1s peak (**g**) CNT-GQDs-TiO$_2$, (**h**) RCNT-GQDs-TiO$_2$, (**i**) CNF-GQDs-TiO$_2$.

2.2. Optical Study and Photoelectrochemical Study

Figure 4a shows the UV-Vis absorption spectra for GQDs-TiO$_2$ nanocomposites. Figure 4c inset represents a UV-Vis plot of the GQD absorption spectra. All the GQDs show characteristic π–π* transitions below 300 nm within the sp^2 C=C hybridized framework of carbon domains [55]. The absorption tail extending to ~350 nm in the visible region was assigned to the n → π* transitions of nonbonding electrons of O, N, and S atoms [56]. Greater absorption, specifically in the case of CNF-GQDs, was due to these extra electrons caused by N, S doping. PL spectra of as-prepared GQDs samples tested in powder form by excitation of a He–Cd laser at 325 nm. For better understanding of GQDs band gap, Tauc plot of respective GQDs was shown in Figure 4c. Bandgap corresponding to CNT-GQDs, RCNT-GQDs, and CNF-GQDs as obtained by the formulation, were 3.3, 2.2, and 2.8 eV, respectively. for composites tailored band gap, Tauc plot of respective GQDs-TiO$_2$ nanocomposites was shown in Figure 4b. The bandgap of the samples can be easily determined by the intersection point of the energy axis with the slope of the curve in graph. Bandgap corresponding to CNT-GQDs-TiO$_2$, RCNT-GQDs-TiO$_2$, and CNF-GQDs-TiO$_2$ nanocomposites as obtained by the formulation, were 2.21, 2.23, and 2.29 eV, respectively. Ti–O–C interfacial bond formation and intimate interaction in nanocomposites could decrease the bandgap value of nanocomposites. Interaction between TiO$_2$ and GQDs was because of chemical coupling and influence of C atoms on the molecular orbitals of TiO$_2$, as a result, the bottom band of the nanocomposite lowers itself as compared with that of TiO$_2$. Incorporation of GQDs will enhance the lifetime of electron-hole pairs and result in faster photogenerated electron transfer in TiO$_2$ due to π-conjugation of the GQDs.

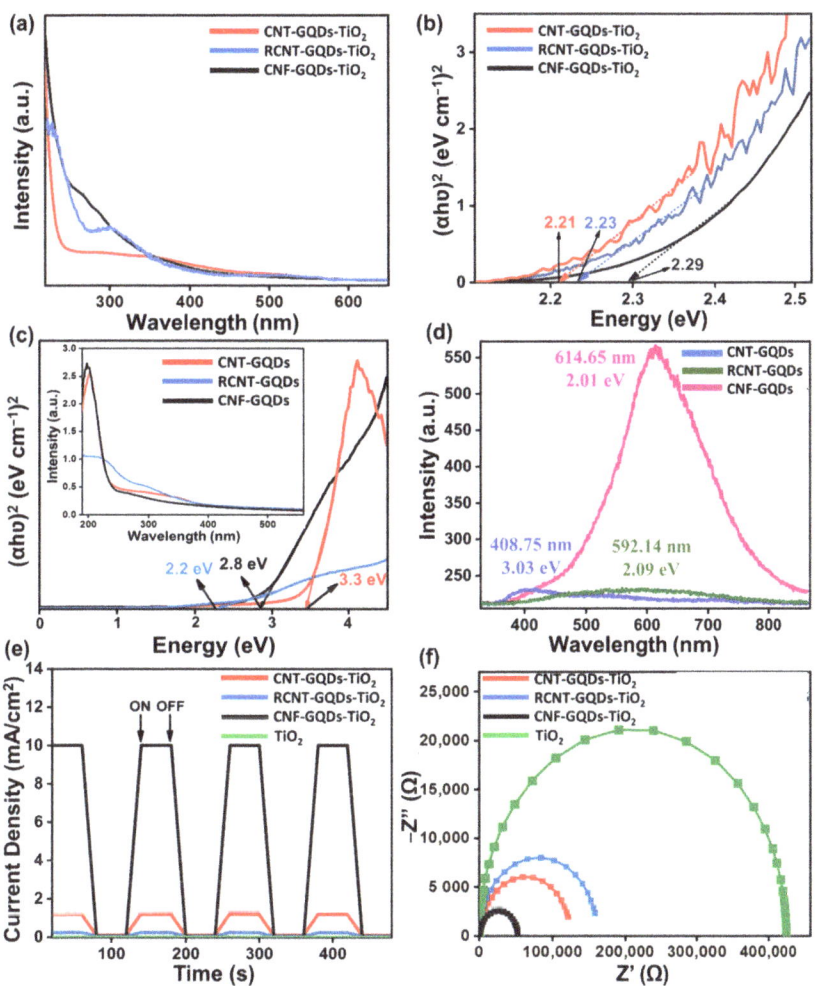

Figure 4. (**a**) UV-vis absorption of as-synthesized GQDs-TiO$_2$ nanocomposites, (**b**) Tauc plot of respective GQDs-TiO$_2$ nanocomposites, (**c**) Tauc plot of as-synthesized GQDs with respective UV-vis absorption as inset image, (**d**) Photoluminescence (PL) spectra of as-synthesized GQDs excited at 325 nm by He-Ne laser. Photoelectrochemical analysis (**e**) Transient photocurrent response of the as-prepared GQDs-TiO$_2$ nanocomposites, (**f**) EIS measurements.

Figure 4d depicts the photoluminescence (PL) spectra of three GQDs, CNF-GQDs clearly show the strongest PL emission in the orange-red region of the visible spectra, with a bandgap energy of 2.01 eV; similarly, the RCNT-GQDs emit in the yellow-orange region, with an equivalent range of photon energy of 2.09 eV but with lesser intensity than the emission of the CNF-GQDs. By contrast, the CNT-GQDs exhibit totally different PL emission at 408.75 nm with a 3.03 eV bandgap energy in the violet region but with the same intensity as the emission of the RCNT-GQDs. PL phenomena were observed in GQDs when the excitation wavelength was comparable to the transition wavelength. Long wavelengths were absorbed and emitted by C=S and C=N chromophores, which resulted in red-yellow emission; by contrast, n → π* excitation of C=O chromophores were assigned to short wavelengths with violet and blue emission [57].

Table 1 denotes the atomic % of elements clearly shows that the CNF-GQDs and RCNT-GQDs were rich in S-group content, whereas the CNT GQDs were rich N-group content. Because of the abundance of S groups and some traces of N in conjugated molecules of CNF-GQDs, the n → π* transition of the C=S group dominates with overlapping of N and O groups in CF-GQDs [58], which led to the synergistic effect of these three groups in lattice formation, as confirmed from the XRD and XPS analysis. Hence, orange–red emission was observed upon excitation of CNF-GQDs. The proposed mechanism responsible for the CNF-GQDs strong PL emission is S π* orbital insertion between C n and N π* orbitals and the formation of a low-electronegative S state. Remarkable difference in PL intensities of GQDs strongly justifies the CNF-GQDs efficiency, these property variations experienced by GQDs was based on synthesis method. On comparing the key parameters of synthesis process, we conclude that simple oxidative cutting with comparatively low temperature range (110–120 °C) using CNF is more feasible and effective for GQDs synthesis because refluxing process and followed hydrothermal reaction at high temperature range proved to be harsh on the surface and also it distorts the edges and structure of the GQDs which hinders the oxygen vacancies, graphitic properties and anchoring properties those can be inherited by GQDs from CNTs. This proves the CNF-GQDs with higher PL intensity as a potential contender for photocatalytic process.

Table 1. Atomic percentage analysis with statistical error using XPS spectrometer for as-prepared GQDs and GQDs-TiO$_2$ nanocomposites.

Elements	CNT-GQDs	CNT-GQDs-TiO$_2$	RCNT-GQDs	RCNT-GQDs-TiO$_2$	CNF-GQDs	CNF-GQDs-TiO$_2$
			Atomic percentage			
C	44.14 ± 2.0	27.41 ± 0.2	62.51 ± 3.2	51.07 ± 5.2	24.51 ± 0.7	37.76 ± 4.7
O	47.61 ± 1.0	53.11 ± 1.0	33.51 ± 3.5	38.11 ± 8.5	62.99 ± 4.1	44.55 ± 0.7
N	6.53 ± 0.8	2.45 ± 0.5	0.21 ± 0.1	0.12 ± 0.5	4.55 ± 2.9	1.46 ± 0.9
S	1.72 ± 0.1	0.15 ± 0.04	3.69 ± 0.09	2.52 ± 0.6	7.95 ± 0.4	1.75 ± 0.3
Ti	-	17.22 ± 0.08	-	8.19 ± 2.2	-	14.49 ± 2.6

Separation of photogenerated charge carriers of as-synthesized GQDs-TiO$_2$ nanocomposites can be investigated by photoelectrochemical study at room temperature using an Ivium electrochemical workstation under solar simulator. The electrolyte used through the experiments was 2 M KOH aqueous solution. A typical three-electrode setup consisting of a working electrode, a Pt coil as a counter electrode, and a saturated calomel electrode (SCE) as the reference electrode was used. Fluorine doped tin oxide coated (FTO) glass electrode was prepared by spin coating technique at 20,000 rpm for 10 s in three runs with exposed area of 1×1 cm^2. Coating slurry was prepared by dispersing 5 mg of photocatalyst in 1 mL ethanol containing 10 μL Nafion. Electrodes were oven dried for 3 h and calcinated at 200 °C in nitrogen atmosphere for 2 h to augment adhesion.

Photoelectrochemical analysis of GQDs-TiO$_2$ nanocomposites as shown in Figure 4e,f transient photocurrent and EIS graphs respectively. While referring to transient current curves, it can be clearly observed that TiO$_2$ has negligible current response as compared to prepared nanocomposites. CNT-GQDs-TiO$_2$ and RCNT-GQDs-TiO$_2$ electrodes exhibit high photocurrent density, well stable and longstanding photoresponse under visible light projection. CNF-GQDs-TiO$_2$ electrode produces much larger photocurrent density than other electrodes, this indicates that CNF-GQDs-TiO$_2$ possess highest charge transfer rate and improved photogenerated charge carriers' separation efficiency. Meanwhile, more effective way to reveal electron transfer efficiency of as prepared sample electrodes is through EIS (electrochemical impedance spectroscopy). In Figure 4f, clear comparative analysis can be done among the nanocomposites based on EIS curve radius. Shorter the radius implies better mobility and well separated photogenerated charge carriers. CNF-GQDs-TiO$_2$ demonstrates the smallest radius curve among all, results in better charge transfer at electrode interface that enhance the separation of carriers. These electrochemical results prove that the GQDs-TiO$_2$ heterojunctions specially CNF-GQDs-TiO$_2$ serves as electron harvester for improved degradation performance and separation of charge carriers, this phenomenon can be thoroughly observed in next sections.

2.3. Visible-Light Photocatalysis and Mechanism Studies

Figure 5 presents the visible-light photocatalytic degradation kinetics for different nanocomposites samples toward the degradation of MB as a model pollutant dye. Figure 5a shows the photocatalytic degradation behavior of different compositions of GQDs-TiO$_2$ nanocomposites. MB degradation first-order-rate-constant behavior was fitted using the Langmuir-Hinshelwood rate formalism, which represents the kinetics of photocatalytic reaction in different catalysts and was expressed as $\ln(C/C_0) = k_a$, where k_a is the pseudo-first-order-reaction rate constant and C and C_0 denote the MB concentrations after and before irradiation time t, respectively [59]. The rate constant for different catalysts was obtained from the slope of the straight-line plot shown in Figure 5b. All the as-prepared nanocomposites exhibit an excellent degradation percentage, as shown in Figure 5e. The CNT-GQDs-TiO$_2$ sample showed a dramatic increase in degradation percentage to 99.78% in 90 min; similar results were obtained with the RCNT-GQDs-TiO$_2$ and CNF-GQDs-TiO$_2$ nanocomposites. Meanwhile, CNF-GQDs-TiO$_2$ exhibited the fastest degradation kinetics, with 99.39% degradation in 60 min of light irradiation. Graphical representation of the rate constants for different nanocomposites is shown in Figure 5f. These data for samples show that GQDs-TiO$_2$ exhibit an approximately 2.3 to 3-fold higher degradation rate constant than the commercial P25 photocatalyst. The highest rate constant recorded for the CNT-GQDs-TiO$_2$ was 6.75×10^{-2} min^{-1}, whereas that of the TiO$_2$ was 2.48×10^{-2} min^{-1}. The characteristic absorption peak of MB rapidly decreased in intensity in the experiments with all of the prepared nanocomposites; also, this decrease to a certain level was also obtained with TiO$_2$ (79% degradation). As-prepared GQDs show strong optical absorption capabilities in a wide light spectrum but with poor photocatalytic properties ascribed to their corresponding large binding energy. The key role of a stable heterojunction interface is evident when the GQDs and TiO$_2$ are blended with each other to form an optimum nanoscale composition containing well-dispersed soluble GQDs on the surface of TiO$_2$ nanoparticles using a simple sonication and stirring method. The content of GQDs cannot be increased further because of their strong tendency to aggregate on the TiO$_2$ surface, which retards the photocatalytic process remarkably. The construction of well-dispersed nanoparticle heterojunctions is critical and can be effectively achieved through bandgap narrowing, as shown by the Tauc plot previously. By contrast, C–O–Ti bond formation ascribed to chemisorption of GQDs on the TiO$_2$ surface plays a strong role in strengthening the interaction between nanocomposites [60]. This strong coupling of GQDs with TiO$_2$ forms the hot-carrier injection pathway, whereas N and S doping causes substantial bandgap reduction. Because of this heterojunction formation, efficient charge separation occurred on the surface, results in a higher photodegradation rate. Comparatively slight variation among the degradation rates of the samples was observed because of the difference in the bandgap value of nanocomposites, where the smallest bandgap value was reported for CNT-GQDs-TiO$_2$, which exhibited the highest degradation rate of 99.78%, confirming that the bandgap is an essential factor in dye degradation. CNT-GQDs-TiO$_2$ exhibits slightly higher degradation from CNF-GQDs-TiO$_2$ because of its smaller bandgap as it can be seen in Tauc plot, but based on TGA analysis and repetitive cyclic runs, it is not as stable as CNF-GQDs-TiO$_2$. Therefore, CNF-GQDs-TiO$_2$ is considered as the best suited composite among all due to strong covalent bonding between CNF-GQDs and TiO$_2$ that indicates stable C–O–Ti bonds formation. Also, CNF-GQDs experienced the N and S dopant synergistic effect that results in better performance of nanocomposite with strong absorption capability and photosensitizer ability.

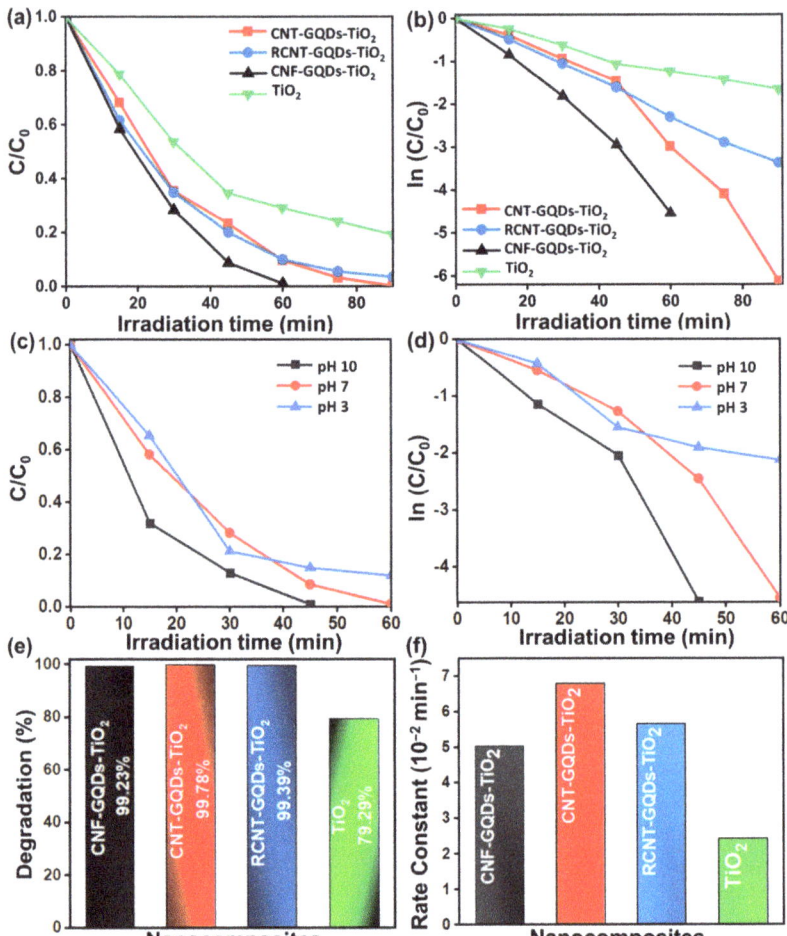

Figure 5. Visible light photocatalytic degradation of MB for as-synthesized GQDs-TiO$_2$ nanocomposites (**a**) Change in relative concentration (C/C0) of MB in GQDs-TiO$_2$ nanocomposites as a function of irradiation time up to 90 min, (**b**) Derived plot from same data to study ln(C/C0) variations with irradiation time. pH variation effect on the (**c**) Change in relative concentration (C/C0) as a function of irradiation time up to 60 min using CNF-GQDs-TiO$_2$ nanocomposites, (**d**) Respective derived plot of ln(C/C0) variations with irradiation time. (**e**) Comparison of MB degradation in different nanocomposites, (**f**) Comparative study of first order rate constants of as-prepared GQDs-TiO$_2$ nanocomposites.

pH effect is an important parameter for the dye degradation. MB degradation was tested in the pH range between 3 to 10, pH values were adjusted using H$_2$SO$_4$ and 0.1 mol/L NaOH. Experimental condition was identical as mentioned in the experimental section. Figure 5c,d shows that photodegradation rate was improved by increasing the pH values. At pH 3, degradation efficiency was dropped to 84%. At pH 10 and pH 7, degradation efficiency is same ~99.23% but with the increase of pH value the process of degradation become fast. At pH 10, 99.23% degradation was achieved in 45 min. This shows that alkaline conditions increase the reaction rate due to the abundance of hydroxyl ions that helps in increasing the degradation reaction. Whereas at low pH, reaction rate reduced due to excess of H^+ ions [61]. Hence, it can be stated that alkaline medium is favorable for MB degradation using CNF-GQDs-TiO_2 nanocomposites.

GQDs-TiO_2 composites were studied by many groups with TiO_2 nanostructures (nanotubes, nanoparticles, and nanosheets) for the degradation of organic dyes to find the best suited one as shown in Table 2. Various chemical composition and formation methods were used to synthesize GQDs and composites with desired physiochemical properties that results in fastest, stable, and complete organic dye degradation. GQDs, N-GQDs, and N, S-GQDs based TiO_2 composites were studied previously but results in slow and lesser degradation efficiency [62–64]. Whereas, Tian et al. synthesized the N, S-GQDs based reduced graphene oxide (rGO)-TiO_2 nanotubes (NTs) using hydrothermal and stirring process for degradation methyl orange (MO). This results in MO degradation in 4 h upto 90% [65]. This work, have presented the three types of GQDs formation with different N, S doping amount due to the different precursors and process of synthesis. Best results were obtained by TiO_2 composites formed using CNF-GQDs due to strong covalent bonding between CNF-GQDs and TiO_2 defining its stability, also it indicates chemisorption and rearrangement of lattices when in-plane (C-O) functional groups were transferred to oxygen-deficient vacancies in the TiO_2 plane to form stable C–O–Ti bonds. This reflects the role of functional groups in GQDs and oxygen vacancies in TiO_2 for determining the configuration of nanocomposites. CNF-GQDs experienced the N and S dopant synergistic effect that leads to the better performance of nanocomposite with strong absorption capability and photosensitizer ability. Based on this, we concluded that CNF-GQDs-TiO_2 has shown the fastest degradation (60 min) of methylene blue (MB) with highest efficiency of 99.23% as compared with the previously reported articles.

Table 2. Comparative data analysis of dye degradation using GQDs based TiO_2 nanocomposites.

Photocatalyst	Pollutant	Irradiation Time (min)	Degradation Efficiency (%)	References
GQD/a-TiO_2	Rhodamine B	35	97	[62]
GQDs/TiO_2	Methylene Blue	120	96.70	[63]
GQDs/TiO_2	Rhodamine B	60	70	[66]
3DGA@CDs-TNs	Rhodamine B	60	97.60	[67]
N-GQDs/TiO_2	Rhodamine B	120	94	[64]
N-GDQs/TiO_2	Methylene Blue	150	100	[68]
N-GDQs/TiO_2	Methylene Blue	70	85	[69]
Ti^{3+}-TiO_2/GQD NSs	Methylene Blue	60	90	[70]
NSTG	Methylene Blue	240	98.40	[40]
N, S-GQDs/rGO/TiO_2 NT	Methyl orange	240	90	[65]
TiO_2/rGO	Methylene Blue	120	91.48	[71]
CNF-GQDs-TiO_2	Methylene Blue	60	99.23	This work

Another essential factor for photocatalysts is reusability and stability. The reusability of the nanocomposites was tested over three repeated cyclic runs of photocatalysis, as shown in Figure 6a. These stability results provide important facts about catalysts. After three cyclic runs, the CNT-GQDs-TiO_2 demonstrated a 0.54% decrement. By contrast, the RCNT-GQDs-TiO_2 exhibited only 55.5% degradation, with a 44.11% decrement compared with its initial degradation results. The CNF-GQDs-TiO_2 catalyst was the most stable for three repeated cyclic runs. Reason behind CNF-GQDs reusability and long-term stability was N, S doping, thermally stable structure and oxygen mediated vacancies which was responsible for C–O–Ti bonding. Poor stability of the RCNT-GQDs-TiO_2 and strong bonding of the CNF-GQDs-TiO_2 were confirmed by the TGA results.

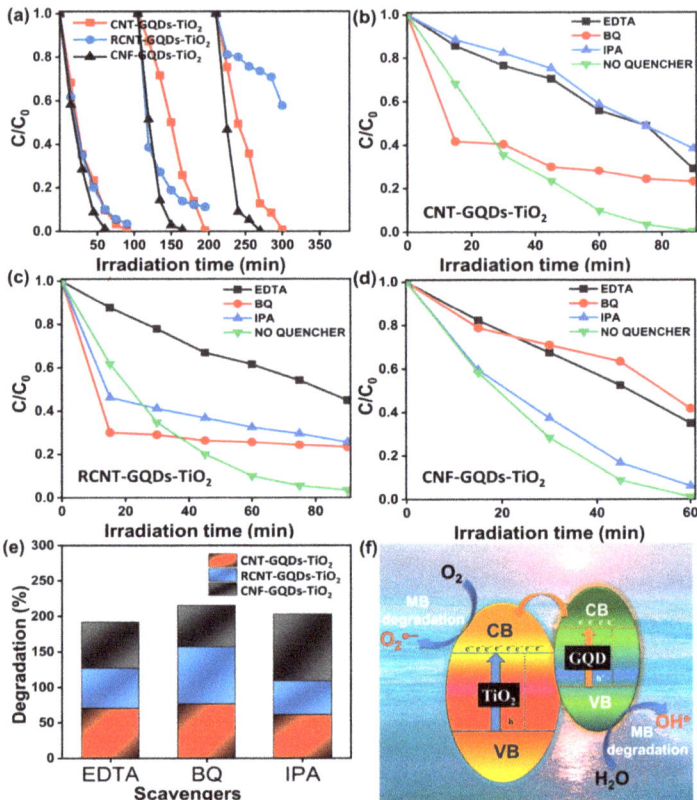

Figure 6. (a) Cyclic runs of MB degradation under visible light exposure for GQDs-TiO$_2$ nanocomposites. MB degradation concentration in absence and presence of different scavengers (scavengers: IPA~•OH, BQ~ O$_2^{\bullet-}$, EDTA~ h$^+$) by (b) CNT-GQDs-TiO$_2$, (c) RCNT-GQDs-TiO$_2$, (d) CNF-GQDs-TiO$_2$. (e) Comparative analysis of effect of scavengers on MB degradation. (f) Schematic illustration of free radical generation and interfacial charge separation at GQDs-TiO$_2$ nanocomposite junction for MB degradation.

To investigate the role of the electron, transfer mechanism and the reaction process in a typical photocatalysis system, active species such as holes (h$^+$), hydroxyl radicals (•OH), and superoxide radicals (O$_2^{\bullet-}$) were studied. Solutions of IPA, BQ, and EDTA scavengers were used to study the roles of •OH, O$_2^{\bullet-}$, and h$^+$, respectively [72]. To investigate and explore their contribution in visible-light photocatalysis, a series of scavenger tests were performed on CNT-GQDs-TiO$_2$ nanocomposites. The dosage addition of scavengers inhibits the degradation efficiency of MB to various degrees, which refers to the involvement of all oxidative species in the photodegradation system. Changes in relative concentration of MB in different scavengers for all of the as-prepared nanocomposite systems are shown in Figure 6b–d. In the case of IPA addition, the MB was degraded by 61.74% for CNT-GQDs-TiO$_2$, implying the vital role of •OH radical, and RCNT-GQDs-TiO$_2$ was quenched to 47.55% in EDTA, indicating the important of h$^+$ species in the degradation process. When BQ was added to the CNF-GQDs-TiO$_2$ system, 58.26% MB degradation was observed, indicating that the major reactive species involved in the process was O$_2^{\circ-}$. Figure 6e shows a comparative analysis of the nanocomposite systems with and without scavenger addition and respective values can be seen in Table 3. All studies of samples show the highest quenching of MB degradation in different scavengers. These results indicate that

•OH, $O_2^{\bullet-}$, and h$^+$ are the major reactive species in MB degradation for CNT-GQDs-TiO$_2$, CNF-GQDs-TiO$_2$, and RCNT-GQDs-TiO$_2$, respectively.

Table 3. Role of quenchers in UV-Vis photocatalysis of MB degradation.

Quencher	CNT-GQDs-TiO$_2$ (% Degradation)	RCNT-GQDs-TiO$_2$ (% Degradation)	CNF-GQDs-TiO$_2$ (% Degradation)
EDTA	70.89	47.55	64.97
IPA	61.74	56.17	93.94
BQ	76.98	80.15	58.26

Based on the aforementioned results, the degradation mechanism can occur through various steps, implying direct electron transfer. Figure 6f shows the proposed mechanism of free radicals involved in the photodegradation process. The phenomena that lead to the formation of the Schottky barrier between photoexcited electrons in GQDs and the TiO$_2$ lattice are thermionic emission and quantum tunneling. Achieving thermionic emission was difficult because the complete experimental procedure was carried out at room temperature. Hence, the most likely factor behind electron transport in the GQD–TiO$_2$ composite was quantum tunneling, which forms the heterojunction type II band alignment.

Degraded methylene blue (MB) results in some intermediate volatile end products of MB solution. When N–S heterocyclic compound of methylene blue was separated 2-Amino-5-dimethylamino-benzenesulfonic acid anion was formed. Degradation intermediate of 2-Amino-5-dimethylamino-benzenesulfonic acid was 4-Amino-benzenesulfonic acid and 2-Amino-5-hydroxy-benzenesulfonic acid. Oxidization of 4-Amino-benzenesulfonic acid results in 4-Nitro-benzenesulfonic acid anion. Products formed after hydroxyl radical attack on the N–S heterocyclic of MB molecules were Dimethyl-(4-nitro-phenyl)-amine and p-Dihydroxybenzene. Tiny molecular products of degradation correspond to succinic acid generated by open-loop benzene [73]. On the basis of these intermediate products, we can deduce the degradation pathway of MB. The photogenerated electrons in the TiO$_2$ with the in situ excited electrons of GQDs reacts with free radicals. Hence, electrons from the GQDs transfer to the conduction band of TiO$_2$ to trap the oxygen (O$_2$) molecule to generate the $O_2^{\bullet-}$ radicals, and h$^+$ react with an electron donor (water molecules or hydroxide ions) to yield oxidizing hydroxyl •OH radicals. These generated radicals of $O_2^{\bullet-}$ and •OH are the essential substances to decompose the dye molecules into CO$_2$ and H$_2$O. Thus, GQDs with high electron mobility are a perfect electron-transfer tank for enhanced electron–hole separation at the interface of TiO$_2$. The high degradation of MB in several scavengers can be interpreted as the conversion of $O_2^{\bullet-}$ radical to •OH radical. The possible pathways for MB degradation are expressed by the following reactions:

$$TiO_2 - GQD + h\upsilon \rightarrow TiO_2\,(e^-) - GQD + TiO_2\,(h^+) - GQD$$

$$TiO_2\,(e^-) - GQD \rightarrow TiO_2 - GQD(e^-)$$

$$GQD\,(e^-) + O_2 \rightarrow GQD + O_2^{\bullet-}$$

$$TiO_2\,(h^+) + H_2O \rightarrow h^+ + OH^{\bullet}$$

$$O_2^{\bullet-} + OH^{\bullet} + MB \rightarrow water + CO_2$$

3. Materials and Experimental

3.1. Chemicals

Vapor-grown carbon nanofibers were purchased from Carbon Nano-material Technology. Co. Ltd., Pohang, Korea. Single-walled carbon nanotubes (SWCNTs (75%)) with an average diameter of 1.8 ± 0.4 nm and length greater than 5 µm were supplied by TuballTM-OCSiAl Pvt. Ltd., Incheon, Korea. TiO$_2$ powder, sulfuric acid (99.99%), nitric acid (ACS reagent, 70%), methylene blue (MB), 2,3-dichloro-5,6-dicyano-p-benzoquinone (98%-BQ), ethylenediaminetetraacetic acid (EDTA, >98%), and isopropyl alcohol solution (70% in

H_2O–IPA) were purchased from Sigma-Aldrich Inc., Seoul, Korea. All reagents were of analytical grade and were used without further purification. Distilled water (18.2 MΩ·cm) was used throughout the experiments.

3.2. Synthesis of GQDs from CNT (CNT-GQDs)

A hydrothermal etching method was used to prepare oxygen-rich GQDs with some traces of N. A total of 300 mg of SWCNTs was refluxed with 150 mL of HNO_3 for 12 h at 100 °C [49]. The suspension was then cooled and filtered to remove the supernatant. The collected residue was dispersed in 50 mL of DI water and heated hydrothermally in a Teflon-lined autoclave at 200 °C for 12 h. The same etching procedure (i.e., refluxing with HNO_3, followed by filtering and hydrothermal heating) was subsequently repeated. After two cycles of the etching process, nitrogen-doped GQDs named as CNT-GQDs were obtained for further use.

3.3. Synthesis of GQDs from RCNT (RCNT-GQDs)

300 mg of SWCNTs was refluxed with H_2SO_4 and HNO_3 (3:1) for 24 h at 120 °C. The mixture was then cooled and diluted with DI water (1:5) while maintaining the pH at ~7. Stepwise gradual heating (100 °C, 150 °C, and 200 °C) of the suspension was conducted at 4 h intervals to obtain sulfur-doped GQDs named as RCNT-GQDs. The as-prepared sample of GQDs was washed with DI water using the filtration technique.

3.4. Synthesis of GQDs from CNF (CNF-GQDs)

Oxygen-rich GQDs doped with nitrogen (N) and sulfur (S) atoms were synthesized using a simple oxidative cutting method. A total of 900 mg of CFs was treated with a mixture of conc. H_2SO_4 and HNO_3 (3:1) under continuous stirring for 24 h at 120 °C. The mixture was diluted with DI water (1:5) after being cooled to room temperature. To maintain the acidity of the mixture, its pH was adjusted to ~7 using NaOH. Finally, a dialysis process of the end product was conducted in a dialysis bag for 3 days and the dialyzed solution was heated at 80 °C to obtain CNF-GQDs.

3.5. Synthesis of TiO_2 and GQD Composite

Different as-prepared samples of GQDs (2 mg/mL) were dispersed in DI water to obtain a uniform suspension using the ultrasonication method. A total of 200 mg of TiO_2 was added to 50 mL of GQD suspension and stirred at 60 °C for 10–12 h, until it dries completely and turned into powder. The GQDs-TiO_2 nanocomposite was obtained in powder form. The nanocomposite samples were named CNT-GQDs-TiO_2, RCNT-GQDs-TiO_2, and CNF-GQDs-TiO_2. The synthesis scheme for respective GQDs is shown in Figure 7.

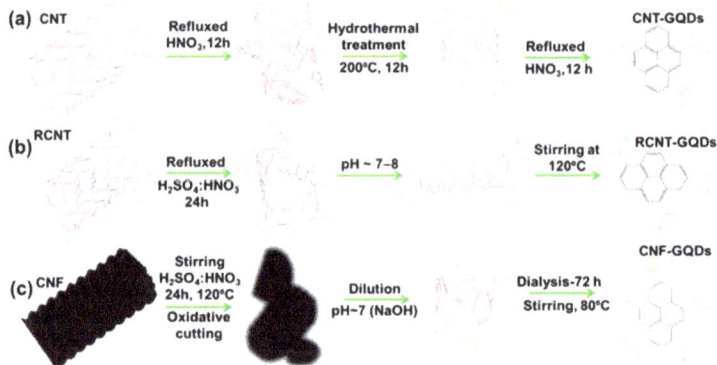

Figure 7. Schematic for the synthesis of N, S co-doped GQDs named as (**a**) CNT-GQDs, (**b**) RCNT-GQDs, (**c**) CNF-GQDs.

3.6. Characterization Techniques

The structure, crystal facets, and phase of the as-prepared samples were examined by powder X-ray diffraction (XRD, Bruker D2 PHASER, Gyeonggi-do, Korea). The morphological features, crystallinity, and chemical composition were studied by field-emission scanning electron microscopy (FE-SEM, Hitachi High-Tech Korea Co., Ltd., S-4300, Seoul, Korea), field-emission transmission electron microscopy (FE-TEM JEM-2100F/JEOL, Seoul, Korea), and energy-dispersive X-ray spectroscopy (EDX) composition analysis. The X-ray photoelectron spectroscopy data were also obtained (XPS, VG Scientific Co., ESCA LAB MK-II, Korea) using a monochromatic Al Kα X-ray beam with a current of 20 mA. Quantitative analysis of the solid organic elements present in the samples was performed with an elemental analyzer (Thermo Fisher Scientific, Inc. EA1112, Korea). The specific surface area measurements were performed using an adsorption analyzer (BEL BELSORP, Inc. Japan), and thermogravimetric analysis (TGA, NETZSCH, TG 209 F3, Germany) was carried out for the as-prepared samples at a heating rate of 10 °C/min and with N_2 gas as a purge gas. Photoluminescence (PL) emission spectra were acquired under excitation at 325 nm (RAM Boss, Maple Dongwoo Optron). The UV-vis absorption was performed using a commercial UV-vis spectrophotometer (Scinco, S-3100, Seoul, Korea). Electrochemical analysis was carried out using an Ivium electrochemical analyzer.

3.7. Photodegradation Test

Photocatalytic studies of the GQDs-TiO$_2$ nanocomposites were conducted using a solar simulator (SUN 2000, Abet Technologies, Inc. Gyeonggi-do, Korea) fitted with a 440 nm cutoff filter. The photocatalytic experiments were carried out in the presence of MB as an organic dye pollutant. A total of 20 mg of catalyst were dispersed in 50 mL of MB aqueous solution with an initial concentration of 10 mg/L. Prior to irradiation, the as-prepared solution was stirred continuously in the dark for 1 h at room temperature to achieve adsorption-desorption equilibrium. At 15 min irradiation intervals, a 5 mL aliquot was withdrawn and analyzed via UV-vis spectrophotometry with a maximum absorption peak at 665 nm.

3.8. Free-Radical Scavenger Test

A free-radical scavenger experiment was conducted to study the involvement of reactive species in the photodegradation of MB. Isopropyl alcohol (IPA) solution, BQ, and EDTA were used as scavengers to analyze the influence of hydroxyl radicals (•OH), superoxide radicals ($O_2^{•-}$), and holes (h^+), respectively. In the scavenging experiments, 10 mg of a scavenger was mixed with the dye–catalyst solution (20 mg, 10 mg/L). Afterward, the same procedure mentioned in previous section was conducted under identical conditions.

4. Conclusions

A novel approach for the synthesis of N, S doped GQDs from 1-D carbon nanomaterial with high photocatalytic and photoelectrochemical response was suggested. The mechanism of interfacial charge transfer in GQDs-TiO$_2$ heterojunction is elucidated through the possible C–O–Ti bonds formation. TGA and XPS analyses revealed that N and S doping with in-plane functional groups of GQDs were responsible for stable interface formation between GQDs and TiO$_2$. Photoelectrochemical measurements provide insights into photocurrent and impedance characteristics of as-prepared GQDs-TiO$_2$ nanocomposites. Remarkably high degradation percentages of MB under visible-light illumination were observed for CNT-GQDs-TiO$_2$ (99.78%), RCNT-GQDs-TiO$_2$ (99.39%), and CNF-GQDs-TiO$_2$ (99.23%) nanocomposites. Rapid interfacial charge transfer from the GQDs to TiO$_2$ is the key phenomenon for the enhanced photodegradation of MB. Consecutive degradation cyclic runs to test the highest efficiency and stability of nanocomposites result that the CNF-GQDs-TiO$_2$ nanocomposites have the fastest photodegradation rate and proved to be the most stable among all. CNT-GQDs-TiO$_2$, RCNT-GQDs-TiO$_2$, and CNF-GQDs-TiO$_2$ nanocomposites show different characteristic radicals •OH, h+, and $O_2^{•-}$, respectively, as revealed by radical quenching experiments. This study is considered to be a platform for nonmetallic GQDs and metal oxide semiconductor nanocomposites for green-environment, sustainable energy, and optoelectronic applications.

Author Contributions: J.R.: Conceptualization, Investigation, Methodology, Validation, Data curation, Writing—original draft, Writing—review and editing. U.K.: Conceptualization, Visualization, Formal analysis, Writing—review and editing. M.P.: Writing—review and editing. B.P.: Writing—review and editing. S.-J.P.: Supervision, Funding acquisition, Writing—review and editing. All authors have read and agreed to the published version of the manuscript.

Funding: This research was supported by Korea Electric Power Corporation (Grant number: R21XO01-5). This work was also supported by Nano-Convergence Foundation (www.nanotech2020.org (accessed on 1 May 2021)) funded by the Ministry of Science and ICT (MSIT, Korea) & the Ministry of Trade, Industry and Energy (MOTIE, Korea). [Project Name: Development of high-efficiency activated carbon filter for removing indoor harmful elements (VOCs, radon, bacteria, etc.)].

Data Availability Statement: The data is contained within the article.

Conflicts of Interest: The authors declare no conflict of interest.

References

1. Ebrahimi, M.; Samadi, M.; Yousefzadeh, S.; Soltani, M.; Rahimi, A.; Chou, T.C.; Chen, L.C.; Chen, K.H.; Moshfegh, A.Z. Improved solar-driven photocatalytic activity of hybrid graphene quantum dots/ZnO nanowires: A direct Z-scheme mechanism. *ACS Sustain. Chem. Eng.* **2017**, *5*, 367–375. [CrossRef]
2. Kamran, U.; Park, S.J. Hybrid biochar supported transition metal doped MnO_2 composites: Efficient contenders for lithium adsorption and recovery from aqueous solutions. *Desalination* **2022**, *522*, 115387. [CrossRef]
3. Kanjwal, M.A.; Barakat, N.A.M.; Sheikh, F.A.; Park, S.J.; Kim, H.Y. Photocatalytic activity of ZnO-TiO_2 hierarchical nanostructure prepared by combined electrospinning and hydrothermal techniques. *Macromol. Res.* **2010**, *18*, 233–240. [CrossRef]
4. Zhang, Y.; Park, M.; Kim, H.Y.; Park, S.J. In-situ synthesis of graphene oxide/BiOCl heterostructured nanofibers for visible-light photocatalytic investigation. *J. Alloy. Compd.* **2016**, *686*, 106–114. [CrossRef]
5. Zhang, Y.; Park, M.; Kim, H.Y.; Ding, B.; Park, S.J. In-situ synthesis of nanofibers with various ratios of $BiOCl_x/BiOBr_y/BiOI_z$ for effective trichloroethylene photocatalytic degradation. *Appl. Surf. Sci.* **2016**, *384*, 192–199. [CrossRef]
6. Kamran, U.; Bhatti, H.N.; Noreen, S.; Tahir, M.A.; Park, S.J. Chemically modified sugarcane bagasse-based biocomposites for efficient removal of acid red 1 dye: Kinetics, isotherms, thermodynamics, and desorption studies. *Chemosphere* **2022**, *291*, 132796. [CrossRef]
7. Kamran, U.; Bhatti, H.N.; Iqbal, M.; Jamil, S.; Zahid, M. Biogenic synthesis, characterization and investigation of photocatalytic and antimicrobial activity of manganese nanoparticles synthesized from Cinnamomum verum bark extract. *J. Mol. Struct.* **2019**, *1179*, 532–539. [CrossRef]
8. Gao, L.; Gan, W.; Qiu, Z.; Zhan, X.; Qiang, T.; Li, J. Preparation of heterostructured WO_3/TiO_2 catalysts from wood fibers and its versatile photodegradation abilities. *Sci. Rep.* **2017**, *7*, 1102. [CrossRef]
9. Nawi, M.A.; Zain, S.M. Enhancing the surface properties of the immobilized Degussa P-25 TiO_2 for the efficient photocatalytic removal of methylene blue from aqueous solution. *Appl. Surf. Sci.* **2012**, *258*, 6148–6157. [CrossRef]
10. Ji, Z.; Wu, J.; Jia, T.; Peng, C.; Xiao, Y.; Liu, Z.; Liu, Q.; Fan, Y.; Han, J.; Hao, L. In-situ growth of TiO_2 phase junction nanorods with Ti^{3+} and oxygen vacancies to enhance photocatalytic activity. *Mater. Res. Bull.* **2021**, *140*, 111291. [CrossRef]
11. Hurum, D.C.; Agrios, A.G.; Gray, K.A.; Rajh, T.; Thurnauer, M.C. Explaining the Enhanced Photocatalytic Activity of Degussa P25 Mixed-Phase TiO_2 Using EPR. *J. Phys. Chem. B* **2003**, *107*, 4545–4549. [CrossRef]
12. Ramakrishnan, V.M.; Natarajan, M.; Santhanam, A.; Asokan, V.; Velauthapillai, D. Size controlled synthesis of TiO_2 nanoparticles by modified solvothermal method towards effective photo catalytic and photovoltaic applications. *Mater. Res. Bull.* **2018**, *97*, 351–360. [CrossRef]
13. Miao, J.; Zhang, R.; Zhang, L. Photocatalytic degradations of three dyes with different chemical structures using ball-milled TiO_2. *Mater. Res. Bull.* **2018**, *97*, 109–114. [CrossRef]
14. Ohtani, B.; Prieto-Mahaney, O.O.; Li, D.; Abe, R. What is Degussa (Evonik) P25? Crystalline composition analysis, reconstruction from isolated pure particles and photocatalytic activity test. *J. Photochem. Photobiol. A Chem.* **2010**, *216*, 179–182. [CrossRef]
15. Luan, Y.; Jing, L.; Meng, Q.; Nan, H.; Luan, P.; Xie, M.; Feng, Y. Synthesis of Efficient Nanosized Rutile TiO_2 and Its Main Factors Determining Its Photodegradation Activity: Roles of Residual Chloride and Adsorbed Oxygen. *J. Phys. Chem. C* **2012**, *116*, 17094–17100. [CrossRef]
16. Asahi, R.Y.O.J.I.; Morikawa, T.A.K.E.S.H.I.; Ohwaki, T.; Aoki, K.; Taga, Y. Visible-light photocatalysis in nitrogen-doped titanium oxides. *Science* **2001**, *293*, 269–271. [CrossRef] [PubMed]

17. Xian, J.; Li, D.; Chen, J.; Li, X.; He, M.; Shao, Y.; Yu, L.; Fang, J. TiO$_2$ nanotube array-graphene-CdS quantum dots composite film in Z-scheme with enhanced photoactivity and photostability. *ACS Appl. Mater. Interfaces* **2014**, *6*, 13157–13166. [CrossRef] [PubMed]
18. Aleksandrzak, M.; Adamski, P.; Kukułka, W.; Zielinska, B.; Mijowska, E. Effect of graphene thickness on photocatalytic activity of TiO$_2$-graphene nanocomposites. *Appl. Surf. Sci.* **2015**, *331*, 193–199. [CrossRef]
19. Yang, Z.; He, Z.; Wu, W.; Zhou, Y. Preparation of graphene-TiO$_2$ photocatalysis materials by laser-induced hydrothermal method. *Colloids Interface Sci. Commun.* **2021**, *42*, 100408. [CrossRef]
20. Ramanathan, S.; Moorthy, S.; Ramasundaram, S.; Rajan, H.K.; Vishwanath, S.; Selvinsimpson, S.; Durairaj, A.; Kim, B.; Vasanthkumar, S. Grape seed extract assisted synthesis of dual-functional anatase TiO$_2$ decorated reduced graphene oxide composite for supercapacitor electrode material and visible light photocatalytic degradation of bromophenol blue dye. *ACS Omega* **2021**, *6*, 14734–14747. [CrossRef]
21. Kamran, U.; Park, S.J. Microwave-assisted acid functionalized carbon nanofibers decorated with Mn doped TNTs nanocomposites: Efficient contenders for lithium adsorption and recovery from aqueous media. *J. Ind. Eng. Chem.* **2020**, *92*, 263–277. [CrossRef]
22. Kamran, U.; Park, S.J. Functionalized titanate nanotubes for efficient lithium adsorption and recovery from aqueous media. *J. Solid State Chem.* **2020**, *283*, 121157. [CrossRef]
23. Qin, X.; Jing, L.; Tian, G.; Qu, Y.; Feng, Y. Enhanced photocatalytic activity for degrading Rhodamine B solution of commercial Degussa P25 TiO$_2$ and its mechanisms. *J. Hazard. Mater.* **2009**, *172*, 1168–1174. [CrossRef] [PubMed]
24. Wang, D.; Zhao, H.; Wu, N.; El Khakani, M.A.; Ma, D. Tuning the Charge-Transfer Property of PbS-Quantum Dot/TiO$_2$-Nanobelt Nanohybrids via Quantum Confinement. *J. Phys. Chem. Lett.* **2010**, *1*, 1030–1035. [CrossRef]
25. Das, G.S.; Shim, J.P.; Bhatnagar, A.; Tripathi, K.M.; Kim, T. Biomass-derived Carbon Quantum Dots for Visible-Light-Induced Photocatalysis and Label-Free Detection of Fe(III) and Ascorbic acid. *Sci. Rep.* **2019**, *9*, 15084. [CrossRef] [PubMed]
26. Xie, Y.; Ali, G.; Yoo, S.H.; Cho, S.O. Sonication-Assisted Synthesis of CdS Quantum-Dot-Sensitized TiO$_2$ Nanotube Arrays with Enhanced Photoelectrochemical and Photocatalytic Activity. *ACS Appl. Mater. Interfaces* **2010**, *2*, 2910–2914. [CrossRef] [PubMed]
27. Wang, Q.; Li, J.; Tu, X.; Liu, H.; Shu, M.; Si, R.; Ferguson, C.T.; Zhang, K.A.; Li, R. Single Atomically Anchored Cobalt on Carbon Quantum Dots as Efficient Photocatalysts for Visible Light-Promoted Oxidation Reactions. *Chem. Mater.* **2019**, *32*, 734–743. [CrossRef]
28. Akbar, K.; Moretti, E.; Vomiero, A. Carbon Dots for Photocatalytic Degradation of Aqueous Pollutants: Recent Advancements. *Adv. Opt. Mater.* **2021**, *9*, 2100532. [CrossRef]
29. Lee, Y.L.; Chang, C.H. Efficient polysulfide electrolyte for CdS quantum dot-sensitized solar cells. *J. Power Sources* **2008**, *185*, 584–588. [CrossRef]
30. Påhlsson, A.M.B. Toxicity of heavy metals (Zn, Cu, Cd, Pb) to vascular plants. *Wat. Air Soil Poll.* **1989**, *47*, 287–319. [CrossRef]
31. Bharathi, G.; Nataraj, D.; Premkumar, S.; Sowmiya, M.; Senthilkumar, K.; Thangadurai, T.D.; Khyzhun, O.Y.; Gupta, M.; Phase, D.; Patra, N.; et al. Graphene Quantum Dot Solid Sheets: Strong blue-light-emitting & photocurrent-producing band-gap-opened nanostructures. *Sci. Rep.* **2017**, *7*, 10850. [PubMed]
32. Mahalingam, S.; Manap, A.; Omar, A.; Low, F.W.; Afandi, N.F.; Chia, C.H.; Abd Rahim, N. Functionalized graphene quantum dots for dye-sensitized solar cell: Key challenges, recent developments and future prospects. *Renew. Sust. Energ. Rev.* **2021**, *144*, 110999. [CrossRef]
33. Wang, L.; Wang, Y.; Xu, T.; Liao, H.; Yao, C.; Liu, Y.; Li, Z.; Chen, Z.; Pan, D.; Sun, L.; et al. Gram-scale synthesis of single-crystalline graphene quantum dots with superior optical properties. *Nat. Commun.* **2014**, *5*, 5357. [CrossRef] [PubMed]
34. Liu, D.; Chen, X.; Hu, Y.; Sun, T.; Song, Z.; Zheng, Y.; Cao, Y.; Cai, Z.; Cao, M.; Peng, L.; et al. Raman enhancement on ultra-clean graphene quantum dots produced by quasi-equilibrium plasma-enhanced chemical vapor deposition. *Nat. Commun.* **2018**, *9*, 193. [CrossRef] [PubMed]
35. Maiti, S.; Kundu, S.; Roy, C.N.; Das, T.K.; Saha, A. Synthesis of Excitation Independent Highly Luminescent Graphene Quantum Dots through Perchloric Acid Oxidation. *Langmuir* **2017**, *33*, 14634–14642. [CrossRef] [PubMed]
36. Yan, Y.; Chen, J.; Li, N.; Tian, J.; Li, K.; Jiang, J.; Liu, J.; Tian, Q.; Chen, P. Systematic Bandgap Engineering of Graphene Quantum Dots and Applications for Photocatalytic Water Splitting and CO$_2$ Reduction. *ACS Nano* **2018**, *12*, 3523–3532. [CrossRef]
37. Li, L.S.; Yan, X. Colloidal Graphene Quantum Dots. *J. Phys. Chem. Lett.* **2010**, *1*, 2572–2576. [CrossRef]
38. Wang, W.S.; Wang, D.H.; Qu, W.G.; Lu, L.Q.; Xu, A.W. Large Ultrathin Anatase TiO$_2$ Nanosheets with Exposed {001} Facets on Graphene for Enhanced Visible Light Photocatalytic Activity. *J. Phys. Chem.* **2012**, *116*, 19893–19901. [CrossRef]
39. Zheng, L.; Su, H.; Zhang, J.; Walekar, L.S.; Molamahmood, H.V.; Zhou, B.; Long, M.; Hu, Y.H. Highly selective photocatalytic production of H$_2$O$_2$ on sulfur and nitrogen co-doped graphene quantum dots tuned TiO$_2$. *Appl. Catal. B* **2018**, *239*, 475–484. [CrossRef]
40. Shen, K.; Xue, X.; Wang, X.; Hu, X.; Tian, H.; Zheng, W. One-step synthesis of band-tunable N, S co-doped commercial TiO$_2$/graphene quantum dots composites with enhanced photocatalytic activity. *RSC Adv.* **2017**, *7*, 23319–23327. [CrossRef]
41. Shaban, M.; Ashraf, A.M.; Abukhadra, M.R. TiO$_2$ Nanoribbons/Carbon Nanotubes Composite with Enhanced Photocatalytic Activity; Fabrication, Characterization, and Application. *Sci. Rep.* **2018**, *8*, 781. [CrossRef] [PubMed]
42. Yi, Z.; Merenda, A.; Kong, L.; Radenovic, A.; Majumder, M.; Dumée, L.F. Single step synthesis of Schottky-like hybrid graphene-titania interfaces for efficient photocatalysis. *Sci. Rep.* **2018**, *8*, 8154. [CrossRef] [PubMed]

43. Zhu, L.; Nguyen, D.C.T.; Woo, J.H.; Zhang, Q.; Cho, K.Y.; Oh, W.C. An eco-friendly synthesized mesoporous-silica particle combined with WSe_2-graphene-TiO_2 by self-assembled method for photocatalytic dye decomposition and hydrogen production. *Sci. Rep.* **2018**, *8*, 12759. [CrossRef] [PubMed]
44. Umrao, S.; Abraham, S.; Theil, F.; Pandey, S.; Ciobota, V.; Shukla, P.K.; Rupp, C.J.; Chakraborty, S.; Ahuja, R.; Popp, J.; et al. A possible mechanism for the emergence of an additional band gap due to a Ti–O–C bond in the TiO_2–graphene hybrid system for enhanced photodegradation of methylene blue under visible light. *RSC Adv.* **2014**, *4*, 59890–59901. [CrossRef]
45. Jin, F.L.; Ma, C.J.; Park, S.J. Thermal and mechanical interfacial properties of epoxy composites based on functionalized carbon nanotubes. *Mater. Sci. Eng. A* **2011**, *528*, 8517–8522. [CrossRef]
46. Yim, Y.J.; Park, S.J. Electromagnetic interference shielding effectiveness of high-density polyethylene composites reinforced with multi-walled carbon nanotubes. *J. Ind. Eng. Chem.* **2015**, *21*, 155–157. [CrossRef]
47. Donnet, J.B.; Park, S.J. Surface characteristics of pitch-based carbon fibers by inverse gas chromatography method. *Carbon* **1991**, *29*, 955–961. [CrossRef]
48. Donnet, J.B.; Park, S.J. Anodic surface treatment on carbon fibers: Determination of acid-base interaction parameter between two unidentical solid surfaces in a composite system. *J. Colloid Interface Sci.* **1998**, *206*, 29–32.
49. Das, R.; Rajender, G.; Giri, P.K. Anomalous fluorescence enhancement and fluorescence quenching of graphene quantum dots by single walled carbon nanotubes. *Phys. Chem. Chem. Phys.* **2018**, *20*, 4527–4537. [CrossRef]
50. Lei, Y.; Hu, J.; Zhang, Z.; Ouyang, Z.; Jiang, Z.; Lin, Y. Photoelectric properties of SnO_2 decorated by graphene quantum dots. *Mater. Sci. Semicond. Process.* **2019**, *102*, 104582. [CrossRef]
51. Futaba, D.N.; Yamada, T.; Kobashi, K.; Yumura, M.; Hata, K. Macroscopic Wall Number Analysis of Single-Walled, Double-Walled, and Few-Walled Carbon Nanotubes by X-ray Diffraction. *J. Am. Chem. Soc.* **2011**, *133*, 5716–5719. [CrossRef] [PubMed]
52. Mo, R.; Lei, Z.; Sun, K.; Rooney, D. Facile Synthesis of Anatase TiO_2 Quantum-Dot/Graphene-Nanosheet Composites with Enhanced Electrochemical Performance for Lithium-Ion Batteries. *J. Adv. Mater.* **2014**, *26*, 2084–2088. [CrossRef] [PubMed]
53. Li, D.; Dai, S.; Li, J.; Zhang, C.; Richard-Plouet, M.; Goullet, A.; Granier, A. Microstructure and Photocatalytic Properties of TiO_2–Reduced Graphene Oxide Nanocomposites Prepared by Solvothermal Method. *J. Electron. Mater.* **2018**, *47*, 7372–7379. [CrossRef]
54. Li, J.; Zhou, S.l.; Hong, G.B.; Chang, C.T. Hydrothermal preparation of P25–graphene composite with enhanced adsorption and photocatalytic degradation of dyes. *J. Chem. Eng.* **2013**, *219*, 486–491. [CrossRef]
55. Yadav, A.; Yadav, M.; Gupta, S.; Popat, Y.; Gangan, A.; Chakraborty, B.; Ramaniah, L.M.; Fernandes, R.; Miotello, A.; Press, M.R.; et al. Effect of graphene oxide loading on TiO_2: Morphological, optical, interfacial charge dynamics-A combined experimental and theoretical study. *Carbon* **2019**, *143*, 51–62. [CrossRef]
56. Peng, J.; Gao, W.; Gupta, B.K.; Liu, Z.; Romero-Aburto, R.; Ge, L.; Song, L.; Alemany, L.B.; Zhan, X.; Gao, G.; et al. Graphene Quantum Dots Derived from Carbon Fibers. *Nano Lett.* **2012**, *12*, 844–849. [CrossRef]
57. Das, R.; Parveen, S.; Bora, A.; Giri, P.K. Origin of high photoluminescence yield and high SERS sensitivity of nitrogen-doped graphene quantum dots. *Carbon* **2020**, *160*, 273–286. [CrossRef]
58. Qu, D.; Sun, Z.; Zheng, M.; Li, J.; Zhang, Y.; Zhang, G.; Zhao, H.; Liu, X.; Xie, Z. Three Colors Emission from S,N Co-doped Graphene Quantum Dots for Visible Light H_2 Production and Bioimaging. *Adv. Opt. Mater.* **2015**, *3*, 360–367. [CrossRef]
59. Zhang, Y.; Yang, H.M.; Park, S.J. Synthesis and characterization of nitrogen-doped TiO_2 coatings on reduced graphene oxide for enhancing the visible light photocatalytic activity. *Curr. Appl. Phys.* **2018**, *18*, 163–169. [CrossRef]
60. Rajender, G.; Kumar, J.; Giri, P.K. Interfacial charge transfer in oxygen deficient TiO_2-graphene quantum dot hybrid and its influence on the enhanced visible light photocatalysis. *Appl. Catal. B* **2018**, *224*, 960–972. [CrossRef]
61. Ali, M.H.H.; Al-Afify, A.D.; Goher, M.E. Preparation and characterization of graphene–TiO_2 nanocomposite for enhanced photodegradation of Rhodamine-B dye. *Egypt. J. Aquat. Res.* **2018**, *44*, 263–270. [CrossRef]
62. Thaweechai, T.; Sirisaksoontorn, W.; Poo-Arporn, Y.; Chanlek, N.; Seraphin, S.; Thachepan, S.; Poo-Arporn, R.P.; Suramitr, S. Transparent graphene quantum dot/amorphous TiO_2 nanocomposite sol as homogeneous-like photocatalyst. *J. Nanopart. Res.* **2021**, *23*, 225. [CrossRef]
63. Sarkar, S.; Raghavan, A.; Giri, A.; Ghosh, S. Graphene Quantum Dots Decorated TiO_2 Nanostructures: Sustainable Approach for Photocatalytic Remediation of an Industrial Pollutant. *ChemistrySelect* **2021**, *6*, 10957–10964. [CrossRef]
64. Sun, X.; Li, H.J.; Ou, N.; Lyu, B.; Gui, B.; Tian, S.; Qian, D.; Wang, X.; Yang, J. Visible-light driven TiO_2 photocatalyst coated with graphene quantum dots of tunable nitrogen doping. *Molecules* **2019**, *24*, 344. [CrossRef] [PubMed]
65. Tian, H.; Shen, K.; Hu, X.; Qiao, L.; Zheng, W. N, S co-doped graphene quantum dots-graphene-TiO_2 nanotubes composite with enhanced photocatalytic activity. *J. Alloy. Compd.* **2017**, *691*, 369–377. [CrossRef]
66. Bai, X.; Ji, Y.; She, M.; Li, Q.; Wan, K.; Li, Z.; Liu, E.; Li, J. Oxygen vacancy mediated charge transfer expediting over GQDs/TiO_2 for enhancing photocatalytic removal of Cr (VI) and RhB synchronously. *J. Alloy. Compd.* **2022**, *891*, 161872. [CrossRef]
67. Shen, S.; Wang, H.; Fu, J. A nanoporous Three-dimensional graphene aerogel doped with a carbon quantum Dot-TiO_2 composite that exhibits superior activity for the catalytic photodegradation of organic pollutants. *Appl. Surf. Sci.* **2021**, *569*, 151116. [CrossRef]
68. Leong, K.H.; Fong, S.Y.; Lim, P.F.; Sim, L.C.; Saravanan, P. Physical mixing of N-doped graphene quantum dots functionalized TiO_2 for sustainable degradation of methylene blue. In *IOP Conference Series: Materials Science and Engineering*; IOP Publishing: Bristol, UK, 2018; Volume 409, p. 012009.

69. Safardoust-Hojaghan, H.; Salavati-Niasari, M. Degradation of methylene blue as a pollutant with N-doped graphene quantum dot/titanium dioxide nanocomposite. *J. Clean. Prod.* **2017**, *148*, 31–36. [CrossRef]
70. Tang, J.; Liu, Y.; Hu, Y.P.; Lv, G.; Yang, C.; Yang, G. Carbothermal Reduction Induced Ti3+ Self-Doped TiO_2/GQD Nanohybrids for High-Performance Visible Light Photocatalysis. *Eur. J. Chem.* **2018**, *24*, 4390–4398. [CrossRef]
71. Kusiak-Nejman, E.; Wanag, A.; Kapica-Kozar, J.; Kowalczyk, Ł.; Zgrzebnicki, M.; Tryba, B.; Przepiórski, J.; Morawski, A.W. Methylene blue decomposition on TiO_2/reduced graphene oxide hybrid photocatalysts obtained by a two-step hydrothermal and calcination synthesis. *Catal. Today* **2020**, *357*, 630–637. [CrossRef]
72. Deng, Y.; Tang, L.; Feng, C.; Zeng, G.; Wang, J.; Lu, Y.; Liu, Y.; Yu, J.; Chen, S.; Zhou, Y. Construction of Plasmonic Ag and Nitrogen-Doped Graphene Quantum Dots Codecorated Ultrathin Graphitic Carbon Nitride Nanosheet Composites with Enhanced Photocatalytic Activity: Full-Spectrum Response Ability and Mechanism Insight. *ACS Appl. Mater. Interfaces* **2017**, *9*, 42816–42828. [CrossRef] [PubMed]
73. Wang, X.Q.; Han, S.F.; Zhang, Q.W.; Zhang, N.; Zhao, D.D. Photocatalytic oxidation degradation mechanism study of methylene blue dye waste water with GR/TiO_2. *MATEC Web Conf.* **2018**, *238*, 03006. [CrossRef]

Article

Insight into the Effect of Anionic–Anionic Co-Doping on BaTiO₃ for Visible Light Photocatalytic Water Splitting: A First-Principles Hybrid Computational Study

Souraya Goumri-Said [1] and Mohammed Benali Kanoun [2,*]

1. Physics Department, College of Science and General Studies, Alfaisal University, P.O. Box 50927, Riyadh 11533, Saudi Arabia
2. Department of Mathematics and Sciences, College of Humanities and Sciences, Prince Sultan University, P.O. Box 66833, Riyadh 11586, Saudi Arabia
* Correspondence: mohammedbenali.kanoun@gmail.com or mkanoun@psu.edu.sa

Abstract: In this research, we thoroughly studied the electronic properties and optical absorption characteristics with double-hole coupling of anions–anion combinations for designing effective photocatalysts for water redox using first-principles methods within the hybrid Heyd–Scuseria–Ernzerhof (HSE06) exchange–correlation formalisms. The findings reveal that the values of formation energy of both the anion mono- and co-doped configurations increase monotonically as the chemical potential of oxygen decreases. The N–N co-doped BaTiO₃ exhibits a more favorable formation energy under an O-poor condition compared with other configurations, indicating that N and N pairs are more likely to be synthesized successfully. Interestingly, all the co-doping configurations give a band gap reduction with suitable position for oxygen production and hydrogen evolution. The obtained results demonstrate that all the co-doped systems constitute a promising candidate for photocatalytic water-splitting reactions. Furthermore, the enhanced ability of the anionic-anionic co-doped BaTiO₃ to absorb visible light and the positions of band edges that closely match the oxidation-reduction potentials of water suggest that these configurations are viable photocatalysts for visible-light water splitting. Therefore, the wide-band gap semiconductor band structures can be tuned by double-hole doping through anionic combinations, and high-efficiency catalysts for water splitting using solar energy can be created as a result.

Keywords: anionic–anionic co-doping; BaTiO₃; Heyd–Scuseria–Ernzerhof (HSE06); photocatalytic water splitting

1. Introduction

The search for new renewable and clean energy sources has drawn considerable attention in recent years, owing to the increasing serious environmental challenges rooted in fossil fuel usage and resource depletion [1]. One of the most competitive methods for providing a cheap and ecologically acceptable energy source is photocatalytic hydrogen synthesis from water reduction, which can be considered a clean and sustainable strategy for resolving energy problems [1–3]. A variety of photocatalysts based on oxide materials have received extensive attention in research because of their strong photocatalytic activity, nontoxicity, and chemical stability [4]. Among them, Barium titanate (BaTiO₃) is a common ferroelectric compound with a perovskite structure and large band gap of 3.0–3.3 eV [5]. The most crucial feature is that it has the appropriate band positions to split water into hydrogen and oxygen [6–10]. This means that the valence band (VB) and conduction band (CB) edges of BaTiO₃ are in the suitable positions, which satisfies the thermodynamic condition for the water-splitting reaction [5]. The ultraviolet part of the solar energy spectrum, which makes up only 5% of it, has fewer practical applications due to the high band gap of this material. It is conventionally known that more than 45% of solar irradiance is visible light.

Therefore, to enhance visible-light photocatalytic performance, it is crucial to design active photocatalysts that operate in this region. Doping impurity elements into crystal lattices is a common and effective method for modifying the electronic structure and band gap position in semiconductors.

Numerous studies [11–21] have recently used the doping of metal and non-metal elements into semiconductors to change the optoelectronic properties. For instance, experimental research by Upadhyay et al. [12] shows that adding 2.0% Fe to $BaTiO_3$ results in a noticeable improvement in photocurrent density and a decreasing in band gap because Fe 3d states exist within the band gap region. Nageri and Kumar [13] suggested that Mn-doped $BaTiO_3$ illustrated higher visible-light-driven photocatalytic activity, which was primarily caused by the energy states connected to the dopants and oxygen vacancies. Demircivi et al. synthesized a W-doped $BaTiO_3$ nanocomposite photocatalyst and observed that it was more effective in degrading tetracycline than both undoped and highly doped $BaTiO_3$ at lower doping levels. This may be because less electron-hole recombination occurs [14]. It has also been demonstrated that Rh-doped $BaTiO_3$, which can produce H_2 from water through a photocatalytic process, has two absorption bands in the visible-light spectrum [15]. Recently, Pie et al. investigated the photocatalytic properties of Mo in $BaTiO_3$, indicating that doping with 4d metal Mo achieved band-to-band visible-light absorption and showed a striking rise in hydrogen-generation efficiency in contrast to the pristine material [16]. In comparison to the undoped $BaTiO_3$, Cao et al. [17] discovered that $BaTiO_3$ doped with anionic N had a lower band gap and increased activity of photo-degradation to Rhodamine-B under irradiation within the visible region, mostly because the valence bandwidth was widened.

On the theoretical side, it has been indicated that doping with various metal impurities modifies the crystal structures and electronic properties of cubic $BaTiO_3$ [18]. It has also been reported that by substituting a metal dopant at Ba or Ti site into a $BaTiO_3$ lattice, the electronic nature and impurity ionic size have a direct effect on the stability and the formation of oxygen vacancies [18]. According to the computed results published by Yang et al. [19], band gap energy of $BaTiO_3$ may be reduced and hole mobility increased by replacing Ti with main group elements including metal (Mo, Tc, Ru, Pd) and transition metal (Cr, Mn, Fe, Co, Ni) atoms. Our previous studies indicated that the electronic structures of $BaTiO_3$ have been modified by doping rare earth into either Ba or Ti sites using first-principles calculations [20,21]. Recently, Huang et al. [22] reported the effect of doping chalcogens at O site into $BaTiO_3$ on optical absorption and photocatalytic water splitting. They discovered that the presence of the p-energy state of dopants at the minimum of CB results in the reducing of the band gap and the improvement of the visible range.

Mono-doping, which is the process of adding cations or anions to semiconductors, typically results in the generation of dopant states within the band gap, which could operate as the recombination site for photo-generation of charge carriers and lower photocatalytic activity. This will destroy the photocatalytic activity of photocatalysts. Co-doping with metal cations and anions in large band gap semiconductors has garnered much interest as a solution to these issues [22–24].

Separating photogenerated holes and electrons is made easier by the compensated nature of co-doping with a cation-anion pair, which also passivates the dopants bands and lowers the band gap to a reasonable value [25,26]. As a result, donor-acceptor pairs such as transition metal (MT = V, Nb, Ta, Mo, and W) and anion (X = N, C) atoms are typically used to adjust the band gap of semiconductor-based photocatalysts like TiO_2 [27–30], $SrTiO_3$ [30,31], $NaNbO_3$ [32,33], $KNbO_3$ [34], $La_2Ti_2O_7$ [35], and $BaTiO_3$ [36,37]. In addition to cation-anion co-doping, double-hole co-doping has become a successful approach for modifying the band gap of photocatalysts based on perovskite such as $SrTiO_3$, $NaTaO_3$, and $KNbO_3$ [38–40]. The band gap is greatly minimized as a result of anionic-anionic co-doping, which also produces completely occupied delocalized intermediate states above the valence region, while increasing the overall stability [41].

In this study, we investigated the impact of double-hole-mediated coupling on the electronic structures and the photocatalytic water-splitting performance of BaTiO$_3$ by performing hybrid density functional theory (DFT) computations. By analyzing the geometrical crystal structures, energies of formation, electronic properties, and optical absorption characteristics of anion mono- and co-doped BaTiO$_3$, we distinctly demonstrate that anion-anion (N–N, C–S, N–P, and P–P) coupling combinations can successfully reduce the band gap by adding an intermediate filled and delocalized state and can also produce attractive band alignments that are suitable for photo-electrochemical water splitting. This type of double-hole-mediated co-doping, for example, is a successful way to modify the band structures of a large band gap photocatalyst and then improve the performance of the photocatalytic ability of photocatalysts, including visible-light water splitting.

2. Computational Details

The Quantum Atomistix ToolKit (QuantumATK) software package, which is based on the local integration of the atomic orbitals method and density functional theory (DFT), was used to perform the reported computations [42]. The exchange and correlation energy functionals have been expressed using the Perdew–Burke–Ernzerhof generalized gradient approximation [43]. The pseudopotential of norm-conserving PseudoDojo was used to define the interaction between ion nuclei and valence electrons [44]. Self-consistent field simulations with a tolerance limit of 10^{-8} Ha were used to determine energy convergence. PseudoDojo-medium was chosen as the basis set, and a mesh cut-off energy of 90 Ha was used. A force on each atom was reduced to 0.05 eV/Å by minimizing the total energy of the system using the Broyden–Fletcher–Goldfarb–Shanno (LBFGS) approach, which led to the stable arrangement of the atoms and the lattice parameters. A $4 \times 4 \times 3$ and $10 \times 10 \times 8$ Monkhorst–Pack [45] k-grid was used for the Brillouin Zone integration in performing geometry optimization and electronic calculations. The accurate band gap energy and optical properties of semiconductors were predicted using the Heyd–Scuseria–Ernzerhof (HSE06) more precise hybrid density functional [46,47]. Short-range (sr) and long-range (lr) components of the exchange and correlation energy in the HSE function were defined as:

$$E_{xc}^{HSE}(\mu, \alpha) = \alpha E_x^{HF,sr,\mu} + (1-\alpha) E_x^{PBE,sr,\mu} + E_x^{PBE,lr,\mu} + E_c^{PBE} \quad (1)$$

where μ denotes the parameter defining the range separation of Coulomb kernel ($\mu = 0.2$/Å), α is the mixing parameter of 0.25, and E_x and E_c represent the exchange and correlation contributions, respectively.

3. Results and Discussion

In the pristine cubic BaTiO$_3$ structure, Ba atoms are situated at the eighth apexes of the cube corner, Ti atoms occupy the center of the octahedron, and O atoms are located at the face centers, forming a symmetric octahedron (TiO$_6$), as illustrated in Figure 1. The computed lattice parameters and Ba–O (2.85) and Ti–O (2.02) bond lengths are in good agreement with the experimental measurements [48] and values of previous theoretical studies [20,21,36,49]. The electronic band structure and density of state (DOS) were computed using HSE06 functional, as shown in Figure 1. The obtained direct band gap of 3.0 eV is more compatible with both experimental results [5] and previous reported results [20,21,36,49]. The analysis of the DOS figure shows that the VB and CB edges principally consist of O 2p states and Ti 3d states, respectively. Nevertheless, the absence of any Ba electronic states around the maximum valence edge and lowest conduction edge suggests that Ba merely contributes to the structure of BaTiO$_3$ and has no impact on the electronic structure close to the Fermi level.

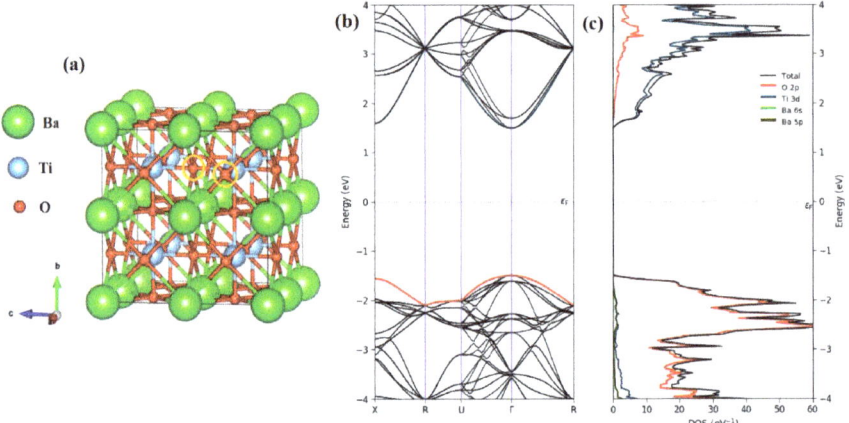

Figure 1. (a) The optimized supercell structure of 2 × 2 × 2 supercell and (b) calculated band structure and total and partial DOS of pristine BaTiO3. (Ba in green, Ti in light blue, O in red, orange circles of doping atoms).

For mono- and co-doped systems, a 2 × 2 × 2 supercell approach is adopted with periodic boundary conditions, consisting of 8 Ba, 8 Ti, and 24 O atoms. The anionic mono-doped configuration is modeled by replacing one of the anion dopants (N, P, C, S) with one of the O atoms in the BaTiO$_3$ supercell. It has been discovered that the anionic–anionic co-doped BaTiO$_3$ system, which is constructed by utilizing two anionic dopants to substitute the two nearby O atoms at the same level, marked with orange circles in Figure 1, is more stable than others [38–40]. In this case, we look only at this particular lowest-energy design. The lattice constants of the modified systems are gathered in Table 1. For all co-doping systems, the lattice constants of the relaxed structures decreased due to the incorporation of an anionic element.

Table 1. The optimized lattice parameters, and the formation of energy for pristine and mono- and co-doped BaTiO$_3$.

	a (Å)	b (Å)	c (Å)	E_b (eV)	E_f (eV)	
					Ti-Rich	O-Rich
Pure	8.005	8.005	8.005	-	-	-
N	8.014	8.014	8.014	-	−10.10	−2.75
C	8.008	8.008	8.073	-	−5.33	−0.08
S	8.070	8.064	8.423	-	−4.31	0.94
P	8.012	8.012	8.232	-	−2.08	3.18
N–N	8.088	8.015	8.088	1.317	−17.53	−6.23
C–S	8.267	8.017	8.107	2.179	−11.89	−1.32
N–P	8.297	8.024	8.086	3.038	−13.12	−2.62
P–P	8.321	7.980	8.321	4.637	−8.79	1.72

To investigate the relative stability of co-doped configurations, we determined the binding energy (E_b) by adopting the following relation:

$$E_b = E_{A1} - E_{A2} - E_{pure} + E_{A1+A2} \qquad (2)$$

where E_{pure}, E_{A1}, E_{A2}, and E_{A1+A2} are the total energies of pure, (first anion, $A1$) mono-doped, (second anion $A2$) mono-doped, and (double-hole, $A1 + A2$) co-doped BaTiO$_3$, respectively. The calculated values of binding energy are gathered in Table 1. The computed binding energy values are positive for the investigated double-hole co-doped systems, which contribute to its stability relative to the analogous mono-doped systems.

To determine the best growth circumstances for different doped configurations, we computed the formation energy, E_f, using the following equation [50,51]

$$E_f = E_{doped} - E_{undoped} - n_X \mu_X + n_O \mu_O \qquad (3)$$

where $E_{undoped}$ and E_{doped} denote the total energies of the pure and doped BaTiO$_3$ supercell. The number n represents the addition or subtraction of dopant and host ions. The chemical potential of the X (N, P, S, C) and O atoms are denoted by the symbols μ_X, and μ_O, respectively, and depend on the parameters of the experiment. The formation energy is dependent on the chemical potentials of the host atoms represented by the environment because defects occur in response to the experimental growth or annealing process. As a stable structure, the chemical potentials of the BaTiO$_3$ and the constituent Ba, Ti, and O atoms should satisfy the following expression:

$$\mu_{Ba} + \mu_{Ti} + 3\mu_O = \mu_{BaTiO_3} \qquad (4)$$

where μ_{Ba} is computed from the total energy of one atom of Ba in bulk crystal structure. For Ti-rich conditions, the μ_{Ti} can be obtained in the ground state with energy of bulk Ti and μ_O estimated according to the Expression (4). Under O-rich conditions, μ_O is determined as the total energy of O$_2$ molecules. The variable μ_X is the energy of a single X atom inside a cube with a side length of 10 Å. The calculated energies of formation for anionic mono- and co-doped are summarized in Table 1 as well as illustrated in Figure 2. It reveals that a larger negative value of formation energy illustrates a thermodynamically more favorable co-doping procedure. The findings imply that the formations of all configurations are conceptually beneficial because of the negative formation energy. The analysis in Figure 2 shows that N mono-doping is proven to be more advantageous than in other mono-doping cases, a result which can be ascribed to the similar radii of atoms. In addition, the values of formation energy of anion mono-doping rise when the μ_O varies from O-poor and O-rich environments. This shows that this type of doping is easily formed under O-poor conditions. It is found that anionic co-doping has much lower formation energy than the comparable mono-doping, which is energetically favorable in O-poor conditions. This is mainly owing to the stronger interaction between anionic and anionic atoms in the former. It should be noted that N–N-co-doped BaTiO$_3$ exhibits a more favorable energy of formation compared with the other configurations, indicating that N and N combinations are more likely to be synthesized successfully.

Next, we investigated the impact of anion mono-doping on the electronic structures of BaTiO$_3$. In the case of anionic mono-doping, the overall electron number may increase or decrease in comparison to that of a pristine BaTiO$_3$, in which spin-polarization calculations are performed to handle these mono-doped systems. For N-mono-doped BaTiO$_3$, as an N-dopant contains one less valence electron compared with an O atom, the N-doping is one electron deficient. This is reflected in asymmetrical DOSs with regard to the electrons in the spin-up and spin-down channels, as displayed Figure S1, exhibiting a magnetic nature with total magnetic moment of +1.0 μ_B. This is principally coming from the N atom (+0.954 μ_B). When the N impurity substitutes the O atom, one acceptor level in the band gap is generated. These unfilled states superior to the Fermi level are extremely unfavorable for photocatalytic activity, as a result of their ability to trap charge carriers and speed up the electron-hole combination procedure. The photocatalytic performance may be impacted by the partially filled impurity levels near the Fermi level, which could serve as an electron-hole recombination junction. For P-mono-doped BaTiO$_3$, replacing one O atom with one P impurity will also result in the formation of a single acceptor, because the P atom has one less valence electron than O atoms. P-doping introduces localized levels that arise in the majority and minority spin channel. Therefore, an additional hole was formed in the system which was followed by a spin-polarization effect. These results show a magnetic behavior. Moreover, the total magnetic moment obtained is 1.0 μ_B, primarily due to the P atom (0.904 μ_B). Below the Fermi level, there are several occupied localized states that mostly result from the combining of P 3p and O 2p states. Additionally, it is discovered that a number of empty dopant states, primarily caused by P-3p localized states, exist close to the minimum of CB. For C-doping, the introduction of two acceptor levels into the BaTiO$_3$ lattice results from the substitution of O with C. The total magnetic moment of C-doping of + 2.002 μ_B is determined by spin-resolved calculations, which originates mainly from the C atom with local moment of (+1.741 μ_B). At the maximum valence band, it is evident that a number of dopant states, largely composed of C-2p orbitals, are present. Surprisingly, above the Fermi level, no unoccupied localized dopant states are visible. The absence of uninhabited localized bands above the Fermi level is mentioned. It also has a narrow effective band gap of 2.22 eV, making it a reasonable photocatalyst for the visible region. In the case of S-doping, the S atom has the same number of valence electrons as O atoms and, as a result, the interaction between S 3p states and Ti 3d states occurs when one S atom replaces one O atom. As can be observed from the DOS shown in Figure S2, there appear to be occupied localized bands above the valence region that are generated by S 3p states and O 2p states, whereas the minimum CB is made up of Ti states. Without empty localized states arising beyond the Fermi level, the band gap is found to be 2.24 eV, which is advantageous for the absorption of visible light.

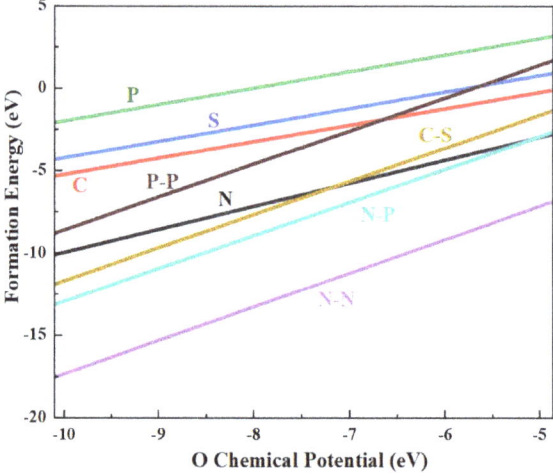

Figure 2. Estimated energies of formation in respect to the oxygen chemical potential for anionic mono- and anionic-anionic co-doped configurations.

BaTiO$_3$ mono-doped with anionic elements, as demonstrated above, produces certain impurity states within the band gap, thus narrowing the gap region. These in-gap states have certain drawbacks for photocatalytic water splitting since they could serve as sites for the capture and recombination of photo-generated electrons and holes, which would reduce the efficiency of photon conversion. In fact, the double-hole-mediated anionic co-doping of BaTiO$_3$ with N–N, C–S, N–P, and P–P atom combinations may compensate for the charge without the formation of unfilled bands above the Fermi level. In N–N co-doped systems, the replacement of two near-neighboring O atoms by two N-dopants introduces two net holes and a new N-N bond is created in which the bond length distance is 1.518 Å. The obtained value is much shorter than that of the distance of 2.830 Å between two O atoms for the pristine BaTiO$_3$, indicating the strong coupling between the two N atoms. The band structure and DOS are calculated to study the impact of N–N co-doping on the electronic characteristics of BaTiO$_3$, as shown in Figure 3a. The HSE06 calculation predicts a band gap of 1.55 eV that is notably lower than the pure compound owing to the existence of the intermediate states located above the valence band. Additionally, since no acceptor states are seen beyond the Fermi level, the recombination effects will be lower than they would be in an N mono-doped system. Note that two completely occupied states are created within the band gap after the N–N co-doping. According to the PDOS, the edge of the VB of N–N co-doped configuration is mainly contributed to by O 2p states, whereas the edge of the CB is dominated by Ti 3d states.

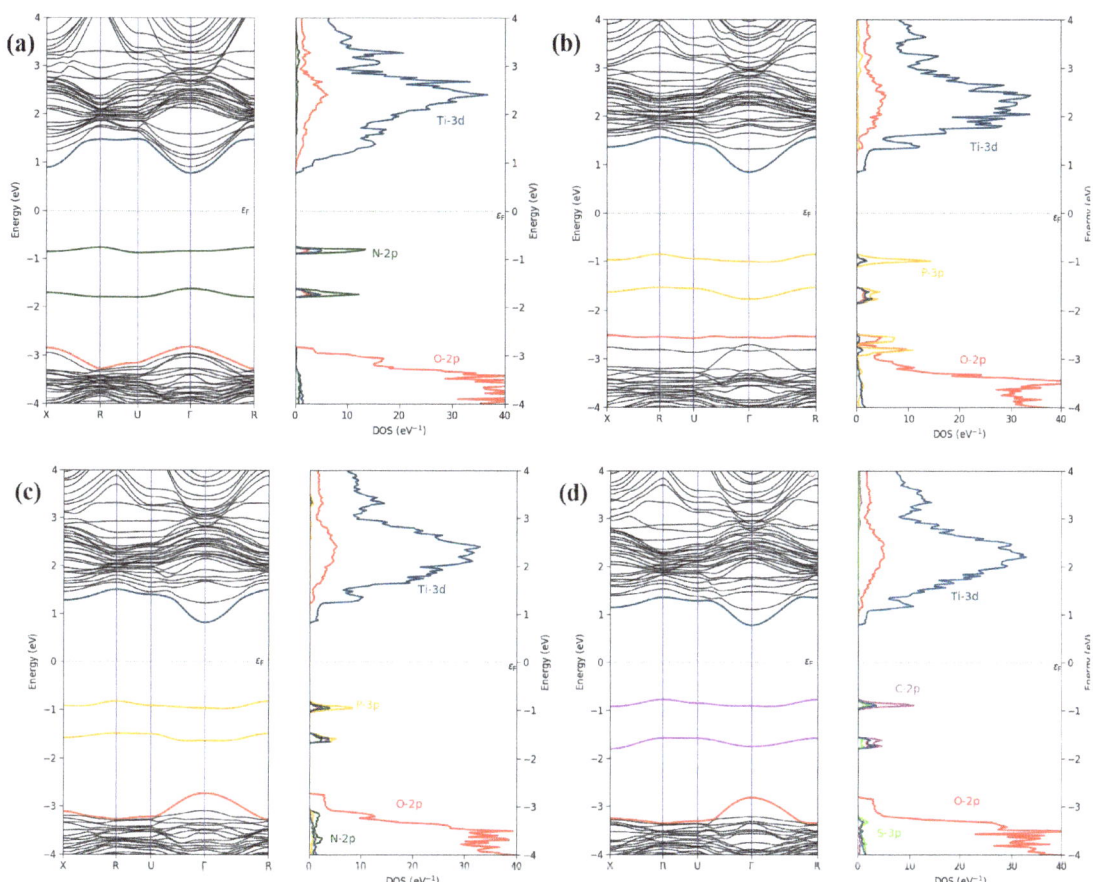

Figure 3. Electronic band structure and partial density of states for (**a**) N–N, (**b**) P–P, (**c**) N–P, and (**d**) C–S co-doped BaTiO$_3$.

Most filled-in gap states are situated above the VB as a result of the strong mixture between the Ti 3d, and N 2p states, as well as the strong coupling between the O 2p states. Additionally, no acceptor states can be seen above the Fermi level, proving that there will be significantly fewer recombination sites than in the case of N mono-doping. Thus, these unoccupied dopants states that manifest in the mono-doping system can be eliminated by such double-hole N–N co-doping. The P–P co-doping into BaTiO$_3$ is nearly identical to the N–N co-doping, because it likewise has two net holes. The P–P bond distance is found to be 2.109 Å. The analysis in Figure 3b shows that the edges of VB and CB are predominantly composed of the P 3p states and Ti 3d states, respectively. When two P atoms are in close proximity to one another, the unpaired 3p electrons of the two neighboring

P atoms hybridize with one another and create a bonding band and an anti-bonding band, since each of the P atoms has one unpaired electron. The latter hybridization introduces three completely occupied sub-bands above the VB edge, while the anti-bonding state is situated near the minimum of CB. Interestingly, there is no unfilled impurity states in the band gap region. The calculated band gap of P–P co-doping is found to be 1.70 eV, which would undoubtedly increase the visible radiation photocatalysis and decrease the recombination losses. For N–P co-doped $BaTiO_3$, co-doping with N and P introduces two net holes, one from the N atom and the other from the P atom. The DOS of the co-doping configuration is shown in Figure 3c, in which fully occupied states appear just above the maximum of VB, so no unoccupied localized states are between the VB and CB edges. Therefore, the band gap is reduced due to 1.64 eV. The N–P bond length obtained is 1.79 Å. Additionally, above the maximum of VB, completely filled dopant states are produced by the admixture of N $2p$, P $3p$, and O $2p$ states. This also represents a reliable link between the N $2p$ and P $3p$ states caused by double holes.

We also investigated the C–S co-doping in a $BaTiO_3$ supercell. The geometry optimization shows that the C–S bond length distance is found to be 1.85 Å, which is much shorter than that of O–O (2.85 Å, proving the strong coupling between C and S atoms). According to the PDOS displayed in Figure 3d, completely occupied states are produced above the upper limit of VB, reducing the band gap to 1.55 eV. Moreover, the maximum of VB consists of O $2p$ states and the lower limit of CB is made up of Ti $3d$ states. Unlike the N–N co-doping, the C atom is responsible for the two holes. All localized states between the borders of VB and CB are occupied as the bonding state moves into the VB and the anti-bonding state moves towards the CB. Moreover, there are two occupied states, primarily originated from C $2p$ and S $3p$ states that are located above the VB edge.

Insightful, and closely connected to the electronic structure of materials are optical functions. For the ongoing design of enhanced semiconductor applications, in-depth knowledge of absorption coefficient and refractive index dispersions is necessary. The depth to which light at a given wavelength (energy) traverses a material before being absorbed is defined by the absorption coefficient, which provides information about the optimal efficiency of energy conversion. In Figure 4, we plotted the absorption spectra versus wavelength (from 0 to 800 nm) for $BaTiO_3$ and the doped derivatives configurations to see the impact of the band gap narrowing on the optical response to photon absorption. According to an investigation of optical absorption characteristics, pure $BaTiO_3$ can accumulate only in the UV area, appearing at one peak as λ = 345 nm, and has no response to visible-light absorption. The anionic-anionic co-doping extends the absorption coefficient leading to the appearance of more peaks compared to that of the pure BaTiO3. These rising peaks are known as absorption edges. The optical absorption spectrum of the P–P doping system shows three peaks, the first one at 290 nm and the second one at λ = 370 nm, in which the most interesting peak appears in the blue zone of the visible spectrum at λ = 470 nm but with less intensity compared to the former peaks. The C–S doping configuration is more interesting as it shows a redshift of the $BaTiO_3$ spectra with four peaks in the visible ranges at 388 nm, 425 nm, 513 nm, and 590 nm, the reasonably highest absorption intensity being 4.25×10^5 (cm^{-1}). The N–P co-doping shows a peak in absorption with an intensity of 6.1×10^5 (cm^{-1}) in the start of the violet wavelength (380 nm) and two more peaks at blue and green wavelengths. The N–N co-doping produces two peaks in blue and green wavelengths with 2.82×10^5 (cm^{-1}) absorption intensity. We can relate this different behavior to the band gap already calculated in Figure 4. In practice, at low frequencies, free carriers are connected to the primary electronic conduction mechanism in a semiconductor. As the photon energy increases and gets closer to the energy gap, a new conduction mechanism may emerge. An inter-band transition occurs when an electron is excited by a photon from an occupied state in the valence band to an unoccupied state in the conduction band. We anticipate that inter-band transition will have a threshold energy at the energy gap. The frequency dependence of the absorption coefficient for direct allowed transitions is governed by the following formula: $\alpha_{abs}(\omega) \propto \frac{1}{\omega}\sqrt{\hbar\omega - E_g}$. As a result, a threshold at the energy gap E_g determines the direct optically permitted inter-band transitions in our materials. As a result, there are many different physical processes with a different frequency dependence of $\alpha_{abs}(\omega)$. In some cases, using the imaginary part of the dielectric function, given by $\varepsilon_2(\omega) = \frac{\tilde{n}c}{\omega}\alpha_{abs}(\omega)$, to describe absorption is preferable, where $\tilde{n}$ denotes the index of refraction. Furthermore, the absorption coefficient of an indirect inter-band transition in which a phonon is absorbed is: $\alpha_{abs}(\omega) = C_a \frac{(\hbar\omega - E_g + \hbar\omega_q)^2}{exp(\hbar\omega_q/k_BT) - 1}$, where C_a is a constant for the phonon absorption process, $\hbar\omega_q$ is phonon energy, and k_B represents the Boltzmann constant (8.380649 J K^{-1} mol^{-1}).

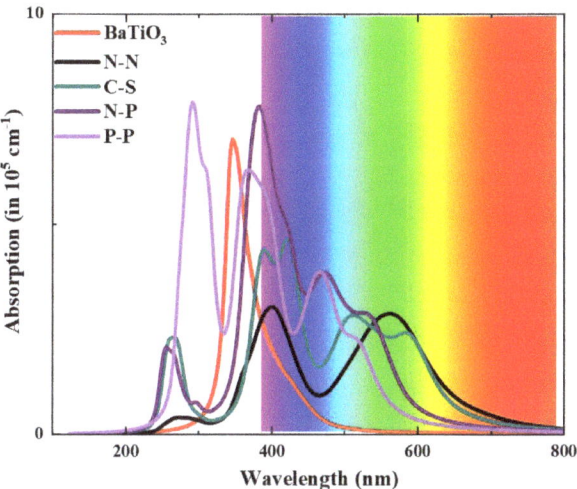

Figure 4. Calculated optical absorption spectra for pristine and anionic-anionic co-doped BaTiO$_3$ at the HSE06 level.

The determination of the energy-band potentials of photocatalysts is crucial for understanding the photocatalytic mechanism of semiconductor materials, since the energy-band potentials of semiconductor-based photocatalysts are closely connected to the redox potential of photogenerated charge carriers. The respective locations of *VB* and *CB* potentials of pristine BaTiO$_3$ can be computed through the expression [5]:

$$E_{CB} = \chi - 0.5E_g + E_0 \quad (5)$$

$$E_{VB} = E_{CB} + E_g \quad (6)$$

where E_g represents the band gap calculated using HSE06 functional, χ is the absolute electronegativity, and E_0 is the hydrogen dimension's free-electron energy (~4.5 eV). For pure BaTiO$_3$, the CB edge is found to be 0.385 eV larger than the reduction potential, and the *VB* edge is 0.89 eV smaller than the oxidation potential, indicating good agreement with previous work [36]. It indicates that while pure BaTiO$_3$ is very effective at splitting water, it can absorb only UV light, which reduces the efficiency of its photocatalytic activity. The alignments of the band edge of pure BaTiO$_3$ are shown in Figure 5 with respect to the redox potentials of water [40]. According to the findings of the band alignment, the designed co-doping systems have the capacity to thermodynamically split water because the CB and VB edge locations of N–N, P–P, N–P, and C–S co-doped BaTiO$_3$ are situated outside the redox potential of water. This demonstrates that both the photo-reduction and the photo-oxidation of water are beneficial processes in this system. Therefore, anionic co-doped BaTiO$_3$ materials would make good water-splitting candidates when exposed to visible light.

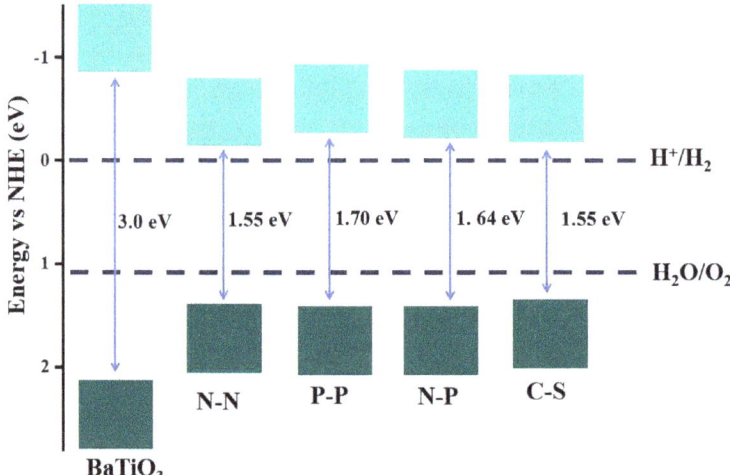

Figure 5. The calculated VB and CB edge energy levels versus normal hydrogen electrode (NHE) using HSE06 functional of N–N, P–P, N–P, and C–S co-doped BaTiO$_3$.

4. Conclusions

We performed hybrid density functional calculations to thoroughly examine the crystal geometry electronic structures and optical properties of anionic mono-doped and anionic-anionic co-doped BaTiO$_3$ in order to generate efficient water redox photocatalysts. According to our findings, most of the co-doping systems can successfully lower the formation energy of the associated mono-doped system. Therefore, co-doping configurations are suitable under O-rich conditions. It was found that the formation energies of all the co-doped systems and anionic mono-doped systems increase monotonically as the chemical potential of O rises. Moreover, the anionic co-doping combinations (N–N, P–P, N–P, and C–S) generate completely filled dopant states in the band gap region, resulting in a reduction in the effective band gaps. It was also revealed that the double-hole anionic co-doped BaTiO$_3$ is proven to be a successful absorber of visible light by the computed optical absorption curves. Additionally, the locations of the band edges in relation to the redox potentials of water demonstrate that anionic co-doping pairs can meet the needs of H$_2$ and O$_2$ generation. According to these findings, combining anionic co-doping pairs via double holes is an efficient method for tuning the band gap of wide–band gap semiconductors, and for creating very effective catalysts for water splitting of solar light.

Supplementary Materials: The following supporting information can be downloaded at: https://www.mdpi.com/article/10.3390/catal12121672/s1. Figure S1. Calculated total and partial DOS of (a) N doped, (b) P doped, (c) C doped and (d) S doped BaTiO$_3$. Here, the vertical dashed line denotes the Fermi level. Figure S2. Side view of the optimized structures of (a) pristine, (b) N–N, (c) C–S, (d) N–P and (e) P–P co-doped BaTiO$_3$.

Author Contributions: Conceptualization, S.G.-S. and M.B.K.; methodology, S.G.-S. and M.B.K.; software, S.G.-S. and M.B.K.; validation, S.G.-S. and M.B.K.; formal analysis, S.G.-S. and M.B.K.; investigation, S.G.-S. and M.B.K.; resources, S.G.-S. and M.B.K.; data curation, S.G.-S. and M.B.K.; writing—original draft preparation, S.G.-S. and M.B.K.; writing—review and editing, S.G.-S. and M.B.K.; visualization, S.G.-S. and M.B.K.; supervision, S.G.-S. and M.B.K.; project administration, S.G.-S. and M.B.K. All authors have read and agreed to the published version of the manuscript.

Funding: This research received no external funding.

Data Availability Statement: Data sharing is not applicable to this article.

Conflicts of Interest: The authors declare no conflict of interest.

References

1. Kudo, A.; Miseki, Y. Heterogeneous photocatalyst materials for water splitting. *Chem. Soc. Rev.* **2009**, *38*, 253–278. [CrossRef] [PubMed]
2. Byrne, C.; Subramanian, G.; Pillai, S.C. Recent advances in photocatalysis for environmental applications. *J. Environ. Chem. Eng.* **2018**, *6*, 3531–3555. [CrossRef]
3. Moniz, S.J.A.; Shevlin, S.A.; Martin, D.J.; Guo, Z.X.; Tang, J.W. Visible-light driven heterojunction photocatalysts for water splitting—A critical review. *Energy Environ. Sci.* **2015**, *8*, 731–759. [CrossRef]
4. Wang, Q.; Domen, K. Particulate Photocatalysts for Light-Driven Water Splitting: Mechanisms, Challenges, and Design Strategies. *Chem. Rev.* **2020**, *120*, 919–985. [CrossRef] [PubMed]
5. Xu, Y.; Schoonen, M.A. The absolute energy positions of conduction and valence bands of selected semiconducting minerals. *Am. Miner.* **2000**, *85*, 543–556. [CrossRef]
6. Maeda, K. Rhodium-Doped Barium Titanate Perovskite as a Stable p-Type Semiconductor Photocatalyst for Hydrogen Evolution under Visible Light. *ACS Appl. Mater. Interfaces* **2014**, *6*, 2167–2173. [CrossRef]
7. Zhang, S.; Chen, D.; Liu, Z.; Ruan, M.; Guo, Z. Novel strategy for efficient water splitting through pyro-electric and pyro-photo-electric catalysis of BaTiO3 by using thermal resource and solar energy. *Appl. Catal. B Environ.* **2021**, *284*, 119686. [CrossRef]
8. Gao, C.; Yu, H.; Zhang, L.; Zhao, Y.; Xie, J.; Li, C.; Cui, K.; Yu, J. Ultrasensitive Paper-Based Photoelectrochemical Sensing Platform Enabled by the Polar Charge Carriers-Created Electric Field. *Anal. Chem.* **2020**, *92*, 2902–2906. [CrossRef]
9. Zhou, Z.; Tang, H.X.; Sodano, A.H. Vertically Aligned Arrays of BaTiO3 Nanowires. *ACS Appl. Mater. Interfaces* **2013**, *5*, 11894–11899. [CrossRef]
10. Fang, T.; Hu, H.; Liu, J.; Jiang, M.; Zhou, S.; Fu, J.; Wang, W.; Yang, Y. Type-II Band Alignment Enhances Unassisted Photoelectrochemical Water-Splitting Performance of the BaTiO3/CdS Ferroelectric Heterostructure Photoanode under Solar Light Irradiation. *J. Phys. Chem. C* **2021**, *125*, 18734–18742. [CrossRef]
11. Fernandes, E.; Gomes, J.; Martins, R.C. Semiconductors Application Forms and Doping Benefits to Wastewater Treatment: A Comparison of TiO2, WO3, and g-C3N4. *Catalysts* **2022**, *12*, 1218. [CrossRef]
12. Upadhyay, S.; Shrivastava, J.; Solanki, A.; Choudhary, S.; Sharma, V.; Kumar, P.; Singh, N.; Satsangi, V.R.; Shrivastav, R.; Waghmare, U.V.; et al. Enhanced photoelectrochemical response of BaTiO3 with Fe doping: Experiments and first-principles analysis. *J. Phys. Chem. C* **2011**, *115*, 24373–24380. [CrossRef]
13. Nageri, M.; Kumar, V. Manganese-doped BaTiO3 nanotube arrays for enhanced visible light photocatalytic applications. *Mater. Chem. Phys.* **2018**, *213*, 400–405. [CrossRef]
14. Demircivi, P.; Simsek, E.B. Visible-light-enhanced photoactivity of perovskite-type W-doped BaTiO3 photocatalyst for photodegradation of tetracycline. *J. Alloys Compd.* **2019**, *774*, 795–802. [CrossRef]
15. Cui, Y.; Sun, H.; Shen, G.; Jing, P.; Pu, Y. Effect of Dual-Cocatalyst Surface Modification on Photodegradation Activity, Pathway, and Mechanisms with Highly Efficient Ag/BaTiO3/MnOx. *Langmuir* **2020**, *36*, 498–509. [CrossRef]
16. Xie, P.; Yang, F.; Li, R.; Ai, C.; Lin, C.; Lin, S. Improving hydrogen evolution activity of perovskite BaTiO3 with Mo doping: Experiments and first-principles analysis. *Int. J. Hydrogen Energy* **2019**, *44*, 11695–11704. [CrossRef]
17. Cao, J.; Ji, Y.; Tian, C.; Yi, Z. Synthesis and enhancement of visible light activities of nitrogen-doped BaTiO3. *J. Alloys Compd.* **2014**, *615*, 243–248. [CrossRef]
18. Sharma, V.; Pilania, G.; Rossetti, G.A.; Slenes, K.; Ramprasad, R. Comprehensive examination of dopants and defects in BaTiO3 from first principles. *Phys. Rev. B Condens. Matter Mater. Phys.* **2013**, *87*, 134109. [CrossRef]
19. Yang, F.; Yang, L.; Ai, C.; Xie, P.; Lin, S.; Wang, C.; Lu, X. Tailoring Bandgap of Perovskite BaTiO3 by Transition Metals Co-Doping for Visible-Light Photoelectrical Applications: A First-Principles Study. *Nanomaterials* **2018**, *8*, 455. [CrossRef]
20. Alshoaibi, A.; Kanoun, M.B.; Haq, B.U.; AlFaify, S.; Goumri-Said, S. Ytterbium doping effects into the Ba and Ti sites of perovskite barium titanate: Electronic structures and optical properties. *Results Phys.* **2020**, *18*, 103257. [CrossRef]
21. Alshoaibi, A.; Kanoun, M.B.; Haq, B.U.; AlFaify, S.; Goumri-Said, S. Insights into the Impact of Yttrium Doping at the Ba and Ti Sites of BaTiO3 on the Electronic Structures and Optical Properties: A First-Principles Study. *ACS Omega* **2020**, *5*, 15502–15509. [CrossRef] [PubMed]
22. Huang, H.-C.; Yang, C.-L.; Wang, M.-S.; Ma, X.-G. Chalcogens doped BaTiO3 for visible light photocatalytic hydrogen production from water splitting. *Spectrochim. Acta Part A* **2019**, *208*, 65–72. [CrossRef]
23. Ohno, T.; Tsubota, T.; Nakamura, Y.; Sayama, K. Preparation of S, C cation-codoped SrTiO3 and its photocatalytic activity under visible light. *Appl. Catal. A Gen.* **2005**, *288*, 74–79. [CrossRef]
24. Niu, M.; Cheng, D.; Cao, D. Enhanced photoelectrochemical performance of anatase TiO2 by metal-assisted SeO coupling for water splitting. *Int. J. Hydrogen Energy* **2013**, *38*, 1251–1257. [CrossRef]
25. Yang, K.; Dai, Y.; Huang, B. Review of First-Principles Studies of TiO2: Nanocluster, Bulk, and Material Interface. *Catalysts* **2020**, *10*, 972. [CrossRef]
26. Zhang, J.; Tse, K.; Wong, M.; Zhang, Y.; Zhu, J. A brief review of co-doping. *Front. Phys.* **2016**, *11*, 117405. [CrossRef]
27. Gai, Y.; Li, J.; Li, S.-S.; Xia, J.-B.; Wei, S.-H. Design of Narrow-Gap TiO2: A Passivated Codoping Approach for Enhanced Photoelectrochemical Activity. *Phys. Rev. Lett.* **2009**, *102*, 036402. [CrossRef]

28. Kanoun, M.B.; Alshoaibi, A.; Goumri-Said, S. Hybrid Density Functional Investigation of Cu Doping Impact on the Electronic Structures and Optical Characteristics of TiO_2 for Improved Visible Light Absorption. *Materials* **2022**, *15*, 5645. [CrossRef]
29. Ikram, M.; Ul-Haq, M.A.; Haider, A.; Haider, J.; Ul-Hamid, A.; Shahzadi, I.; Bari, M.A.; Ali, S.; Goumri-Said, S.; Kanoun, M.B. The enhanced photocatalytic performance and first-principles computational insights of Ba doping-dependent TiO_2 quantum dots. *Nanoscale Adv.* **2022**, *4*, 3996–4008. [CrossRef]
30. Niishiro, R.; Kato, H.; Kudo, A. Nickel and either tantalum or niobium-codoped TiO_2 and $SrTiO_3$ photocatalysts with visible-light response for H_2 or O_2 evolution from aqueous solutions. *Phys. Chem. Chem. Phys.* **2005**, *7*, 2241–2245. [CrossRef]
31. Wang, J.; Wang, Y.; Wang, Y.; Zhang, X.; Fan, Y.; Liu, Y.; Yi, Z. Role of P in improving V-doped $SrTiO_3$ visible light photocatalytic activity for water splitting: A first—Principles study. *Int. J. Hydrogen Energy* **2021**, *46*, 20492–20502. [CrossRef]
32. Wang, B.; Kanhere, P.D.; Chen, Z.; Nisar, J.; Pathak, B.; Ahuja, R. Anion-doped $NaTaO_3$ for visible light photocatalysis. *J. Phys. Chem. C* **2013**, *44*, 22518–22524. [CrossRef]
33. Wang, G.-Z.; Chen, H.; Wu, G.; Kuang, A.-L.; Yuan, H.-K. Hybrid Density Functional Study on Mono- and Codoped $NaNbO_3$ for Visible-Light Photocatalysis. *ChemPhysChem* **2016**, *17*, 489. [CrossRef]
34. Maarouf, A.A.; Gogova, D.; Fadlallah, M.M. Metal-doped $KNbO_3$ for visible light photocatalytic water splitting: A first principles investigation. *Appl. Phys. Lett.* **2021**, *119*, 063901. [CrossRef]
35. Liu, P.; Nisar, J.; Pathak, B.; Ahuja, R. Cationic–anionic mediated charge compensation on $La_2Ti_2O_7$ for visible light photocatalysis. *Phys. Chem. Chem. Phys.* **2013**, *15*, 17150–17157. [CrossRef]
36. Fo, Y.; Ma, Y.; Dong, H.; Zhou, X. Tuning the electronic structure of $BaTiO_3$ for an enhanced photocatalytic performance using cation–anion codoping: A first-principles study. *New J. Chem.* **2021**, *45*, 8228. [CrossRef]
37. Fo, Y.; Zhou, X. A theoretical study on tetragonal $BaTiO_3$ modified by surface co-doping for photocatalytic overall water splitting. *Int. J. Hydrogen Energy* **2022**, *47*, 19073–19085. [CrossRef]
38. Liu, P.; Nisar, J.; Pathak, B.; Ahuja, R. Hybrid density functional study on $SrTiO_3$ for visible light photocatalysis. *Int. J. Hydrogen Energy* **2012**, *16*, 11611–11617. [CrossRef]
39. Wang, J.; Teng, J.; Pu, L.; Huang, J.; Wang, Y.; Li, Q. Double-hole-mediated coupling of anionic dopants in perovskite $NaNbO_3$ for efficient solar water splitting. *Int. J. Quantum Chem.* **2019**, *119*, e25930. [CrossRef]
40. Wang, G.; Huang, Y.; Kuang, A.; Yuan, H.; Li, Y.; Chen, H. Double-Hole-Mediated Codoping on $KNbO_3$ for Visible Light Photocatalysis. *Inorg. Chem.* **2016**, *55*, 9620–9631. [CrossRef]
41. Wang, J.; Huang, J.; Meng, J.; Li, Q.; Yang, J. Single-and few-layer BiOI as promising photocatalysts for solar water splitting. *Phys. Chem. Chem. Phys.* **2016**, *18*, 17517. [CrossRef]
42. Smidstrup, S.; Markussen, T.; Vancraeyveld, P.; Wellendorff, J.; Schneider, J.; Gunst, T.; Verstichel, B.; Stradi, D.; Khomyakov, P.A.; Vej-Hansen, U.G.; et al. QuantumATK: An integrated platform of electronic and atomic-scale modelling tools. *J. Phys. Condens. Matter* **2020**, *32*, 015901. [CrossRef] [PubMed]
43. Perdew, J.P.; Burke, K.; Ernzerhof, M. Generalized gradient approximation made simple. *Phys. Rev. Lett.* **1996**, *77*, 3865. [CrossRef] [PubMed]
44. Van Setten, M.; Giantomassi, M.; Bousquet, E.; Verstraete, M.; Hamann, D.; Gonze, X.; Rignanese, G.-M. The PseudoDojo: Training and grading a 85 element optimized norm-conserving pseudopotential table. *Comput. Phys. Commun.* **2018**, *226*, 39–54. [CrossRef]
45. Monkhorst, H.J.; Pack, J.D. Special points for Brillouin-zone integrations. *Phys. Rev. B* **1976**, *13*, 5188. [CrossRef]
46. Ferreira, L.G.; Marques, M.; Teles, L.K. Approximation to density functional theory for the calculation of band gaps of semiconductors. *Phys. Rev. B* **2008**, *78*, 125116. [CrossRef]
47. Kanoun, M.B.; Goumri-Said, S.; Schwingenschlögl, U.; Manchon, A. Magnetism in Sc-doped ZnO with zinc vacancies: A hybrid density functional and GGA+U approaches. *Chem. Phys. Lett.* **2012**, *532*, 96–99. [CrossRef]
48. Ravel, B.; Stern, E.A.; Vedrinskii, R.I.; Kraizman, V. Local structure and the phase transitions of $BaTiO_3$. *Ferroelectrics* **1998**, *206*, 407–430. [CrossRef]
49. Chakraborty, A.; Liton, M.N.H.; Sarker, M.S.I.; Rahman, M.M.; Khan, M.K.R. A comprehensive DFT evaluation of catalytic and optoelectronic properties of $BaTiO_3$ polymorphs. *Phys. B Condens. Matter* **2023**, *648*, 414418. [CrossRef]
50. Janotti, A.; van de Walle, C.G. Native point defects in ZnO. *Phys. Rev. B* **2007**, *76*, 165202. [CrossRef]
51. Kanoun, M.B. Vacancy defects- and strain-tunable electronic structures and magnetism in two-dimensional $MoTe_2$: Insight from first-principles calculations *Surf. Interfaces* **2021**, *27*, 101442. [CrossRef]

Review

CuS-Based Nanostructures as Catalysts for Organic Pollutants Photodegradation

Luminita Isac [1,2], Cristina Cazan [1,2], Luminita Andronic [1] and Alexandru Enesca [1,*]

[1] Product Design, Mechatronics and Environmental Department, Transilvania University of Brasov, 500036 Brasov, Romania
[2] Renewable Energy Systems and Recycling Research Center, Transilvania University of Brasov, 500036 Brasov, Romania
* Correspondence: aenesca@unitbv.ro

Abstract: The direct or indirect discharge of toxic and non-biodegradable organic pollutants into water represents a huge threat that affects human health and the environment. Therefore, the treatment of wastewater, using sustainable technologies, is absolutely necessary for reusability. Photocatalysis is considered one of the most innovative advanced techniques used for pollutant removal from wastewater, due to its high efficiency, ease of process, low-cost, and the environmentally friendly secondary compounds that occur. The key of photocatalysis technology is the careful selection of catalysts, usually semiconductor materials with high absorption capacity for solar light, and conductivity for photogenerated charge carriers. Among copper sulfides, CuS (covellite), a semiconductor with different morphologies and bandgap values, is recognized as an important photocatalyst used for the removal of organic pollutants (dyes, pesticides, pharmaceutics etc.) from wastewater. This review deals with recent developments in organic pollutant photodegradation, using as catalysts various CuS nanostructures, consisting of CuS NPs, CuS QDs, and heterojunctions (CuS/carbon-based materials, CuS/organic semiconductor, CuS/metal oxide). The effects of different synthesis parameters (Cu:S molar ratios, surfactant concentration etc.) and properties (particle size, morphology, bandgap energy, and surface properties) on the photocatalytic performance of CuS-based catalysts for the degradation of various organic pollutants are extensively discussed.

Keywords: CuS nanostructures; heterojunctions; photocatalysis; organic pollutants; wastewater treatment

Citation: Isac, L.; Cazan, C.; Andronic, L.; Enesca, A. CuS-Based Nanostructures as Catalysts for Organic Pollutants Photodegradation. *Catalysts* **2022**, *12*, 1135. https://doi.org/10.3390/catal12101135

Academic Editor: John Vakros

Received: 1 September 2022
Accepted: 25 September 2022
Published: 28 September 2022

Publisher's Note: MDPI stays neutral with regard to jurisdictional claims in published maps and institutional affiliations.

Copyright: © 2022 by the authors. Licensee MDPI, Basel, Switzerland. This article is an open access article distributed under the terms and conditions of the Creative Commons Attribution (CC BY) license (https://creativecommons.org/licenses/by/4.0/).

1. Introduction

The global population growth has resulted in constantly increasing environmental pollution, with serious consequences for human health. Harmful impacts on humans, animals, and the environment are due to the water resource contamination with inorganic, i.e., heavy metals [1–3], and organic, pollutants. Toxic, non-biodegradable, and recalcitrant organic pollutants, such as dyes [4–7], pesticides [8], pharmaceutical active compounds (PhACs) [9,10], and other organic pollutants [11,12], are discharged into water from various industries: textile, leather, food, pharmaceutical, cosmetic, agriculture, plastic, etc. Consequently, the removal of these pollutants from wastewater, via appropriate technology, still remains a great challenge for worldwide researchers.

In this context, many technologies, such as sedimentation, reverse osmosis (RO), coagulation, flotation, solvent extraction, electrolysis, biodegradation, ozonation, sonolysis, sonophotocatalysis, gammaeradiolysis, photo-Fenton, photo-, electro-, and photoelectrocatalysis and chemical catalysis, and combined anaerobic photocatalysis with membrane techniques, have been developed [10,13–15]. Among these technologies, photocatalysis, an advanced oxidation process (AOP), has attracted considerable attention as a green and sustainable technology with promising prospects in global environmental issues remediation [16–18]. Although the photocatalytic degradation of organic pollutants has been proven to have significant results, its application at the industrial level faces certain deficiencies related to the efficient use of solar energy, mainly due to the high photo-generated carriers recombination rate [19,20]. Thus, the design and development of low-cost and highly efficient catalysts is the main topic in photocatalysis research. Semiconductor materials, with wide solar selective spectral response, high activity, and increased chemical and physical stability, are the most frequently used photocatalysts. As wastewater treatment technology, semiconductor-based photocatalysis has numerous advantages, such as simplicity, ease of handling, good reproducibility, high efficiency, and low costs. Moreover, it is an ecological, non-toxic and energy-free technology [5].

An important class of semiconductor photocatalysts is that of metal sulfides. For these materials, the band gap energy can be easily tuned by simply controlling the particle size without changing the chemical composition of the metal sulfide.

Copper sulfides (Cu_xS, x = 0.5–2), one of the most significant metal chalcogenide representatives, have been intensively studied in recent decades, due to their particular properties (optical and electrical), that result from their various chemical compositions and morphologies. Considering the chemical composition (x variation), at least eight crystalline phases of Cu_xS have been reported to date, ranging from the "copper low" sulfide CuS_2 (copper disulfide, x = 0.5) to the copper-rich phase Cu_2S (chalcocite, x = 2).

Adjusting the chemical composition (x value), optoelectronic properties are modified, affecting the photocatalytic performance of Cu_xS catalyst. Generally, the increase of x causes a decrease in the Cu_xS photocatalytic activity. Copper sulfides with x = 1.8–2 act as more efficient materials for solar cells and optoelectronic devices than as photocatalysts [21–23]. Accordingly, it was reported [24] that Cu_2S NPs (0.08 g/L pollutant solution), obtained by a template free polyol reaction, degraded only 51% of dye ABRX-3B (100 mg/L) under Vis light (300W Xe lamp), after 100 min. More recently, the photocatalytic activity of Cu_2S-metal oxide(s) heterostructures, prepared by a simple two-step sol–gel procedure, was studied under UV and UV–Vis irradiation scenarios using herbicide S-MCh (30 mg/L) as the reference pollutant [25]. The results showed that the three-component heterostructure of $Cu_2S/TiO_2/WO_3$ had higher photocatalytic efficiency (61%) compared with two-component heterostructures, 30% for Cu_2S/TiO_2, and 28% for Cu_2S/WO_3, respectively, after 8 h in UV-Vis light irradiation. A lower photocatalytic efficiency (~10%) was reported for Cu_9S_5 ($Cu_{1.8}S$) MCs in the degradation of MB solution (5.8 mg/L), under Vis light irradiation, after 175 min [26].

Another potential non-stoichiometric copper sulfide catalyst is Cu_7S_4 ($Cu_{1.75}S$). However, its photocatalytic application is reduced, due to both the high recombination rate of the photogenerated electron/holes, and also to the powder recycling issues. Thus, as an efficient alternative to powders, flexible Cu mash/Cu_7S_4 films were developed via a facile in situ anodization technique [16]. Using Cu/Cu_7S_4 films of 4 cm^2 area as catalysts, the highest efficiency in the photocatalytic Fenton-like dye MB (10 mg/L) degradation was 98.4%, under simulated sunlight, after 140 min.

Over recent years, the CuS semiconductor has been shown to be a promising candidate for visible light photocatalysis, due to its narrow band gap and good optical absorption properties in the Vis-NIR region. Based on the increased photocatalytic performances of CuS, compared with Cu_xS (x = 1.8–2) and $Cu_{1.75}S$ (more efficient and more environmentally friendly as films), therefore, a large amount of research was conducted for this review, which is focused on the use of CuS nanostructured powders as photocatalysts for organic pollutant degradation in wastewater.

Due to its special properties, e.g., increased conductivity at high temperature, superconductivity at 1.6 K, (electro)chemical-sensing capabilities, low-toxicity, nonlinear optical and ideal solar energy absorption capacity, CuS shows a versatile range of applications, including solar thermal collectors [27], artificial photosynthesis [28], gas sensors [29–32], biosensors [29,33,34], photodynamic therapy of cancer [29,32,34], lithium-ion batteries [29,35], supercapacitors [29,36], photoconductors [37], water disinfection [38–41], hydrogen production via water splitting [37,42], thermoelectric materials [29], and photocatalysis [4,37].

According to literature [10,14,43,44], the lack of chemical stability, fast recombination of photogenerated charge carriers, and particle aggregation in solution are still significant drawbacks of a single semiconductor photocatalyst, such as CuS. An effective solution to these impediments is heterojunction construction with additional advantages, such as improving solar energy absorption and carriers charge transfer rate. The photocatalytic performance of CuS nanostructures (NSs) and CuS-based hetero-nanostructures depends on particles size, surface area, morphology, and the interface properties, that can be tailored by tuning the synthesis methods and/or conditions.

In recent years, many studies have been developed on CuS NSs with applications in environmental issues, especially in wastewater treatment. In this review, the photocatalytic activity of various CuS-based nanostructures for degradation of different organic pollutants, such as organic dyes, pesticides, phenol and phenolic compounds, pharmaceutical active compounds, etc., under simulated (UV, Vis) and sunlight irradiation, is extensively discussed. This study aims to identify the photocatalytic performances of recently reported CuS-based catalysts, for possible future improvements so as to ensure it is as efficient as possible for the degradation of industrial organic pollutants in wastewater effluents. In Figure 1 is illustrated the schematic representation of this review regarding CuS-based nanostructures photocatalysts obtaining, their specific properties (morphology, band gap energy, specific surface area), and application in organic pollutants degradation from wastewater.

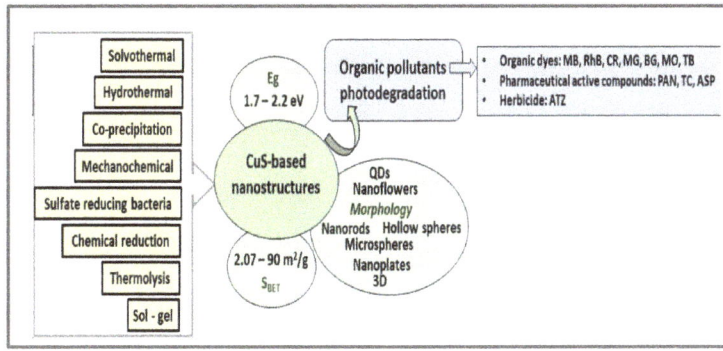

Figure 1. Synthesis methods and specific properties of CuS-based nanostructures photocatalysts used in organic pollutants degradation from wastewater.

2. CuS Nanostructures with Different Morphologies as Catalysts for Organic Pollutant Photodegradation

At room temperature, CuS (covellite) has a hexagonal crystalline structure (space group P63/mmc, a = 3.8020 Å, c = 16.430 Å) consisting of unit cells, in which layers of planar CuS_3 (triangles) and CuS_4 (tetrahedrons) with S–S bonds alternate. At 55 K, the CuS hexagonal structure changes to orthorhombic, due to second-order phase transition resulting in orthorhombic distortion of the Cu-S and S-S bond lengths [45,46].

The energy band alignments and total–partial density of states, calculated with the generalized gradient approximation (GGA) of the density functional theory, showed that CuS has significant metallic behavior due to p(S)–d(Cu) orbital interactions up to Fermi level [45]. The Fermi level, an empty energy band induced by both copper vacancy and sulfur vacancy on the top of VB, is an important factor in the evaluation of CuS photoelectronic properties, mainly related to the potential local surface plasmon resonance (LSPR) [47]. Due to its deficiency in Cu atoms (comparing with Cu_2S), CuS exhibits the highest concentration of free carriers (holes) in the VB, resulting in LSPR bands in the NIR region, hence having extended light absorption [48].

As a copper sulfides class representative, CuS is a p-type semiconductor with a bandgap energy ranging from 1.7 eV to 3.46 eV [33,49] but, for most CuS, the Eg values are in the range of 1, 7–2.2 eV. Controlling the CuS semiconductor morphology (shape and particle size), the band gap can be tailored without changing the chemical composition of the material [50]. For example, the band gap energy for CuS nanoparticles (NPs) can vary from 1.7 to 2.14 eV, depending on morphology: 1.7 eV for hollow spheres [51], 1.87 eV for nanoflowers [52], 1.97 eV for microspheres [53], 2.1 eV for nanoplates [54] and 2.12 eV for flake-like nanostructures [55].

Thus, many works have reported the preparation of CuS with different morphologies, from quantum dots (QDs) to 3D hierarchical CuS architectures consisting of 1D nanotubes, using methods such as hydrothermal [50,56,57], solvothermal [58–63], co-precipitation [64,65], photochemical precipitation [66], mechanochemical [67,68], thermolysis [49], chemical reduction [69], microwave assisted growth [70] and solid-state reaction routes [71]. However, the preparation of uniform CuS nanostructures, through simple, fast, ecological and low-cost technologies, still remains a significant challenge for all researchers in the field.

Covellite nanoparticles (CuS NPs) with remarkable chemical, structural and surface properties, significantly different from those of bulk, are considered promising photocatalytic materials [55]. The photocatalytic properties of CuS catalyst can be tailored by changing preparation method parameters. In addition, the photocatalyst dosage, the pollutant and its concentration, the type and intensity of the irradiation source are other important factors to be considered in the photodegradation process. The photocatalytic activity of CuS-based nanostructures for the degradation of different organic pollutants is selectively presented in Table 1.

Table 1. Representative studies on organic pollutants photodegradation using nanostructured CuS-based catalysts.

Catalyst	Synthesis Method	Pollutant Conc. (mg/L)	Catalyst Dosage (g/L)	Light Source	η* (%)	t (min)	Ref.
CuS NPs	ST	MB (500)	0.5–2	UV (90 W Xe lamp) Vis (160 W Hg -W lamp)	98.26 100	30	[72]
CuS NPs	ST	MB (6.4) CR (13.9) RhB (9.6) EY (13)	0.03	UV (12 W G8 T5 Philips) Vis (160 W Hg lamp) Solar	70.79, 85.13, 90.29 75.47, 63.21, 60.35 50.04, 60.37, 69.23 79.49, 56.02, 91.97	240	[61]
CuS NPs	Aqueous solution route	MB (20) MB + H_2O_2	0.1	Vis (150 W Xe arc lamp)	39 92	90	[52]
CuS NPs	Sulfate reducing bacteria (SRB)	MB (16) RhB (24) + H_2O_2	0.3–0.5	Vis (600 W halogen lamp)	94 61.2	5.5 25	[55]
CuS NPs	PVP assisted ST	MB (400) + H_2O_2	0.1	Solar	96.5	48	[53]
CuS 3D NSs	ST	MB (20) + H_2O_2	0.2	Vis (10 kW/m^2, CEL HXF 300)	90	30	[62]
CuS NPs	Solution aerosol thermolysis	MB (12.8) RhB (9.6) MO (9.8) + H_2O_2	1	Vis (300 W Xe lamp)	98 98 50	15 50 45	[27]
CuS MCs	Subcritical & supercritical methanol reaction	MB (5.8)	0.2	Vis (25 W day-light lamp)	85.4	300	[26]
CuS NPs	CoPp	RhB (20) RhB + H_2O_2	0.05	Vis (150 W Xe lamp)	99 99	120 60	[51]
CuS QDs	Mechano-chemical	RhB (10) RhB + H_2O_2	0.4	Vis (150 W Xe lamp)	60 95	30	[67]

Table 1. Cont.

Catalyst	Synthesis Method	Pollutant Conc. (mg/L)	Catalyst Dosage (g/L)	Light Source	η* (%)	t (min)	Ref.
CuS NPs	Mixed solvent route	RhB (50) + H_2O_2	0.2	Vis (1000 W halide lamp)	94	60	[73]
3D CuS NSs	ST & self-assembly	RhB (50) + H_2O_2	0.2	Vis (150 W Xe lamp)	99	45	[74]
3D CuS NSs	One-step in situ heating sulfuration route	MB (10) RhB (10) MB/RhB + H_2O_2	1.25	Vis (300 W xenon lamp)	99 99 99	25	[75]
CuS NPs	One pot synthesis from Cu(II) dithio-carbamate	CR (100)	0.5	Solar	100	40	[54]
CuS NPs	Solid-state reaction	MoO (60) SO (60) AO (60)	0.1	Vis (500 W Hg lamp)	51 22 45	180	[71]
3D CuS NSs	HT	4-CP (100) 4-CP + H_2O_2	1	Vis (49.700 lux)	62 83	300	[57]
CuS NPs Cu_2S NPs CuS-Cu_2S NPS	One-step HT (tuning Cu:S molar ratios)	RhB (10) + H_2O_2	0.2	Vis (250 W cold Xe lamp)	96 87 92	5	[56]
CuS-Cu_2S NPs	Chemical reduction	MB (3.2) MG (3.65) MO (3.27) MV (3.6) RhB (4.79)	0.4	Solar	61.95 90.25 9.4 85.03 70.16	100 40 180 60 80	[69]
CuS/KCC1	ST	HA (2)	0.1	UV-C (18 W lamp)	89.5	90	[76]
CuS/ZnO	HT	MB (9.6) TB (16.2)	0.7	Vis (52 W renewable household Philips LEDs)	93 87.5	16 18	[77]
$Cu_{1-x}Ag_xS$ (0.0 ≤ x ≤ 0.1)	CoPp	MB (6)	0.25	Vis (indigenous light reactor)	75.4	90	[33]
CuS/rGO	CoPp	MG (10)	0.1	Solar	99.2	90	[14]
CuS/CQDs	Carbonization of water hyacinth weed + SG (CuS)	BG (50)	0.09	Vis (200 W tungsten lamp)	96	90	[78]
CuS/CQDs	Carbonization of peanut shells + HT (CuS)	PAN (20)	0.2	Vis (400 W Hg lamp)	96.5	150	[50]
CuS/PDI	Two-step self-assembly	TC (50)	0.6	Vis (300 W Xe lamp)	90	120	[79]
CuS/rGO	SG	ATZ (50)	0.8	Vis (300 W Xe lamp)	100	20	[43]
CuS/WO_3-AC	HT	ASP (10)	-	Vis (400 W metal halide lamp)	97.6	150	[80]

η* is the efficiency degradation of pollutant after t min of irradiation.

The Vis and natural light-driven photocatalytic degradation of cationic azo dye methylene blue (MB) using CuS nanoparticles (NPs) with different morphologies, has been demonstrated to be an attractive topic for many research. Flower-like [52], hollow microspheres [53], and flake- and spherical-like CuS NPs [55], prepared by simple aqueous solution route, facile PVP assisted solvothermal process and biological sulfate reduction, were investigated as photocatalysts in the degradation of MB, in the absence/presence of H_2O_2.

The CuS nanostructured flowers, with diameters of about 800–1200 nm, Eg = 1.87 eV and BET surface areas of 61.55 m^2/g, obtained by using a simple aqueous solution route without any surfactant addition, removed only 39% MB after 90 min, under Vis light irradiation [52]. The addition of H_2O_2 in the photocatalytic system significantly increased the degradation efficiency to 92%, after the same irradiation time. This increase was due to the presence of H_2O_2 molecules, which enhanced the dye degradation by accelerating the formation of hydroxyl radicals (·OH), the active species which promote the oxidation of MB into smaller, non-toxic molecules [52,81].

An enhanced photocatalytic efficiency (94%) in the degradation of MB under natural light (sunlight), in the presence of H_2O_2, was reported for CuS hollow microspheres mesoporous structures with a bandgap of ~1.97 eV and surface area of 36 m²/g [53]. The CuS hollow microspheres nanostructures were obtained by the solvothermal process, varying the amounts of PVP surfactant (0–2 g), the copper/sulfur precursors, and solvents, while the precursor ratios were maintained at constant during the experiments. The average diameter of microspheres in the nanosheets-based hierarchical structure was about 2.3 μm. The studied photocatalysts showed excellent stability and recyclability, with 96.5% of the dye removed after 6th cycle. The photocatalytic performance of CuS catalyst was attributed to its hollow microsphere morphology, which favors the absorption of more MB molecules, promotes the light generated charge carriers transfer to the reactive surface and allows rapid diffusion of the reactants and products during the oxidation/reduction reaction [53].

CuS NPs with two different morphologies were synthesized via the sulfate reducing bacteria (RBS) method by changing the copper precursor concentration: lower concentrations favored the obtaining of flake-like nanoparticles (Eg = 2.12 eV) with an average width of 25 nm and average length of 130 nm, while spherical-like CuS NPs (Eg = 2.14 eV), with average diameter in the range 30–50 nm, were synthesized at higher concentrations [55].

Both the flake-like and spherical-like CuS NPS showed high photodegradation efficiency for the MB + H_2O_2 system, respectively 94% after 5.5 min and 93% after 1 min illumination with a halogen lamp (600 W). Although these photocatalysts have the advantage of being used in applications that require short irradiation time or short catalyst–pollutant contact time, their reusability is still an open-issue for further studies.

The studies mentioned above confirm that the photocatalytic activity of CuS NPs with hollow microspheres or spherical-like morphologies, with consistent shape and size of spherical particles, was higher than that of CuS NPs with non-spherical structures.

The MB dye photodegradation mechanism, through oxidation/reduction reactions, using CuS NPs as the photocatalyst, in the absence/presence of H_2O_2 molecules, is illustrated in Figure 2.

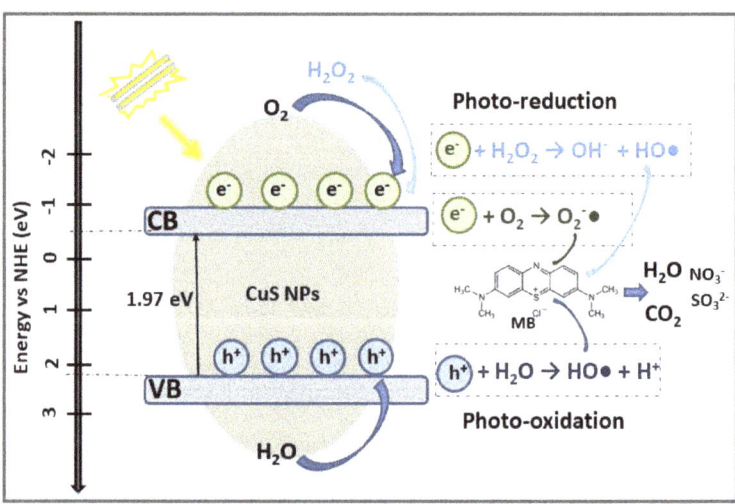

Figure 2. The photocatalytic degradation mechanism of MB dye using CuS NPs photocatalyst under sunlight irradiation.

The adsorption and photocatalysis performance of CuS nanostructures with different morphologies of particles, prepared by a simple co-precipitation method using $Cu(NO_3)_2$ and thioacetamide as precursors, for degradation of the dye Rhodamine B (RhB) aqueous solutions (20 mg/L) under Vis light irradiation (150 W Xe lamp equipped with a glass filter) were evaluated by Li and Wang [51]. It was reported that CuS NPs with different morphologies (flakes, rods and hollow spheres), and Eg = 1.7 eV had good photodegradation efficiency for RhB, with removal rates higher than 85% within 120 min. As was expected, CuS hollow spheres, with 400–3500 nm outer diameter and 10–400 nm thickness, formed by self-assembly of nanoparticles with sizes in the range 1–2 µm, showed a RhB photodegradation of 99% in 120 min. The addition of a small amount of H_2O_2 (1mL/100 mL dye solution) had the result of halving the time necessary for photodegradation of 99% RhB with high concentration (50 mg/L) in wastewater. The increased photocatalytic activity of the CuS NPs, attributed to the combination of the adsorption and photocatalysis processes, demonstrates the potential application of CuS NPs photocatalyst in high dye concentration wastewater treatment.

The photocatalytic activity and stability of CuS QDs in degradation of RhB solutions under visible light illumination (150 W Xe lamp) were studied by Li et al. [67]. Using a low-cost and simple mechanochemical ball milling method, CuS powders containing ultrafine crystals with uniform size (1–5 nm) and an average diameter of 2.9 nm, were prepared. The calculated BET surface area of about 90.0 m^2/g, which was significantly higher than that of CuS NPs (30.6 m^2/g, [51]), was expected to improve the photocatalytic activity of the CuS QDs. Higher specific surface area, with more reactive sites and shorter migration distance, reduced the recombination rate of photogenerated electron/hole pairs. Accordingly, the H_2O_2 addition increased the degradation efficiency of RhB from about 60% to 95%, after 30 min exposure in Vis light, due to the quantum effect of CuS QDs which favors ultra-fast transfer of charge carriers from CuS QDs to RhB dye.

Based on previous studies, both CuS NPs and CuS QDs are excellent photocatalysts for RhB degradation, depending on the dye concentration in wastewater: CuS QDs is more efficient in wastewater with a low content of dye, while CuS NPs shows good efficiency in dye-concentrated wastewater as well.

3D nanostructured CuS have been remarked on as materials with high solar catalytic efficiency due to their large surface areas with sufficiently active sites, which improve CuS surface-reactants contact [62].

The hierarchical 3D CuS nanostructures, prepared by low-temperature solvothermal grow of 1D CuS nanotubes on a self-assembled 3D $Cu(MAA)_2$ precursor, not only provided the advantage of a large number of active sites on the surfaces, but also the advantages of improved molecular/ionic transport (through the 1D nanotubes), and of good mechanical stability (due to the 3D structure) [74]. The as-prepared 3D hierarchical CuS architectures showed a degradation of almost 99% of RhB within 45 min, in the presence of visible light and oxidizing agent (H_2O_2). The stability and reusability of the photocatalysts proved to be quite good, the removal rate of RhB reaching 70% after 5 cycles.

A simple and one-step in situ heating sulfuration procedure was used to prepare the CuS photocatalyst with a 3D hierarchical nanostructure with CuS nanoplates formed on the copper foam precursor structure in [75]. According to this study's results, the excellent photocatalytic performance of 3D nanostructured CuS catalyst resulting in 99% degradation of RhB + H_2O_2 solution when exposed in simulated solar light for 25 min, was attributed to the synergistic effects of high optical absorption (Eg = 1.58 eV), and large specific surface area (12.06 m_2/g), resulting in sufficient reaction active sites. Moreover, the photocatalytic activity and stability of the studied catalysts did not alter after 4 photocatalytic cycles, which made them ideal recyclable catalysts.

Cu_xS nanoparticles with different compositions (x = 1, 2) and morphologies were synthesized via simple and environmentally friendly methods, e.g., one-step hydrothermal and thermal chemical reduction of S precursor with/without surfactant, in [56,69].

By tuning the molar ratios Cu:S (1:0.1–1:3) of copper acetate and sublimed sulfur precursors in polyethylene (PEG-400) surfactant, CuS, Cu_2S, and CuS–Cu_2S NPs with various morphologies and particle sizes, and, therefore, different band gap energies, were obtained:

Cu_2S spherical and irregular nanoflakes (Eg = 3.5 eV) for Cu:S = 1:0.25

Cu_2S-CuS irregular flakes with 30 nm thickness (Eg = 2.72 eV) for Cu:S = 1:0.75

CuS irregular nanoflakes with particle sizes between 200 and 300 nm and thickness less than 30 nm (Eg = 2.01 eV) for Cu:S = 1:1.

The photocatalytic experiments showed that photocatalyst efficiency in RhB degradation under Vis light (250W cold xenon lamp with cut-off wavelength of 420 nm), in the presence of H_2O_2, increases with band gap energy decrease, thus CuS NPs degraded 96% RhB in 5 min, while Cu_2S and Cu_2S-CuS NPs degraded 87% and 92% RhB respectively [56].

The RhB photodegradation reaction mechanism in presence of CuS NPs under visible light irradiation is schematically presented in Figure 3.

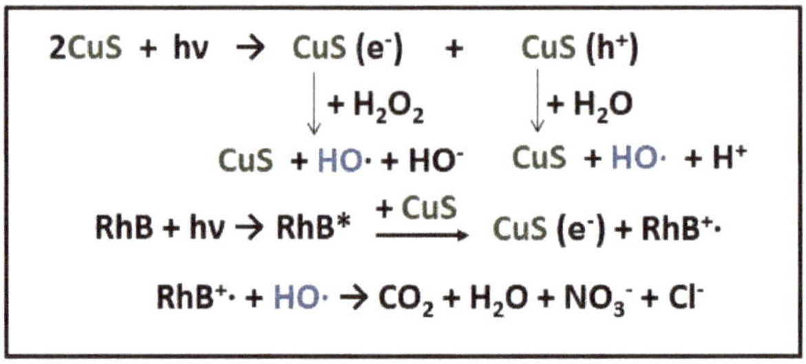

$$2CuS + h\nu \rightarrow CuS\,(e^-) + CuS\,(h^+)$$
$$\downarrow + H_2O_2 \qquad \downarrow + H_2O$$
$$CuS + HO\cdot + HO^- \quad CuS + HO\cdot + H^+$$
$$RhB + h\nu \rightarrow RhB^* \xrightarrow{+CuS} CuS\,(e^-) + RhB^{+\cdot}$$
$$RhB^{+\cdot} + HO\cdot \rightarrow CO_2 + H_2O + NO_3^- + Cl^-$$

Figure 3. The reaction mechanism proposed for RhB photodegradation by CuS NPs catalyst.

When CuS photocatalyst absorbs photons from solar light or an irradiation source, with energy equal or higher than its band-gap, the electrons from CB are transferred to VB, resulting in photogenerated electrons CuS (e^-) and holes CuS (h^+). The H_2O_2 molecules capture these photogenerated electrons and rapidly generate hydroxyl radicals (HO·) and hydroxide ions (OH^-), while the photogenerated holes react with H_2O forming HO·. In the meantime, RhB dye absorbs Vis light and undergoes a transition to its excited state (RhB*), which transfers electrons to CuS, resulting in CuS(e^-) and RhB^+. Then, the highly reactive radical HO· oxidizes and decomposes RhB^+ to CO_2, H_2O and other salt ions [4,56].

Shamraiz et al. synthesized CuS–Cu_2S NPs, with an average size less than 30 nm, by chemical reduction of copper thiourea complex, without any surfactant, at moderate temperature [69]. The CuS–Cu_2S NPs were tested as photocatalysts in the degradation of different dyes under direct sunlight (outdoor lightening), without H_2O_2 addition. The results showed that the photodegradation process was faster for cationic dyes, such as MV (85.03% in 60 min), MG (90,25% in 40 min) and RhB (70.16% in 80 min), but almost negligible for the anionic dye MO (9.4% in 3 h). This behavior of the CuS–Cu_2S catalyst in cationic dye photodegradation was attributed to the presence of active negative charges (OH^-) on the catalyst surface, which were electrostatically attracted by cationic dye molecules, thus facilitating the electron transfer under direct sunlight irradiation [69].

The photocatalytic performance of CuS nanoparticles, with size < 20 nm and specific surface area of 34.37 m^2/g, in the degradation of organic dye pollutants, MB, RhB, EY and CR, under various light (UV, Vis and solar) irradiations, was evaluated by Ayodhya et al. [61]. For CuS NPs synthesis, a simple, relatively fast and green (using xanthan gum as a capping agent) solvothermal method was proposed. The photodegradation of the MB, RhB, EY and CR dyes in the absence and presence of CuS NPs were studied under similar experimental conditions, using UV, Vis and solar light sources. The photocatalytic performances of CuS NPs are shown in Figure 4.

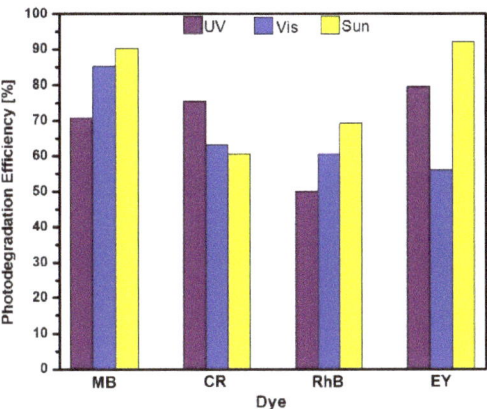

Figure 4. The photocatalytic activity of CuS NPs in dyes degradation under different light irradiations, for 4 h.

It can be observed that CuS NPs showed good photocatalytic activity for dye degradation in sunlight irradiation, ranging from 69.2% for RhB to 92% for EY. The exception was CR which degraded better in UV light (75.5% photodegradation efficiency) in the presence of CuS catalyst.

3. CuS-Based Heterostructures as Catalysts for Organic Pollutant Photodegradation

Although previous studies confirmed the successful applicability of CuS NP photocatalysts in the degradation of organic contaminants (especially organic dyes), the photocatalytic efficiency of single CuS is quite low for the complete degradation of persistent, much more toxic, organic compounds (dyes, pharmaceuticals, pesticides) from wastewater.

3.1. CuS/Carbon-Based Materials Heterostructures as Catalysts for Organic Pollutant Photodegradation

Carbon quantum dots (CQDs), with sizes under 10 nm and fluorescent properties, have large surface areas, high porosity, excellent electrical conductivity, relatively low cost and toxicity, and high aqueous stability. Due to these properties, CQDs can act as absorbents for water impurities, but also as supporting catalyst (co-catalyst) for the main photocatalyst (CuS) [50,78].

Another promising supporting catalyst for CuS is reduced graphene oxide (rGO) with high surface area, efficient electron transfer, and superior conductivity [39,52].

Therefore, the CuS/CQD and CuS/rGO composites could be attractive solutions for photocatalytic degradation of persistent organic contaminants under Vis light irradiation.

Recent investigations on the degradation of brilliant green (BG) highly toxic dye, usually used in textiles and paper industries but also in poultry feed to avoid fungi and parasite contagion, and panadol (PAN), one of the top three most commonly prescribed drugs in the world, were done using CuS/CQD photocatalysts [50,73]. In both studies, CQDs were obtained by carbonization of vegetative wastes (water hyacinth weed, peanut shells), while CuS NPs were prepared via sol–gel [78], respectively, hydrothermal [50] methods. The structural, optical, surface and photocatalytic properties of CuS/CQD composites and CuS NPs, together with organic pollutant details, are given in Table 2.

Table 2. The structural, optical, surface and photocatalytic properties of CuS NPs and CuS/carbon-based materials composites.

Catalyst	Morphology	Eg (eV)	S_{BET} (m^2/g)	Photodegradation				Ref.
				Dye				
				Chemical Formula and Structure	λ_{max} (nm)	η^* (%)	t (min)	
CuS NPs	spherical (10–12 nm)	3.46	9.36	BG, $C_{27}H_{33}N_2 \cdot HO_4S$		38		
CuS/CQDs	compact loading of carbon dots (4–8 nm) over CuS NPs	2.96	15.42		625	91.8	90	[78]
CuS NPs	nano-flower-like structure	-	850.5	PAN, $C_8H_9NO_2$		76.9		
CuS/CQDs	small nano-flowers with nano-petals	-			243	96.5	100	[50]
CuS NPs	urchin-like structure and some irregular hexagonal NPs	2.08	20.25	MG, $C_{23}H_{25}ClN_2$		92		
CuS/rGO	uniform CuS NPs distributed on rGO nanosheets	1.9	34.4		621	99.2	90	[14]
CuS NPs	hexagonal	2.07	130	ATZ, $C_8H_{14}ClN_5$		60	50	
CuS/rGO	separated hexagons of CuS assembled on rGO	1.76	155		222.5	100	20	[43]

η^* is the efficiency degradation of pollutant after t min of irradiation.

Due to improved morphologies, large surface areas and reduced band gap energies, and therefore, superior capacity for light absorption, these composites were shown to be the most effective catalysts for dyes and active pharmaceutical photodegradation. The CuS/CQDs catalyst act as a scavenger during the photocatalytic process, reducing the electron-hole recombination rate, and the possible redox reactions are summarized in Figure 5.

Both CuS/CQDsq photocatalysts were prepared via green techniques, representing an easy and low-cost way to remove organic pollutants and plant waste from the environment (i.e., wastewater).

The photocatalytic performance of CuS/rGO nanocomposites was evaluated for the degradation of toxic persistent organic pollutants, malachite green (MG) and atrazine (ATZ), under direct/simulated sunlight radiance [14,43]. Malachite green is a carcinogenic and non-biodegradable dye used in numerous industrial applications. Atrazine, a commonly used herbicide in the agriculture and food industries, is one of the most stable toxic contaminants in polluted water. The molecular formulae and structures of the organic pollutants, together with the structural, optical, surface and photocatalytic properties of CuS/rGO composites and CuS NPs are presented in Table 2.

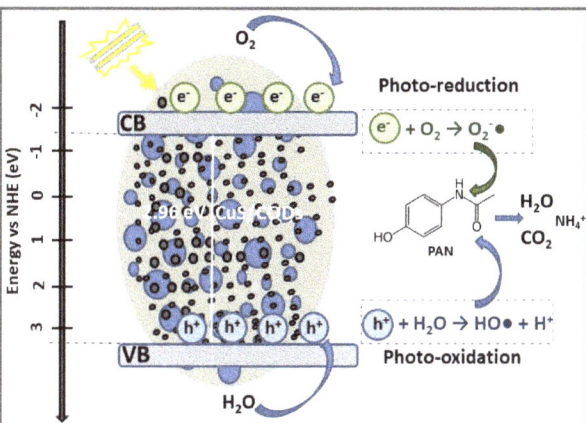

Figure 5. The photocatalytic degradation mechanism of PAN drug using CuS/CQDs photocatalyst under Vis light irradiation (400 W mercury lamp).

The experimental results revealed that the addition of various wt% of rGO in CuS had significant contributions to persistent organic pollutant photodegradation with CuS/rGO composite, such as the following: (a) enhanced light absorption and decreased bandgap energy values from 2.08 eV to 1.76–1.9 eV; (b) ensured relatively large specific surface area which provided more active reaction sites exhibiting strong photo-absorption of pollutant molecules under solar/Vis light irradiations; (c) allowed strong adsorption of pollutant molecules due to the oxygen-containing functional groups on the surface of rGO. The mechanism of ATZ degradation by CuS/rGO photocatalyst is illustrated in Figure 6. The absorption of light photons with energy (hυ) higher than the band gap energy of CuS (E_g = 2.07 eV) generated photo-excited electrons (e^-) on CB, and holes (h^+) in the VB. The rGO nanosheets, which were anchored to the surface of the assembled CuS NPs, could receive the photo-excited e^-, favoring the separation of photogenerated charge carriers and oxidized species (HO·, ·O_2^-) formation. These active radicals reacted with the ATZ molecules adsorbed on the photocatalyst active sites, resulting in CO_2, H_2O and other non-harmful ions [43].

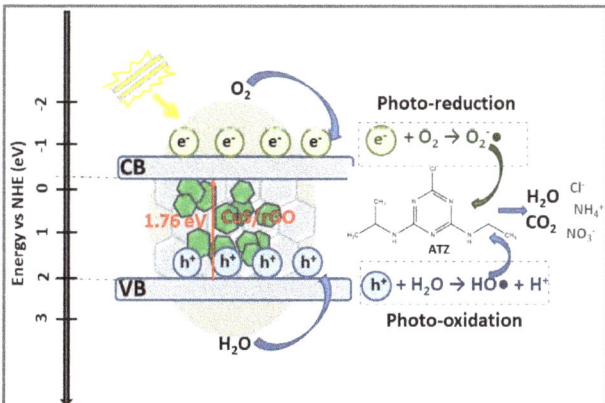

Figure 6. The photocatalytic degradation mechanism of ATZ herbicide using CuS/rGO photocatalyst under Vis light irradiation (300 W Xe lamp with λ < 410 nm cut-off sieve).

The reusability and photo-stability studies showed that CuS/rGO heterojunction nanostructure had the capacity to regenerate several times (e.g., 5 times), without a significant loss in its photodegradation efficiency, therefore exhibiting a high stability under light irradiation [14,43].

3.2. CuS/Organic Semiconductor Heterostructures as Catalysts for Organic Pollutant Photodegradation

PDI, one of the most studied organic fluorescent dyes, has excellent optoelectronic properties, specific to an n-type semiconductor, and, therefore, is used in solar cells and sensor applications. Supramolecular architectures (NS 1D) formed by self-assembled PDI are of particular interest in photocatalytic materials development due to their high photo-thermal stability, charge mobility, and electron affinity [79,82].

A recent study [79] reported the synthesis of a novel CuS/PDI p-n heterojunction photocatalyst for removal of tetracycline (TC) from wastewater. The CuS/PDI composites were prepared using a two-step self-assembly procedure, varying the CuS concentration (5%, 10%, 15%) in self-assembled PDI. By comparison with pure self-assembled PDI and CuS nanosheets, higher efficiency in photodegradation of TC, under simulated visible light, after 2 h, was obtained for the CuS10%/PDI catalyst. Thus, before coupling, both pure CuS (Eg = 2.01 eV, SBET = 1.87 m^2/g) and PDI (Eg = 1.67 eV, SBET = 0.76 m^2/g) catalysts showed poor photocatalytic performance, meaning about 40% and 60%, respectively, within 120 min. After coupling, and p-n heterojunction formation, the CuS/PDI composite (Eg = 1.71 eV, SBET = 39.43 m^2/g) exhibited higher efficiency (about 90%) for TC degradation. The enhancement of CuS/PDI photocatalytic activity was mainly attributed to the highly ordered H-type p-p stacking in the PDI structure and the p-n junction, which allowed the fast separation of the photo-generated charge carrier pairs. After the formation of a p-n junction, the photo-electrons were transferred from CuS CB to self-assembled PDI CB, under the action of the internal electric field, while photo-holes migrated in the opposite direction, from the self-assembled PDI VB to that of CuS. The transferred electrons further reacted with O_2, resulting in peroxide ion radicals ·O_2^- and holes, which produced active radicals ·OH after the reaction with H_2O. Both active oxidized radicals acted as reactants in further decomposition of TC on the catalyst surface (Figure 7).

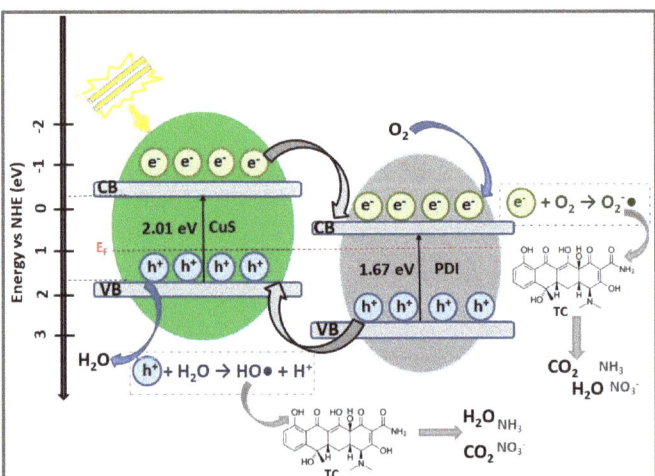

Figure 7. The photocatalytic degradation mechanism of TC antibiotic using CuS/PDI catalyst under Vis light irradiation (300 W Xe lamp with a cut filter with λ > 420 nm).

The stability and reusability experiments encountered some problems with the powder catalyst, therefore PDI/CuS composite was coupled with the modified cotton fibers through electrostatic adsorption. The newly-designed photocatalytic fabric was tested under simulated actual water quality conditions and, after 5 cycles, the degradation rate was maintained at about 80%, which indicated good reusability [79].

3.3. CuS/Metal Oxide Heterostructures as Catalysts for Organic Pollutant Photodegradation

Wide band gap semiconductors, such as TiO_2 (Eg = 3.25 eV [83]), SnO_2 (Eg = 3.4 eV [84]) and ZnO (Eg = 3.2 eV [85]), perform better as photocatalysts in the UV region, which causes their limited use in industrial applications, and, thus, a significant increase in process costs. Another factor that limits their photocatalytic efficiency is the rapid recombination of charge carriers. To extend the photocatalytic response in the Vis spectral region, and to reduce the recombination processes, many heterojunctions have been developed in recent years by coupling a wide band gap semiconductor with a suitable narrow band gap semiconductor, such as CuS. Most of the previously published review articles on CuS nanostructured materials have focused on synthesis methods, special properties, and prospective applications [28,29,36]. More recently, in a mini-review, we published the developments on dye photodegradation using various copper sulfide-based heterojunctions (copper sulfide/metal oxide, copper sulfide/metal sulfide, copper sulfide/graphene, copper sulfide/organic semiconductors) as catalysts [4].

Recently, Sudhaik et al. [46] published an extensive literature study (review) on CuS and different CuS-based heterostructured materials as photocatalysts for wastewater treatment. Thus, in this part of the article, approaches related to recently developed CuS-based heterojunctions as catalysts for organic pollutant photodegradation are highlighted, issues that were not considered in the previously mentioned publications.

The enhanced visible light photocatalytic efficiency of type II heterojunction CuS/ZnO in MB and toluidine blue (TB) degradation was reported by Khausik B et al. [80]. The CuS/ZnO photocatalyst was obtained by assembling p-type CuS NPs on n-type ZnO heterostructures, using hydrothermal method. For the photocatalysis experiments, 35 mg of CuS/ZnO catalyst and 50 mL aqueous solution of MB (3×10^{-5} M), respectively TB (6×10^{-5} M), were used. The photocatalytic experiments showed that the tandem structure of CuS/ZnO had excellent photocatalytic efficiency for MB and TB, with 93% and 87.5% dyes degradation within 16 respectively 18 min, under visible light irradiation. The remarkable photocatalytic activity was attributed to the CuS/ZnO p-n heterojunction formation, which favored the efficient separation of photoinduced charge carriers.

Based on scavenging experiments using different trappers, the mechanism proposed for the degradation of MB and TB under Vis light corresponds to type II heterojunction CuS-based photocatalyst (Figure 8), allowing the transfer of electrons and holes to the other semiconductor from heterojunction (ZnO), when spatial charge carriers separation occurred.

To summarize, in the photocatalytic degradation of dyes by CuS/ZnO catalyst, the photo-excited electrons and holes generate highly reactive hydroxyl radicals (HO·) which decompose pollutants into CO_2, H_2O and other environmentally friendly compounds that do not require any other chemical or physical treatment.

Recently, WO_3 with activated carbon (AC, 1% and 2%) and co-doped with CuS (10%, 15%) composites, WO_3-AC/CuS, were prepared via a facile hydrothermal method [80].

The hydrothermal method was used because it allows the control of the composite's structural, morphological and optical properties, in order to increase photocatalytic activity, by varying the synthesis parameters, such as temperature and time. The nano-rod-like structure (500 nm with size of nano-rods of 80–89 nm) of WO_3 became sharper and clear after doping with AC and was covered with small nanoflowers of hexagonal CuS in WO_3-AC/CuS composites. This morphology demonstrated itself to be suitable for the degradation of organic pollutants, in this case aspirin. In addition, the UV–Vis analysis results showed that increasing the AC and CuS amounts in WO_3 caused the bandgap energy to decrease from 2.49 eV (WO_3) to 2.3 eV (WO_3-2% AC), and to 1.92 eV for WO_3-2% AC/15% CuS. This reduction of band-gap energy confirmed that the addition of CuS enhanced the optical properties, and, therefore, the photocatalytic performance of WO_3-Ac material.

The photocatalytic experiments were focused on the comparative degradation of aspirin (10 mg/L), as an active pharmaceutical contaminant model, under Vis light (homemade photocatalytic reactor with a 400 W metal halide lamp) in the presence of catalysts WO_3, WO_3-AC and WO_3-AC/CuS.

The results showed that ASP photodegradation increased from 40% (WO_3) to 97.6% when WO_3-AC/CuS heterostructure was used as catalyst, after 150 min of Vis light illumination. This high photocatalytic activity was due to the increase of the photo-generated charges recombination ratio and the decrease of the photon depth penetration. The proposed mechanism for the degradation of ASP by WO_3-AC/CuS photocatalyst is schematically presented in Figure 9.

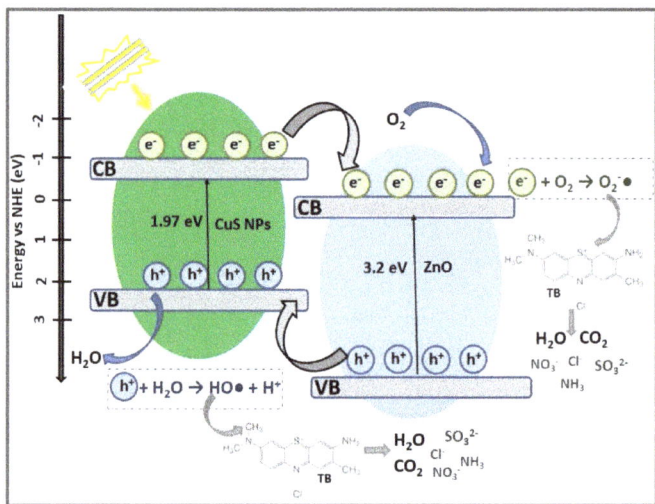

Figure 8. The photocatalytic degradation mechanism of toluidine blue (TB) dye using CuS/ZnO catalyst under Vis light irradiation.

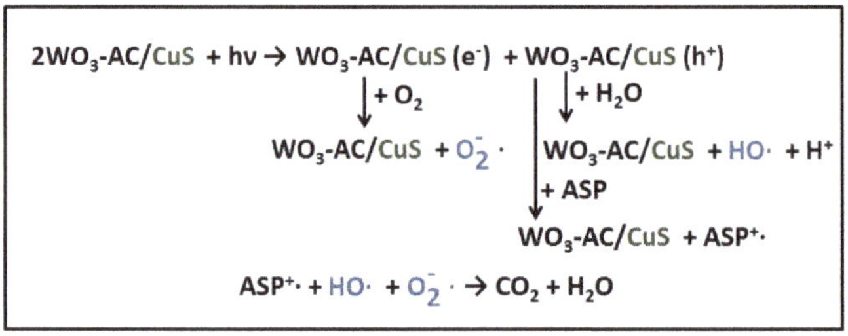

$$2WO_3\text{-AC/CuS} + hv \rightarrow WO_3\text{-AC/CuS (e}^-) + WO_3\text{-AC/CuS (h}^+)$$
$$\downarrow + O_2 \qquad \downarrow + H_2O$$
$$WO_3\text{-AC/CuS} + O_2^- \cdot \qquad WO_3\text{-AC/CuS} + HO\cdot + H^+$$
$$\downarrow + ASP$$
$$WO_3\text{-AC/CuS} + ASP^{+\cdot}$$
$$ASP^{+\cdot} + HO\cdot + O_2^- \cdot \rightarrow CO_2 + H_2O$$

Figure 9. The reaction mechanism for ASP photodegradation by WO_3-AC/CuS catalyst.

4. Conclusions

Covellite (CuS), a p-type semiconductor, is a promising Vis light-responsive photocatalyst due to its narrow bandgap (Eg = 1.7–2.2 eV) and good optical absorption properties in the Vis to NIR region. In this review, the influence of different CuS properties (morphology, particle size, bandgap energy, and surface properties) and synthesis parameters (Cu:S molar ratios, precursor concentration, surfactant amount in precursors solutions, etc.) on the photocatalytic activity of CuS-based catalysts (CuS nanostructures and CuS-based heterostructures) for various toxic, non-biodegradable, and recalcitrant organic pollutants (dyes, herbicides, active pharmaceuticals compounds) degradation was discussed.

Photocatalytic performances of CuS nanostructures and CuS-based heterostructures in organic pollutants removal from wastewater contribute to the further development of various CuS-based photocatalysts with enhanced solar energy conversion efficiency for environmental remediation and green energy production.

Author Contributions: Conceptualization, L.I; methodology, L.I.; literature investigation, L.I., C.C., L.A. and A.E.; writing—original draft preparation, L.I.; writing—review and editing, L.I. and A.E.; visualization, L.I., C.C., L.A. and A.E.; supervision, A.E. All authors have read and agreed to the published version of the manuscript.

Funding: This work was supported by a grant of the Ministry of Research, Innovation, and Digitization, CNCS–UEFISCDI, project number PN-III-P4-PCE-2021-1020 (PCE87), within PNCDI III.

Conflicts of Interest: The authors declare no conflict of interest.

Abbreviations

MCs	microcrystals
HT	hydrothermal
ST	solvothermal
SG	sol–gel
CoPp	Co-precipitation
MV	methyl violet
MG	malachite green
MO	methyl orange
EY	eosin Y
CR	congo red
MoO	mordant orange
SO	safranine orange
AO	acridine orange
ABRX-3B	active brilliant red X-3B
4-CP	4-chlorophenol
HA	humic acid
ASP	aspirin
S-MCh	S-metolachlor
PVP	polyvinylpyrrolidone
KCC1	Fibrous silica KAUST Catalysis Centre
MAA	methacrylic acid
PDI	perylene diimide
CB	conduction band
VB	valence band

References

1. Xu, G.; Du, M.; Zhang, J.; Li, T.; Guan, Y.; Guo, C. Facile fabrication of magnetically recyclable $Fe_3O_4/BiVO_4/CuS$ heterojunction photocatalyst for boosting simultaneous Cr(VI) reduction. *J. Alloys Compd.* **2022**, *895*, 162631. [CrossRef]
2. Li, Z.; Menga, X.; Zhang, Z. Recent development on MoS_2-based photocatalysis: A review. *J. Photochem. Photobiol. C Photochem. Rev.* **2018**, *35*, 39–55. [CrossRef]
3. Jiang, N.; Du, Y.; Ji, P.; Liu, S.; He, B.; Qu, B.; Wang, J.; Suna, X. Enhanced photocatalytic activity of novel $TiO_2/Ag/MoS_2/Ag$ nanocomposites for water-treatment. *Ceram. Int.* **2020**, *46*, 4889–4896. [CrossRef]
4. Isac, L.; Cazan, C.; Enesca, A.; Andronic, A. Copper Sulfide Based Heterojunctions as Photocatalysts for Dyes Photodegradation. *Front. Chem.* **2019**, *7*, 694. [CrossRef]
5. Jabbar, Z.H.; Graimed, B.H. Recent developments in industrial organic degradation via semiconductor heterojunctions and the parameters affecting the photocatalytic process: A review study. *J. Water Process. Eng.* **2022**, *47*, 102671. [CrossRef]
6. El-Hakam, S.A.; Alshorifi, F.T.; Salama, R.S.; Gamal, S.; Abo El-Yazeed, W.S.; Ibrahim, A.A.; Ahmed, A.I. Application of nanostructured mesoporous silica/bismuth vanadate composite catalysts for the degradation of methylene blue and brilliant green. *J. Mater. Res. Technol.* **2022**, *18*, 1963–1976. [CrossRef]
7. Alshorifi, F.T.; Abduliah, A.A.; Salama, R.S. Gold-selenide quantum dots supported onto cesium ferrite nanocomposites for the efficient degradation of rhodamine B. *Helyon* **2022**, *8*, E09652. [CrossRef] [PubMed]

8. Huang, X.; Chen, C.; Tsai, H.; Shaya, J.; Lu, C. Photocatalytic degradation of thiobencarb by a visible light-driven MoS_2 photocatalyst. *Sep. Purif. Technol.* **2018**, *197*, 147–155. [CrossRef]
9. Prabhakar Vattikuti, S.V.; Shim, J. Synthesis, characterization and photocatalytic performance of chemically exfoliated MoS_2. *IOP Conf. Ser. Mater. Sci. Eng.* **2018**, *317*, 012025. [CrossRef]
10. Sivaranjani, P.R.; Janani, B.; Thomas, A.M.; Raju, L.L.; Khan, S.S. Recent development in MoS_2-based nano-photocatalyst for the degradation of pharmaceutically active compounds. *J. Clean. Prod.* **2022**, *352*, 131506. [CrossRef]
11. Wang, W.; Liu, Y.; Yu, S.; Wen, X.; Wu, D. Highly efficient solar-light-driven photocatalytic degradation of pollutants in petroleum refinery wastewater on hierarchically-structured copper sulfide (CuS) hollow nanocatalysts. *Sep. Purif. Technol.* **2022**, *284*, 120254. [CrossRef]
12. Vaiano, V.; Sacco, O.; Barba, D.; Palma, V. Zinc Sulfide Prepared through ZnO Sulfuration: Characterization and Photocatalytic Activity. *Chem. Eng. Trans.* **2019**, *74*, 1159–1164. [CrossRef]
13. Ikram, M.; Khan, M.I.; Raza, A.; Imran, M.; Ul-Hamid, A.; Ali, S. Outstanding performance of silver-decorated MoS_2 nanopetals used as nanocatalyst for synthetic dye degradation. *Phys. E Low-Dimens. Syst. Nanostruct.* **2020**, *124*, 114126. [CrossRef]
14. El-Hout, S.I.; El-Sheikh, S.M.; Gaber, A.; Shawky, A.; Ahmed, A.I. Highly efficient sunlight-driven photocatalytic degradation of malachite green dye over reduced graphene oxide-supported CuS nanoparticles. *J. Alloys Compd.* **2020**, *849*, 156573. [CrossRef]
15. Sabbah, A.; Shown, I.; Qorbani, M.; Fu, F.-Y.; Lin, T.-Y.; Wu, H.-L.; Chung, P.-W.; Wu, C.-Y.; Santiago, S.R.M.; Shen, J.-L.; et al. Boosting photocatalytic CO_2 reduction in a $ZnS/ZnIn_2S_4$ heterostructure through strain-induced direct Z-scheme and a mechanistic study of molecular CO_2 interaction thereon. *Nano Energy* **2022**, *93*, 106809. [CrossRef]
16. He, R.-M.; Yang, Y.-L.; Chen, H.-J.; Liu, J.-J.; Sun, Y.-M.; Guo, W.-N.; Li, D.-H.; Hou, X.-J.; Suo, G.-Q.; Ye, X.-H.; et al. In situ controllable growth of Cu_7S_4 nanosheets on copper mesh for catalysis: The synergistic effect of photocatalytic Fenton-like process. *Colloids Surf. A Physicochem. Eng. Asp.* **2022**, *642*, 128651. [CrossRef]
17. Jiang, G.; Zhu, B.; Sun, J.; Liu, F.; Wang, Y.; Zhao, C. Enhanced activity of ZnS (111) by N/Cu co-doping: Accelerated degradation of organic pollutants under visible light. *J. Environ. Sci.* **2023**, *125*, 244–257. [CrossRef]
18. Mouchaal, Y.; Enesca, A.; Mihoreanu, C.; Khelil, A.; Duta, A. Tuning the opto-electrical properties of SnO_2 thin films by Ag^{+1} and In^{+3} co-doping. *Mater. Sci. Eng. B* **2015**, *199*, 22–29. [CrossRef]
19. Yang, F.; Zhang, Z.; Wang, Y.; Xu, M.; Zhao, W.; Yan, J.; Chen, C. Facile synthesis of nano-MoS_2 and its visible light photocatalytic property. *Mater. Res. Bull.* **2017**, *87*, 119–122. [CrossRef]
20. Dudita, M.; Bogatu, C.; Enesca, A.; Duta, A. The influence of the additives composition and concentration on the properties of SnO_x thin films used in photocatalysis. *Mater. Lett.* **2011**, *65*, 2185–2189. [CrossRef]
21. Hamida, A.A.; Al-Maiyaly, B.K.H. Synthesis and characterization of Cu_2S:Al thin films for solar cell applications. *Chalcogenide Lett.* **2022**, *19*, 579–590. [CrossRef]
22. Wong, A.B.; Brittman, S.; Yu, Y.; Dasgupta, N.P.; Yang, P. Core-Shell CdS-Cu_2S Nanorod Array Solar Cells. *Nano Lett.* **2015**, *15*, 4096–4101. [CrossRef] [PubMed]
23. Wen, L.; Chen, Q.; Zhou, L.; Huang, M.; Lu, J.; Zhong, X.; Fan, H.; Ou, Y. $Cu_{1.8}S$@CuS composite material improved the photoelectric efficiency of cadmium-series QDSSCs. *Opt. Mater.* **2020**, *108*, 110172. [CrossRef]
24. Peng, M.; Ma, L.-L.; Zhang, Y.-G.; Tan, M.; Yu, Y. Controllable synthesis of self-assembled Cu_2S nanostructures through a template-free polyol process for the degradation of organic pollutant under visible light. *Mater. Res. Bull.* **2009**, *44*, 1834–1841. [CrossRef]
25. Enesca, A.; Isac, L. Tuned S-Scheme $Cu_2S\_TiO_2\_WO_3$ Heterostructure Photocatalyst toward S-Metolachlor (S-MCh) Herbicide Removal. *Materials* **2021**, *14*, 2231. [CrossRef]
26. Li, S.; Zhang, Z.; Yan, L.; Jiang, S.; Zhu, N.; Li, J.; Li, W.; Yu, S. Fast synthesis of CuS and Cu_9S_5 microcrystal using subcritical and supercritical methanol and their application in photocatalytic degradation of dye in water. *J. Supercrit. Fluids* **2017**, *123*, 11–17. [CrossRef]
27. Isac, L.; Nicoara, L.; Panait, R.; Enesca, A.; Perniu, D.; Duta, A. Alumina matrix with controlled morphology for colored spectrally selective coatings. *Environ. Eng. Manag. J.* **2017**, *16*, 715–724. [CrossRef]
28. Nemade, K.R.; Waghuley, S.A. Band gap engineering of CuS nanoparticles for artificial photosynthesis. *Mater. Sci. Semicond. Process.* **2015**, *39*, 781–785. [CrossRef]
29. Shamraiz, U.; Hussain, R.A.; Badshah, A. Fabrication and applications of copper sulfide (CuS) nanostructures. *J. Solid State Chem.* **2016**, *238*, 25–40. [CrossRef]
30. Mohamed, E.F.; Dob, T.-O. Synthesis of New Hollow Nanocomposite Photocatalysts: Sunlight Applications for Removal of Gaseous Organic Pollutants. *J. Taiwan Inst. Chem. Eng.* **2020**, *111*, 181–190. [CrossRef]
31. Borthakur, P.; Boruah, P.K.; Das, M.R. Facile synthesis of CuS nanoparticles on two-dimensional nanosheets as efficient artificial nanozyme for detection of Ibuprofen in water. *J. Environ. Chem. Eng.* **2021**, *9*, 104635. [CrossRef]
32. Arshad, M.; Wang, Z.; Nasir, J.A.; Amador, E.; Jin, M.; Li, H.; Chen, Z.; Rehman, Z.; Chen, W. Single source precursor synthesized CuS nanoparticles for NIR phototherapy of cancer and photodegradation of organic carcinogen. *J. Photochem. Photobiol. B Biol.* **2021**, *214*, 112084. [CrossRef] [PubMed]
33. Kaur, A.; Singh, K. Investigation of crystallographic, morphological and photocatalytic behaviour of Ag doped CuS nanostructures. *Mater. Today Proc.* **2022**, *60*, 1090–1098. [CrossRef]

34. Goel, S.; Chen, F.; Cai, W. Synthesis and Biomedical Applications of Copper Sulfide Nanoparticles: From Sensors to Theranostics. *Small* **2014**, *10*, 631–645. [CrossRef] [PubMed]
35. Jiang, K.; Chen, Z.; Meng, X. CuS and Cu$_2$S as Cathode Materials for Lithium Batteries: A Review. *ChemElectroChem* **2019**, *6*, 28243. [CrossRef]
36. Majumdar, D. Recent progress in copper sulfide based nanomaterials for high energy supercapacitor applications. *J. Electroanal. Chem.* **2021**, *880*, 114825. [CrossRef]
37. Yu, B.; Meng, F.; Zhou, T.; Fan, A.; Khan, M.W.; Wu, H.; Liu, X. Construction of hollow TiO$_2$/CuS nanoboxes for boosting full-spectrum driven photocatalytic hydrogen evolution and environmental remediation. *Ceram. Int.* **2021**, *47*, 8849–8858. [CrossRef]
38. Mutalik, C.; Okoro, G.; Krisnawati, D.I.; Jazidie, A.; Rahmawati, E.Q.; Rahayu, D.; Hsu, W.-T.; Kuo, T.-R. Copper sulfide with morphology-dependent photodynamic and photothermal antibacterial activities. *J. Colloid Interface Sci.* **2022**, *607*, 1825–1835. [CrossRef]
39. Ding, H.; Han, D.; Hana, Y.; Liang, Y.; Liu, X.; Li, Z.; Zhu, S.; Wu, S. Visible light responsive CuS/protonated g-C3N4 heterostructure for rapid sterilization. *J. Hazard. Mater.* **2020**, *393*, 122423. [CrossRef]
40. Zhang, Z.; Sun, Y.; Mo, S.; Kim, J.; Guo, D.; Ju, J.; Yu, J.; Liu, M. Constructing a highly efficient CuS/Cu$_9$S$_5$ heterojunction with boosted interfacial charge transfer for near-infrared photocatalytic disinfection. *Chem. Eng. J.* **2022**, *431*, 134287. [CrossRef]
41. Agboola, P.O.; Haider, S.; Shakir, I. Copper sulfide and their hybrids with carbon nanotubes for photocatalysis and antibacterial studies. *Ceram. Int.* **2022**, *48*, 10136–10143. [CrossRef]
42. Chang, C.-J.; Lin, Y.-G.; Chen, J.; Huang, C.-Y.; Hsieh, S.-C.; Wu, S.-Y. Ionic liquid/surfactant-hydrothermal synthesis of dendritic PbS@CuS core-shell photocatalysts with improved photocatalytic performance. *Appl. Surf. Sci.* **2021**, *546*, 149106. [CrossRef]
43. Alhaddad, M.; Shawky, A. CuS assembled rGO heterojunctions for superior photooxidation of atrazine under visible light. *J. Mol. Liq.* **2020**, *318*, 114377. [CrossRef]
44. Enesca, A.; Duta, A. Tailoring WO$_3$ thin layers using spray pyrolysis technique. *Phys. Status Solidi C* **2008**, *5*, 3499–3502. [CrossRef]
45. Morales-García, A.; Soares, A.L., Jr.; Dos Santos, E.C.; de Abreu, H.A.; Duarte, H.A. First-Principles Calculations and Electron Density Topological Analysis of Covellite (CuS). *J. Phys. Chem. A* **2014**, *118*, 5823–5831. [CrossRef] [PubMed]
46. Sudhaik, A.; Raizada, P.; Rangabhashiyam, S.; Singh, A.; Nguyen, V.-H.; Le, Q.V.; Khan, A.A.P.; Hu, C.; Huang, C.-W.; Ahamad, T.; et al. Copper sulfides based photocatalysts for degradation of environmental pollution hazards: A review on the recent catalyst design concepts and future perspectives. *Surf. Interfaces* **2022**, *33*, 102182. [CrossRef]
47. Deng, J.; Zhao, Z.-Y. Effects of non-stoichiometry on electronic structure of Cu$_x$S$_y$ compounds studied by first-principle calculations. *Mater. Res. Express* **2019**, *6*, 105513. [CrossRef]
48. Kalanur, S.S.; Seo, H. Tuning plasmonic properties of CuS thin films via valence band filling. *RSC Adv.* **2017**, *7*, 11118–11122. [CrossRef]
49. Yooa, J.-H.; Ji, M.; Kim, J.H.; Ryu, C.-H.; Lee, Y.-I. Facile synthesis of hierarchical CuS microspheres with high visible-light driven photocatalytic activity. *J. Photochem. Photobiol. A Chem.* **2020**, *401*, 112782. [CrossRef]
50. Iqbal, T.; Ashraf, M.; Afsheen, S.; Masood, A.; Qureshi, M.T.; Obediat, S.T.; Hamed, M.F.; Othman, M.S. Copper sulfide (CuS) doped with carbon quantum dots (CQD) as an efficient photocatalyst. *Opt. Mater.* **2022**, *125*, 112116. [CrossRef]
51. Li, Y.-H.; Wang, Z. Green synthesis of multifunctional copper sulfide for efficient adsorption and photocatalysis. *Chem. Zvesti.* **2019**, *73*, 2297–2308. [CrossRef]
52. Wu, H.; Li, Y.; Li, Q. Facile synthesis of CuS nanostructured flowers and their visible light photocatalytic properties. *Appl. Phys. A* **2017**, *123*, 196. [CrossRef]
53. Tanveer, M.; Cao, C.; Ali, Z.; Aslam, I.; Idrees, F.; Khan, W.S.; But, F.K.; Tahir, M.; Mahmood, N. Template free synthesis of CuS nanosheet-based hierarchical microspheres: An efficient natural light driven photocatalyst. *CrystEngComm* **2014**, *16*, 5290–5300. [CrossRef]
54. Ain, N.; Rehman, Z.; Aamir, A.; Khan, Y.; Rehman, M.; Lin, D.-J. Catalytic and photocatalytic efficacy of hexagonal CuS nanoplates derived from copper(II) dithiocarbamate. *Mater. Chem. Phys.* **2020**, *242*, 1224078. [CrossRef]
55. Nancucheo, A.; Segura, A.; Hernandez, J.; Hernandez-Montelongo, J.; Pesenti, H.; Arranz, A.; Benito, N.; Romero-Saez, M.; Contreras, B.; Díaz, V.; et al. Covellite nanoparticles with high photocatalytic activity bioproduced by using H$_2$S generated from a sulfidogenic bioreactor. *J. Environ. Chem. Eng.* **2022**, *10*, 107409. [CrossRef]
56. Jiang, J.; Jiang, Q.; Deng, R.; Xie, X.; Meng, J. Controllable preparation, formation mechanism and photocatalytic performance of copper base sulfide nanoparticles. *Mater. Chem. Phys.* **2020**, *254*, 123504. [CrossRef]
57. Adhikari, S.; Sarkar, D.; Madras, G. Hierarchical Design of CuS Architectures for Visible Light Photocatalysis of 4-Chlorophenol. *ACS Omega* **2017**, *2*, 4009–4021. [CrossRef]
58. Bhatt, V.; Kumar, M.; Yun, J.-H. Unraveling the photoconduction characteristics of single-step synthesized CuS and Cu$_9$S$_5$ micro-flowers. *J. Alloys Compd.* **2021**, *891*, 161940. [CrossRef]
59. Fang, J.; Zhang, P.; Chang, Z.; Wang, X. Hydrothermal synthesis of nanostructured CuS for broadband efficient optical absorption and high-performance photo-thermal conversion. *Sol. Energy Mater. Sol. Cells* **2018**, *185*, 456–463. [CrossRef]
60. Yan, X.; Michael, E.; Komarneni, S.; Brownson, J.R.; Yan, Z.-F. Microwave- and conventional-hydrothermal synthesis of CuS, SnS and ZnS: Optical properties. *Ceram. Int.* **2013**, *39*, 4757–4763. [CrossRef]

61. Ayodhya, D.; Venkatesham, M.; Kumari, A.S.; Reddy, G.B.; Ramakrishna, D.; Veerabhadram, G. Photocatalytic degradation of dye pollutants under solar, visible and UV lights using green synthesised CuS nanoparticles. *J. Exp. Nanosci.* **2016**, *11*, 418–432. [CrossRef]
62. Wang, X.; He, Y.; Hu, Y.; Jin, G.; Jiang, B.; Huang, Y. Photothermal-conversion-enhanced photocatalytic activity of flower-like CuS superparticles under solar light irradiation. *Sol. Energy* **2018**, *170*, 586–593. [CrossRef]
63. Saranya, M.; Ramachandran, R.; Samuel, E.J.; Jeong, S.K.; Grace, A.N. Enhanced visible light photocatalytic reduction of organic pollutant and electrochemical properties of CuS catalyst. *Powder Technol.* **2014**, *252*, 25–32. [CrossRef]
64. Iqbal, S.; Shaid, N.A.; Sajid, M.M.; Javed, Y.; Fakhar-e-Alam, M.; Mahmood, A.; Ahmad, G.; Afzal, A.M.; Hussain, S.Z.; Ali, F.; et al. Extensive evaluation of changes in structural, chemical and thermal properties of copper sulfide nanoparticles at different calcination temperature. *J. Cryst. Growth* **2020**, *547*, 125823. [CrossRef]
65. Pejjai, B.; Reddivari, M.; Kotte, T.R.R. Phase controllable synthesis of CuS nanoparticles by chemical co-precipitation method: Effect of copper precursors on the properties of CuS. *Mater. Chem. Phys.* **2020**, *239*, 122030. [CrossRef]
66. Andronic, L.; Isac, L.; Cazan, C.; Enesca, A. Simultaneous Adsorption and Photocatalysis Processes Based on Ternary TiO_2–Cu_xS–Fly Ash Hetero-Structures. *Appl. Sci.* **2020**, *10*, 8070. [CrossRef]
67. Li, S.; Ge, Z.-G.; Zhang, Z.-P.; Yao, Y.; Wang, H.-C.; Yang, J.; Li, Y.; Gao, G.; Lin, Y.-H. Mechanochemically synthesized sub-5 nm sized CuS quantum dots with high visible-light-driven photocatalytic activity. *Appl. Surf. Sci.* **2016**, *384*, 272–278. [CrossRef]
68. Balaz, M.; Dutkov, E.; Bujnakova, Z.; Tothova, E.; Kostova, N.G.; Karakirova, Y.; Briancin, J.; Kanuchov, M. Mechanochemistry of copper sulfides: Characterization, surface oxidation and photocatalytic activity. *J. Alloys Compd.* **2018**, *746*, 576–582. [CrossRef]
69. Shamraiz, U.; Badshah, A.; Hussain, R.A.; Nadeem, M.A.; Sab, S. Surfactant free fabrication of copper sulphide (CuS–Cu_2S) nanoparticles from single source precursor for photocatalytic applications. *J. Saudi Chem. Soc.* **2017**, *21*, 390–398. [CrossRef]
70. Cruz, J.S.; Hernández, S.A.M.; Delgado, F.P.; Angel, O.Z.; Pérez, R.C.; Delgado, G.T. Optical and Electrical Properties of Thin Films of CuS Nanodisks Ensembles Annealed in a Vacuum and Their Photocatalytic Activity. *Int. J. Photoenergy* **2013**, *2013*, 178017. [CrossRef]
71. Siddique, F.; Rafiq, M.A.; Afsar, M.F.; Hasan, M.M.; Chaudhry, M.M. Enhancement of degradation of mordant orange, safranin-O and acridine orange by CuS nanoparticles in the presence of H_2O_2 in dark and in ambient light. *J. Mater. Sci. Mater. Electron.* **2018**, *29*, 19180–19191. [CrossRef]
72. Hu, X.-S.; Shen, Y.; Xu, L.-H.; Wang, L.-M.; Xing, Y.-J. Preparation of flower-like CuS by solvothermal method and its photodegradation and UV protection. *J. Alloys Compd.* **2016**, *674*, 289–294. [CrossRef]
73. Nabi, G.; Tanveer, M.; Tahir, M.B.; Kiran, M.; Rafique, M.; Khalid, N.R.; Alzaid, M.; Fatima, N.; Nawaz, T. Mixed solvent based surface modification of CuS nanostructures for an excellent photocatalytic application. *Inorg. Chem. Commun.* **2020**, *121*, 108205. [CrossRef]
74. Chai, Z.; Pan, X.; Cui, F.; Ma, Q.; Zhang, J.; Liu, M.; Chen, Y.; Li, L.; Cui, T. Synthesis of 3D hierarchical CuS architectures consisting of 1D nanotubes for efficient photocatalysts. *Mater. Lett.* **2020**, *275*, 128168. [CrossRef]
75. Qin, N.; Wei, W.; Huang, C.; Mi, L. An efficient strategy for the fabrication of CuS as a highly excellent and recyclable photocatalyst for the degradation of organic dyes. *Catalysts* **2020**, *10*, 40. [CrossRef]
76. Mohammadi, N.; Allahresani, A.; Naghizadeh, A. Enhanced photo-catalytic degradation of natural organic matters (NOMs) with a novel fibrous silica-copper sulfide nanocomposite (KCC1-CuS). *J. Mol. Struct.* **2022**, *1249*, 131624. [CrossRef]
77. Kaushik, B.; Yadav, S.; Rana, P.; Solanki, K.; Rawat, D.; Sharma, R.K. Precisely engineered type II ZnO-CuS based heterostructure: A visible light driven photocatalyst for efficient mineralization of organic dyes. *Appl. Surf. Sci.* **2022**, *590*, 153053. [CrossRef]
78. Vyas, Y.; Chundawat, P.; Dharmendra, D.; Jain, A.; Punjabi, P.B.; Ameta, C. Biosynthesis and characterization of carbon quantum Dots@CuS composite using water hyacinth leaves and its usage in photocatalytic dilapidation of Brilliant Green dye. *Mater. Chem. Phys.* **2022**, *281*, 125921. [CrossRef]
79. Yan, L.; Wang, W.; Zhao, Q.; Zhu, Z.; Liu, B.; Hua, C. Construction of perylene diimide/CuS supramolecular heterojunction for the highly efficient visible light-driven environmental remediation. *J. Colloid Interface Sci.* **2022**, *606*, 898–911. [CrossRef]
80. Iqbal, T.; Ashraf, M.; Masood, A. Simple synthesis of WO_3 based activated carbon co-doped CuS composites for photocatalytic applications. *Inorg. Chem. Commun.* **2022**, *139*, 109322. [CrossRef]
81. Baneto, M.; Enesca, A.; Mihoreanu, C.; Lare, Y.; Jondo, K.; Napo, K.; Duta, A. Effects of the growth temperature on the properties of spray deposited $CuInS_2$ thin films for photovoltaic applications. *Ceram. Int.* **2015**, *41*, 4742–4749. [CrossRef]
82. Zhang, F.; Ma, Y.; Chi, y.; Yu, H.; Li, Y.; Jiang, T.; Wei, X.; Shi, J. Self-assembly, optical and electrical properties of perylene diimide dyes bearing unsymmetrical substituents at bay position. *Sci. Rep.* **2018**, *8*, 8208. [CrossRef]
83. Tuckute, S.; Varnagiris, S.; Urbonavicius, M.; Lelis, M.; Sakalauskaite, S. Tailoring of TiO_2 film crystal texture for higher photocatalysis efficiency. *Appl. Surf. Sci.* **2019**, *489*, 576–583. [CrossRef]
84. Aslam, M.; Qamar, M.T.; Ali, S.; Rehman, A.U.; Soomro, M.T.; Ahmed, I.; Ismail, I.M.I.; Hameed, A. Evaluation of SnO_2 for sunlight photocatalytic decontamination of Water. *J. Environ. Manag.* **2018**, *217*, 805–814. [CrossRef]
85. Chen, W.; Liu, Q.; Tian, S.; Zhao, X. Exposed facet dependent stability of ZnO micro/nano crystals as a photocatalyst. *Appl. Surf. Sci.* **2019**, *470*, 807–816. [CrossRef]

Article

Direct One-Step Seedless Hydrothermal Growth of ZnO Nanostructures on Zinc: Primary Study for Photocatalytic Roof Development for Rainwater Purification

Marie Le Pivert [1,2], Aurélie Piebourg [1], Stéphane Bastide [3], Myriam Duc [4] and Yamin Leprince-Wang [1,*]

1. ESYCOM Lab, CNRS, Université Gustave Eiffel, F-77454 Marne-la-Vallée, France
2. COSYS/ISME Lab, IFSTTAR, Université Gustave Eiffel, F-77447 Marne-la-Vallée, France
3. ICMPE, UMR 7182 CNRS, Université Paris-Est Créteil, F-94320 Thiais, France
4. GERS/SRO, IFSTTAR, Université Gustave Eiffel, F-77447 Marne-la-Vallée, France
* Correspondence: yamin.leprince@univ-eiffel.fr

Abstract: To shift towards the greener city, photocatalytic urban infrastructures have emerged as a promising solution for pollution remediation. To reach this goal, the large bandgap semiconductors, such as nontoxic Zinc Oxide (ZnO), already proved their excellent photocatalytic performances. However, integrating and developing cost-effective and greener photocatalytic surfaces with an easily scaled-up synthesis method and without energy and chemical product overconsumption is still challenging. Therefore, this work proposes to develop a depolluting Zinc (Zn) roof covered by ZnO nanostructures (NSs) using a one-step seedless hydrothermal growth method in 2 h. The feasibility of this synthesis was firstly studied on small areas of Zn (1.25 cm^2) before being scaled up to medium-sized areas (25 cm^2). The efficiency of this functionalization route for ZnO NSs grown without seed layer was attributed to the presence of Zn^{2+} sites and the native oxide film on the Zn surface. Their photocatalytic efficiency was demonstrated by removing in less than 3 h the Methylene Blue (MB) and Acid Red 14 (AR14) in both DI water and rainwater under UV-light. Promising results were also recorded under solar light. Therefore, the photocatalytic Zn roof functionalized by ZnO NSs is a promising route for rainwater purification by photocatalysis.

Keywords: photocatalysis; water purification; ZnO nanostructures; hydrothermal synthesis; seedless; building materials

1. Introduction

Access to water poses a growing risk to the economy as well as to the communities and ecosystems that depend on it. Fresh water is indeed a limited resource whose direct consumption is restricted and could be affected by pollution and contamination. It is therefore more important than ever to preserve easily accessible fresh water sources such as rainwater. Nevertheless, the quality of rainwater is affected by polluted air [1], i.e., it naturally contains some pollutants present in the atmosphere [2]. Hence, some pollutants are drawn from the atmosphere to the ground during rainy days. At this stage, the concentration of pollutants is extremely low, but it continues to increase as the water passes through the urban environment to the wastewater treatment plants. As a result, rainwater affects at the same time, the soil, the groundwater, the human health and the ecosystem. It is therefore essential to reflect upon direct rainwater treatment in urban areas.

As a green and sustainable method of environmental pollution remediation, photocatalytic processes are being studied extensively as possible air and water treatments due to their ability to remove refractory pollutants at different scales and places without requiring complex and expensive infrastructures [3,4]. They are particularly attractive and suitable for this type of treatment because they could use only solar energy (inexhaustible) and a photocatalyst, which is mostly a wide bandgap metal oxide semiconductor. Moreover, they do not require the handling of expensive chemicals and can be operated autonomously to mineralize micropollutants into CO_2, H_2O, and other light by-products [4].

Therefore, the development of photocatalytic urban infrastructures has emerged as a promising solution to address environmental pollution directly near emission sources, thereby avoiding pollutant accumulation and dispersion in urban areas [5–7]. To develop these infrastructures, large bandgap semiconductors, such as nanostructured dioxide titanium (TiO_2) and zinc oxide (ZnO), were mainly used and already proved their excellent photocatalytic performance at the laboratory scale [2,5–9]. For these materials to be highly efficient, they must be elaborated as nanostructures (NSs) owing a high volume-to-surface ratio, in order to develop a very large free active area [10]. The production pathway should be short (few hours) and easily scale-up, using the less energy and chemical products to be the as green as possible. Efforts are still being made to develop this kind of synthesis method to produce large-scale surface. To the best of our knowledge, photocatalytic urban infrastructures are currently produced by incorporating nanoparticles in civil engineering materials [11,12], such as concrete or paint; by depositing a semiconductor layer on the surface [12]; or recently by growing ZnO NSs on the urban infrastructures by two-step hydrothermal synthesis [13,14]. This last process has the advantage to functionalize the civil engineering material surface in less than 2 h and consumes few energy and chemical products. Nevertheless, it requires a ZnO seed layer deposition and double annealing at 350 °C. In order to optimize the synthesis process and to make it greener, an improved ZnO NSs synthesis method will be studied by choosing the appropriate substrate allowing a surface functionalization with only one step instead of two steps.

Therefore, to manage rainwater directly at the first source of runoff with ecofriendly photocatalytic urban infrastructures, this work proposes to develop a depolluting Zinc (Zn) roof covered with ZnO NSs using a one-step seedless hydrothermal growth method in 2 h to produce a large surface of 5 cm × 5 cm. Zinc is indeed a material commonly used in new construction buildings, which allows the use of hydrothermal synthesis in the absence of the seed layer deposition step, thus avoiding the overconsumption of energy and chemical products. Moreover, ZnO is nontoxic, biocompatible, and inexpensive in terms of raw materials, making it a remarkable large bandgap photocatalyst. First, the feasibility of a cost-effective and environmentally compatible production of this photocatalytic surface was studied, as well as the use of a post-annealing treatment. Samples were thoroughly characterized by scanning electron microscopy (SEM FEG ZEISS Merlin, Oberkochen, Germany) to investigate the morphology of the ZnO NSs, by X-ray diffraction (XRD, Brucker Advance) using a cobalt anticathode for the study of the microstructures, by UV–visible spectrophotometry for ZnO bandgap determination (UV-visible, Maya2000 Pro from Ocean Optics, Duiven, Netherlands) and for absorbance measurements of the organic dyes (Perkin Elmer Lambda 35, Waltham, MA, USA) to monitor their degradation, and by photoluminescence (PL) to obtain more information concerning defects both in as-grown and annealed ZnO samples. Then, the photocatalyst properties of both sample types were evaluated with respect to the removal of Methylene Blue (MB) and Acid Red 14 (AR14) in DI water and in rainwater under UV-light and natural solar light. Finally, the synthesis was scaled-up to produce a 5 cm × 5 cm ZnO/Zn roof surface in order to simulate on a small-scale the behavior of rainwater on this surface under artificial solar light illumination.

2. Results

2.1. Characterization of ZnO Nanostructures Grown on Zn Surface

To evaluate the production of ZnO NSs based photocatalytic surfaces by one-step seedless hydrothermal synthesis, the functionalization process was firstly developed on a small scale flattened Zn wires (2.5 cm × 0.5 cm), as shown in Figure 1a. Functionalized Zn samples, both as grown (ZnO/Zn(AG)) and post-annealed (ZnO/Zn, 30 min at 350 °C), exhibited ZnO bandgap values similar to those grown on silicon or on tiling substrates by the classical two-step hydrothermal growth (Figure 1b) [13,14]. No apparent difference was recorded between the UV-Vis spectra of the as-grown and annealed sample, with an average

ZnO bandgap of 3.20 ± 0.02 eV. Only a slight variation in intensity was recorded depending on the location of the measurement. Indeed, the surface of mechanically flattened Zn wire is not smooth, thereby affecting the light path during the acquisition of the UV-vis spectrum.

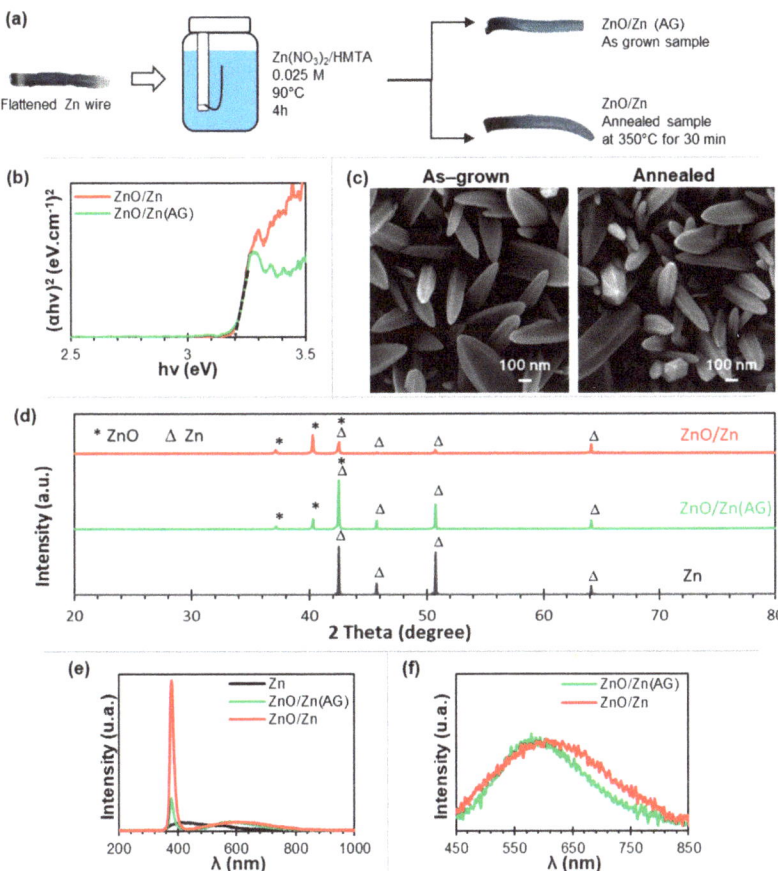

Figure 1. (a) Schematic of the experimental set-up for the synthesis of ZnO NSs on flattened Zn wires; (b) UV–visible spectral plots according to the Tauc-Lorentz model, (c) SEM images and (d) XRD patterns of ZnO/Zn(AG) and ZnO/Zn samples, (e) PL spectrum of ZnO/Zn(AG) and ZnO/Zn samples, (f) PL visible emission band of ZnO/Zn(AG) and ZnO/Zn samples.

The non-perfect homogeneity and the roughness of the Zn substrate surface resulted in some defects on the grown ZnO nanostructures. Nevertheless, the entire surface of the Zn wire was covered by the ZnO NSs (Figure 1c). Indeed, SEM images of ZnO NSs grown on Zn showed a high density of nanowires (NWs) and nanorods (NRs), as traditionally obtained by the hydrothermal route [15,16]. It should be noted that here the substrate is not the typical laboratory substrate, such as glass or silicon, with a well-defined planar surface. Therefore, the NWs/NRs are not perfectly vertical to the substrate and some other punctual NSs appeared.

Figure 1d shows the XRD diffractograms of the ZnO/Zn(AG) and ZnO/Zn sampleswhere the two hexagonal crystal structures of Zn and ZnO can be clearly identified. The four peaks at 42.5°, 45.7°, 50.5°, and 64.2° correspond respectively to the (002), (100), (101), and (102) plan of Zn (reported values from ICDD N° 01-085-5877). Thus, these four peaks are related to the substrate used as clearly identified on the Figure 1d. The ZnO XRD pattern correspond to the three main peaks of Würtzite ZnO, which are (100) at 37.2°, (002) at 40.3°, and (101) at 42.5° (reported values from ICDD N° 98-002-9272) [17,18]. An improvement in crystallinity related to the annealing post-treatment (350 °C) can be observed from the XRD patterns: the intensity of the ZnO peaks increased while the intensity of the Zn peaks decreased. The XRD patterns and SEM images show a preferential growth direction of the ZnO NSs along the c-axis ([0001] direction of the hexagonal structure).

The crystallinity improvement induced the post-treatment at 350 °C during 30 min were also observed by PL measurements (Figure 1e). Indeed, an increase of the UV emission band; related to near band-edge transition, namely the free exciton recombination through an excition-excition collision process; and a slight decrease of visible emission band; related to defect levels, such as Zn_i, V_O, and O_i; were conjointly recorded on ZnO/Zn compared to ZnO/Zn(AG) spectrum. This trend was attributed to the improvement in crystallinity. Moreover, by examine the visible emission band, a red-shift in the peak position was recorded after the annealing treatment (Figure 1f). This shift may due to the deep level defects, such as V_O and O_i, which were usually observed in the oxygen-rich sample giving orange-red emission [19].

After obtaining on small Zn samples with ZnO NSs showing good morphological and textural characteristics despite the absence of both pretreatment and seedlayer coating, the suitability of large-scale synthesis was evaluated by functionalizing a flat Zn sheet (5 cm × 5 cm, sampled from a commercial Zn roof—Leroy Merlin, Collégien, France) (Figure 2a). At this scale, the presence of ZnO NSs on the substrate is accompanied by a slight whitish coloration of the sheet. As expected, the as-grown functionalized Zn sheet, ZnO/Zn(AG), exhibits similar ZnO bandgap values to those obtained at a smaller scale, with an average at 3.21 ± 0.01 eV (Figure 2b).

SEM observations revealed a sligthly different ZnO nanostructure with longer and thinner NWs (Figure 2c), which could be caused by the different nature of the substrate: the Zn wires used in the small-sample experiments has only a native oxide film, while the Zn roof material has been laminated and treated chemically by the manufacturer to form a protective layer containing alkaline Zn carbonate ($2\,ZnCO_3 \cdot 3\,Zn(OH)_2$) on the Zn surface. Moreover, the Zn sheet has the advantage to be more smooth than above used wires avoiding irregularity during the growth. This result is consistent with the literature, where hydrothermal scaled-up synthesis on silicon wafer has already been demonstrated [20,21]. Furthermore, a shorter growth time, 2 h instead of 4 h, was also tested to determine its influence on the characteristics of ZnO NWs. SEM images showed that smaller NWs are obtained but with a lager areal density for 2 h: ~35 NWs/μm^2 vs. ~15 NWs/μm^2, for 4 h-sample.

Figure 2d shows the XRD diffractograms of the ZnO/Zn(AG) samples obtained after 4 h and 2 h synthesis, as well as that of the bare Zn roof substrate. On the ZnO/Zn(AG) roof samples, the two hexagonal crystal structures of Zn and ZnO are clearly identified with no significant difference between the 4 h and 2 h samples. As previously, on the wire samples, the Zn XRD pattern is related to the Zn substrate used. Meanwhile, the ZnO XRD patterns correspond to the typical XRD patterns of the ZnO Würtzite phase.

The similarity between ZnO/Zn(AG) 4 h and ZnO/Zn(AG) 2 h were also observed on the PL spectrum with no shift of UV or visible emission bands (Figure 2e,f). Only a slight difference of intensity, with more intense peaks for ZnO/Zn(AG) 4 h, were recorded. The increase of intensity may be due to the growing amount of ZnO with the hydrothermal duration. Moreover, in accordance with our results, previous PL analysis works showed an increasing intensity in the visible emission band with the growth time, which was mainly attributed to the change of the concentration of the structural defects in the ZnO nanowires [22].

It should be emphasized that the absence of post-treatment (as-grown samples) is tested with the aim of developing a future application. In this way, it is in principle possible to save the energy consumed by a post-annealing (350 °C, 30 min), while a shorter growth time would be advantageous to increase the production rate of photocatalytic roofs. Of course, these savings will be meaningful only if good photocatalytic performances are verified.

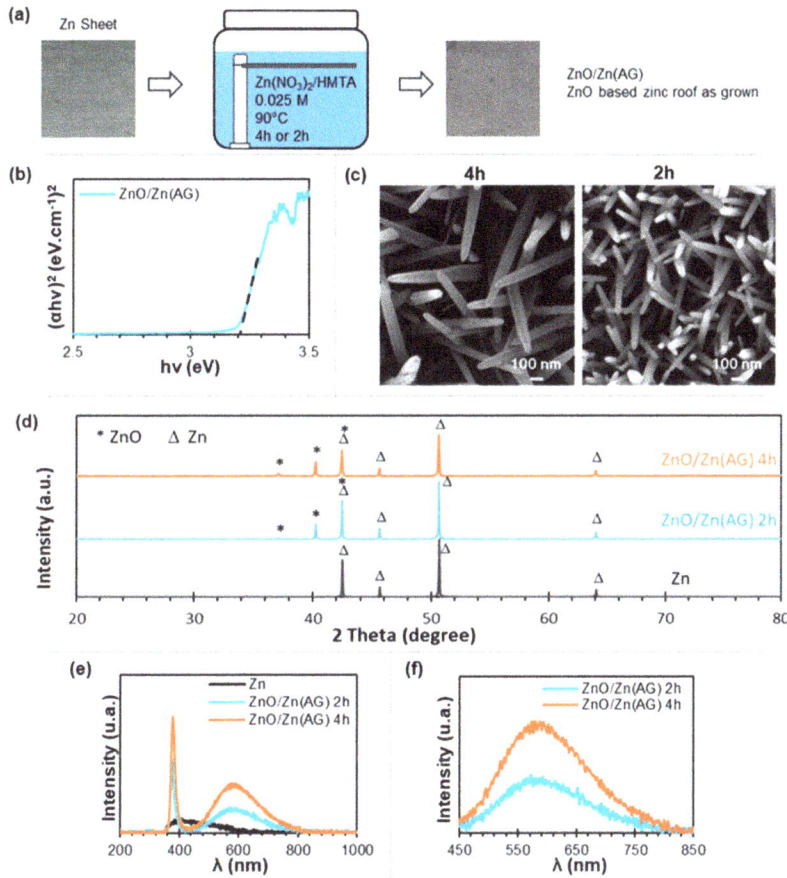

Figure 2. (**a**) Schematic of the experimental set-up for the synthesis of ZnO NSs on Zn roof sheets; (**b**) UV–visible spectral plots according to the Tauc-Lorentz model, (**c**) SEM images and (**d**) XRD patterns of ZnO/Zn(AG) samples grown for 4 h or 2 h, (**e**) PL spectrum of ZnO/Zn(AG) samples grown for 4 h or 2 h, (**f**) PL visible emission band of ZnO/Zn(AG) samples grown for 4 h or 2 h.

2.2. Photocatalytic Activity Evaluation

2.2.1. Methylene Blue Photodegradation under UV-Light

To test the photocatalytic activity of ZnO NSs on flattened Zn wires under reproducible conditions, the photodegradation of MB in DI water was first studied under a powerful UV-light source (Hamamatsu LC-8, ~365 nm, ~35 mW/cm^2). Results demonstrated that ZnO NSs lead to a total degradation in less than 120 min against 71% after 180 min for photolysis performed without ZnO (Figure 3a). Comparing these results with those obtained previously with larger substrates [13], ZnO NSs grown on flattened Zn wires seem to provide a good surface area/efficiency ratio. This means that good photocatalytic activity was recorded for ZnO/Zn(AG) and ZnO/Zn. No significant difference was observed between the as-grown and post-annealed samples, with only a slight improvement in the efficiency of ZnO/Zn compared to ZnO/Zn(AG). Taking this into account, the ZnO/Zn(AG) photocatalyst (the most eco-compatible option) appears to be the best choice for treating polluted rainwater via a functionalized Zn roof.

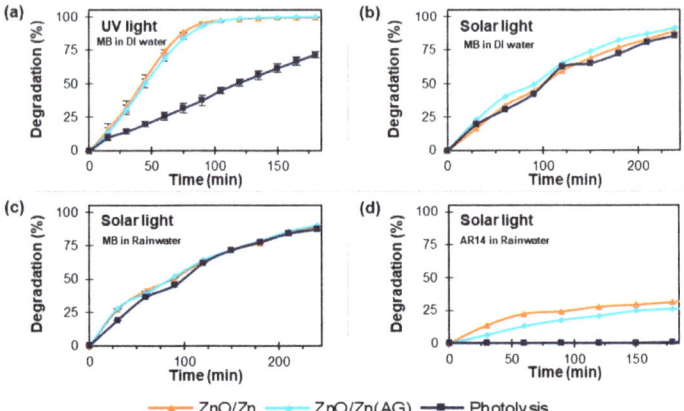

Figure 3. MB photodegradagtion with and without ZnO NSs under (**a**) UV light source (~365 nm, 35 mW/cm^2) in DI Water, (**b**) natural solar light (April/May, France, I_{UV} = 2.3 mW/cm^2) in DI Water, and (**c**) natural solar light (April/May, France, I_{UV} = 2.3 mW/cm^2) in rainwater; (**d**) AR14 photodegradation with and without ZnO NSs under natural solar light (April/May, France, I_{UV} = 2.9 mW/cm^2) in rainwater.

2.2.2. Photodegradation of Methylene Blue under Natural Solar Light

In order to evaluate the photocatalytic activity of nanostructured ZnO under conditions closed to the reality, tests were performed in April/May in Paris, France under natural solar light, whose UV part of the spectrum corresponds to ~2.3 mW/cm^2. The results (Figure 3b) showed that the use of ZnO NSs grown on flattened Zn wires improves the photodegradation of MB very slightly compared to natural photolysis. The same observation was made in a previous work and explained in that MB is very easy to degrade under solar light [14].

Chemical stability and possible competition in terms of photodegradation with other media-specific species were evaluated with the same experiments done in rainwater instead of DI water. Rainwater is known to have a slightly acidic pH and to contain a number of micro-pollutants whose presence could affect the MB removal process due to side reactions, competition at the surface of the photocatalyst, and partial poisoning of the active sites by chemicals present in rainwater. A comparison between Figure 3b,c, however, shows that none of these effects are observed, as the degradation rates are similar in both media. In fact, the easy photolysis degradation of MB can mask the photocatalytic contribution.

2.2.3. Photodegradation of Acid Red 14 in Rainwater under Natural Solar Light

To extend the evaluation of the photocatalytic performance of nanostructured ZnO, another organic dye, AR14, which is more resistant than MB to natural solar light was chosen for a photodegradation test [14]. This test was carried out in rainwater under cloudy weather, as previously, the efficiency of ZnO/Zn(AG) and ZnO/Zn was evaluated and compared to the photolysis in the same conditions. No photolysis was recorded under natural solar light, thereby allowing a better estimation of the photocatalytic activity of ZnO NSs (Figure 3d). Indeed, the photodegradation observed is only due to the photocatalytic activity of the roof functionalization. In contrast to the results under UV light, ZnO/Zn appears to be slightly more efficient than ZnO/Zn(AG) under natural solar light (Figure 3d).

2.2.4. Photodegration of Polluted Runoff Waters on ZnO/Zn Sheets under Solar Light

This section is a preliminary study to evaluate the relevance of ZnO/Zn(AG) development at large scale for the depollution of runoff waters using functionalized Zn roofs. An AR14 at a concentration of 10 µM in a rainwater solution was flowed over a ZnO/Zn(AG) sheet (5 cm × 5 cm, sampled from commercial Zn roof) under a solar lamp (Zolix, Sirius-

300P, Beijing, China, 300 W, 320–2500 nm) with a flow rate of 0.17–1.03 mL/min (Figure 4a). Only a single pass was carried out without recirculation of the treated water in order to simulate roof runoff conditions during rainy weather. Same experiments were carried out with a Zn sheet without functionalization treatment. Results with and without ZnO NSs were then compared to calculated an average gain corresponding to the deviation between the photolysis rate and the photocatalysis rate.

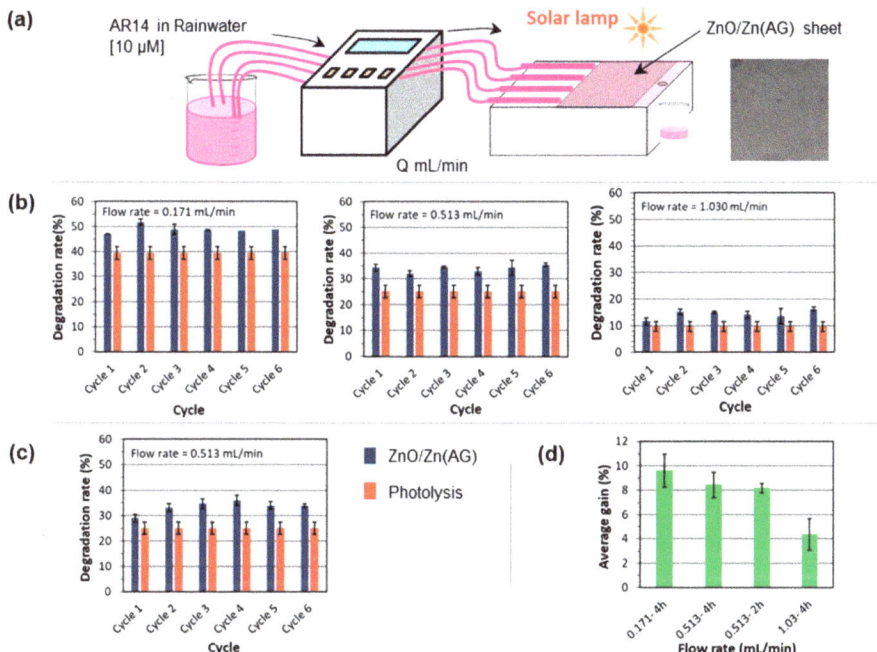

Figure 4. (**a**) Schematic of the experimental set-up of runoff water photodegradation with ZnO NSs on Zn roof sheets. Photodegradation rate of AR14 under artificial solar light with or without the presence of ZnO NSs: (**b**) at different flow rate: 0.171, 0.513, and 1.03 mL/min for 4 h-sample; (**c**) for 2 h-sample with a flow rate of 0.513 mL/min. (**d**) Comparison of the photocatalytic efficiencies for the different cases under studied.

The ZnO NSs average gain varied between 7% and 12% for a flow rate of 0.171 mL/min and between 6% and 9% for a flow rate of 0.513 mL/min, whereas the high flow rate of 1.03 mL/min resulted in a very low average gain of 2% to 6%. Coupled to this decrease of the average gain, the photodegradation rate decreases as the flow rate increases, which can be explained by the shorter residence time of the dye solution under the solar lamp.

For all flow rates, the ZnO/Zn(AG) stability was studied. The results shown in Figure 4b correspond to successive cycles of photocatalysis with a DI water rinse step followed by drying with a hot airflow stream and 15 min under solar lamp irradiation. These results show that there is no loss of efficiency during cycles 1 to 4. No loss of photocatalytic activity was also recorded between cycles 4 to 6, corresponding to consecutive cycles of photodegradation with only a short DI water rinse step followed by a drying with hot airflow. This observation was made in the case of all flow rates tested and supports the stability of the samples.

A comparative study between samples synthesized in 4 h and in 2 h was carried out. It aims at reducing the manufacturing cost for future photocatalytic roof applications. An intermediate flow rate of 0.513 mL/min was chosen for this test. At this flow rate, the degradation rates reported in Figure 4b (4 h-sample) and Figure 4c (2 h-sample) are very close. The Figure 4d gives a comparative illustration of the four presented cases, which allows to conclude that the hydrothermal synthesis time of ZnO NSs can be reduced from 4 h to 2 h without loss in efficiency.

3. Discussion

3.1. ZnO Nanostructures Growth on Zinc Substrate

The characterization work (Figures 1 and 2) demonstrated that ZnO NSs could be grown on a Zn substrate by a one-step seedless hydrothermal growth method, although the route generally used to produce ZnO NSs on a substrate is a seeded growth on ZnO nanoparticles or film-coated substrate [23,24]. However, for some substrates, self-seeded technology or local surface properties allow the direct growth of ZnO NSs, as is the case for Zn substrates.

A first hypothesis that could explain this observation is the formation of a local seed layer on the substrate surface at the initial stage preceding the crystal growth. The Zn substrate can react with hydroxide ions produced in the synthesis solution to form soluble zincate ions ZnO_2^{2-}, which in turn can react with water to deposit a ZnO (solid phase) seed layer according to the following reactions [23–25]:

$$Zn + 2\,OH^- \rightarrow ZnO_2^{2-} + H_2 \tag{1}$$

$$ZnO_2^{2-} + H_2O \rightarrow ZnO + 2\,OH^- \tag{2}$$

This preformed ZnO layer on the Zn substrate has proven to be a crucial step for the successful nucleation and subsequent growth of ZnO NSs [24,25]. Therefore, after the local oxidation of Zn, ZnO NSs can grow directly on the substrate without phase matching problems, according to the classic hydrothermal mechanisms:

$$(CH_2)_6 N_4 + 6\,H_2O \rightarrow 6\,HCHO + 4\,NH_3 \tag{3}$$

$$NH_3 + H_2O \rightarrow OH^- + NH_4^+ \tag{4}$$

$$Zn^{2+} + 4\,OH^- \rightarrow Zn(OH)_4^{2-} \rightarrow ZnO + 2\,OH^- + H_2O \tag{5}$$

$$Zn(NH_3)_4^{2+} + 2\,OH^- \rightarrow ZnO + 4\,NH_3 + H_2O \tag{6}$$

$$Zn^{2+} + 2\,OH^- \rightarrow Zn(OH)_2 \xrightarrow{\Delta} ZnO + H_2O \tag{7}$$

The second hypothesis, which can explain the growth of ZnO NSs without a seed layer, is also directly related to the species generated by the hydrothermal synthesis from reactions (3) and (4). Zn can react with NH_4^+ to form Zn^{2+} and other species (Equation (8)) [26]. Then Zn^{2+} can react with OH^- following Equation (7) to form ZnO. Therefore, at the initial growth stage, the Zn substrate acts as both a source of Zn^{2+} and of nucleation sites. Finally, after this initial stage, the hydrothermal growth process will follow the classic mechanisms for 4 h (reactions (3) to (7)). This process happens preferentially along the c-axis [0001] due to the minimum value of the (002) surface free energy (9.9 eV·nm^{-2}) compared to ones of different plans, such as (110) and (100) (12.3 and 20.9 eV·nm^{-2}, respectively), justifying the ZnO synthesis as NWs and NRs instead nanosheets or other morphologies.

$$Zn + 2\,NH_4^+ \rightarrow Zn^{2+} + 2\,NH_3 + H_2 \tag{8}$$

According to these two hypotheses, which may be valid conjointly, the participation of Zn on the substrate surface as a driving force occurs only during the initial phase. Indeed, after the formation of the first ZnO layers, Zn is protected and can no longer react directly with the solution [26], thus explaining the ZnO NSs well-synthesized as nanowires and nanorods after 4 h.

The results in Figure 2, demonstrating efficient up-scaling, imply that this method is applicable to the mass production of ZnO NSs-based roofs. This is in good agreement with the literature, where hydrothermal synthesis has been shown to be effective in functionalizing large silicon wafers [20,21] and in producing a few square meters of ZnO NSs-based pavers for tiles and bitumen roads [27]. The one-step functionalization of Zn sheets is also consistent with previous work where Zn substrates were functionalized in a chemical bath solution at low temperature [25,28].

3.2. ZnO/Zn Photocatalytic Activity

The total photodegradation of MB within 3 h of light exposure demonstrates an excellent photocatalytic activity under UV light (Figure 3a) of both types of sample (as grown and annealed). By comparison, with the results obtained on large substrates tested under the same experimental conditions [13], it appears that a 1.25 cm^2 sample of ZnO NSs grown on Zn by the present method (one-step) develops the same photocatalytic activity as a 1.55 cm^2 sample of ZnO NSs grown onto silicon by the classical method (two-step). No significant difference was observed between as grown and annealed samples, due to the fact that the morphologies of ZnO NSs are similar (Figure 1c). Only a slight variation in kinetics could be recorded between the two types of samples due to the better crystallinity after post-annealing at 350 °C during 30 min (diffractograms in Figure 1d).

Under natural solar light, there is no real difference between photolysis and photocatalysis for MB removal due to its fairly easy degradation under illumination. The study of AR14 degradation reveals that the annealed sample performed slightly better than the as grown sample. The degradation rates under solar illumination are overall much lower than those under UV lamp illumination because natural solar light contains much less UV (~5%).

The ease of scale-up of this synthesis is already proved (Figure 2) and the suitability of ZnO/Zn(AG) to clean-up runoff water is promising given the stability of the photocatalytic activity over several water cycles (Figure 4). It is mentioned in the literature that after long rainfall (which corresponds to the case of successive cycles in our study), the photocatalytic surface tends to be saturated and a limit in the purification capacity can be reached [7]. Under these conditions, incomplete photodegradation leads to the adsorption of by-product on the ZnO surface, which results in an inhibition of the photocatalyst surface by blocking all active sites. Nevertheless, this was not the case in this work, even in the absence of sample regeneration. Comparing the degradation rates by photocatalysis and photolysis, it is possible to observe a small gain related to the presence of ZnO NSs under this condition, which could be largely improved by increasing the residence time under irradiation, but it will be against the real working conditions of the Zn roof application.

The photocatalytic degradation rate depends largely on the experimental conditions and the type of photocatalyst. Nevertheless, it is therefore important to put our results into perspective. An efficiency close to 70% for the degradation of MB (10 mM) was obtained with ZnO nanorods substrate (3 cm × 1 cm) in a PMMA cuvette under visible light irradiation (AM 1.5G, 1 kW·m^2) from a solar simulator (SS1.6 kW from science tech, Canada) after 90 min [29]. Quartz cuvettes with 10 mL of MB (10 mM) and various ZnO NSs under four visible light lamps (20 W each) lead to a maximal degradation rate of 36.3% [30]. In borosilicate beakers, under solar light, MB at 100 μM and pH 3 was photodegraded in the presence of TiO_2 (0.5 g/L) and H_2O_2 (10 mM) at 92% after 3 h [31]. 48 mg of $ZnCo_2O_4/Co_3O_4$ was poured in AR14 solution (1.2 mg) leading to a photodegradation rate from 29% to 95% under solar illumination (125 W) depending on the photocatalyst synthesis method used [32]. TiO_2 nanopowders demonstrated a photodegradation rate range from 90% to 100% in 60 min under simulated solar light (7.7 μW/cm^2) for AR14 at 20 ppm [33].

4. Materials and Methods

4.1. ZnO Nanostructures Based Zinc Roof Synthesis and Characterization

Flattened Zn wire (2.5 cm × 0.5 cm) and Zn roof sheet (5 cm × 5 cm) were used as substrate for ZnO NSs growth by one-stage seedless hydrothermal synthesis. The samples were functionalized by the two different routes described below, before being carefully characterized by UV–visible spectroscopy (Maya2000 Pro from Ocean Optics,

IDIL, Orsay, France) and by field-emission scanning electron microscopy (SEM FEG ZEISS Merlin, Oberkochen, Germany) to determine their bandgap value and to investigate the morphology of the obtained ZnO nanostructures. Samples were also characterized by X-ray diffraction (Bruker Advance, Champs-sur-Marne, France) using a cobalt anticathode (λ = 1.78897 Å, 35 kV, 40 mA) and by photoluminescence (He-Cd Laser, IK Series, KIMMON, Laser Components, Meudon, France, λ = 325 nm, P = 10 mW).

4.1.1. ZnO/Zn Wires Production

Without any special washing process, the Zn wire substrates were directly functionalized by a one-step seedless hydrothermal route in a small sealed autoclave containing 50 mL of equimolar solution of hexamethylenetetramine (HMTA, $\geq$99%, VWR, CAS-No 100-97-0) and zinc nitrate hexahydrate ($Zn(NO_3)_2 \cdot 6H_2O$, 98%, Sigma-Aldrich, CAS-No 10196-18-6) at 25 mM and 90 °C for 4 h (Figure 1a). The flattened Zn wire substrate was kept vertically in the growth precursor solution during the growth duration. The Zn flat wire deposited was removed from the hydrothermal growth solution and thoroughly washed with distilled water before to be dried by a hot air flow (~30 s at ~53 °C). As-grown ZnO NSs were produced and compared to ZnO NSs post-annealed samples for 30 min at 350 °C. Samples were named ZnO/Zn(AG) and ZnO/Zn for as grown and post-annealed samples, respectively.

4.1.2. ZnO/Zn Sheets Production

The Zn roof sheet (5 cm × 5 cm) substrates were directly, without any washing process, functionalized by a one-step seedless hydrothermal route in sealed autoclave containing 300 mL of equimolar solution of HMTA and $Zn(NO_3)_2$ at 25 mM and 90 °C for 4 h (Figure 2a). The Zn roof sheet substrate was kept horizontally in the growth precursor solution during the growth duration. After 4 h, the Zn flat sheet functionalized was removed from the hydrothermal growth solution and thoroughly washed with DI water before to be dried by a hot air flow (~30 s at ~53 °C). Samples were named ZnO/Zn(AG).

4.2. ZnO/Zn Photocatalytic Activity Evaluation

4.2.1. Photodegradation under UV-Light

Photodegradation of MB was carried out under UV light irradiation (Hamamatsu LC-8, ~365 nm, $I_{received}$ by the sample ~35 mW/cm^2) with the aim to evaluate the best photocatalytic wire samples. Samples were immerged in 30 mL of an MB in DI water solution with an initial concentration of 10 µM placed under the UV source. After the light and the agitation were turned on, the photocatalysis process was monitored by UV–visible spectrophotometry (Perkin Elmer, Lambda 35) every 15 min for 3 h and the degradation efficiency X(%) was estimated using Equation (9):

$$X(\%) = \left(\frac{A_0 - A}{A_0} \right) \times 100 \tag{9}$$

where A_0 and A, respectively, stand for the initial and actual absorption peak values at the wavelength of the maximum absorption of MB (λ_{max} = 664 nm).

4.2.2. Photodegradation under Natural Solar Light

To prove the relevance of samples to answer to rainwater pollution under natural solar light, MB photocatalysis under natural solar light was carried out with ZnO NSs grown onto zinc wire samples. Experiments were realized at ambient temperature between April and May in Paris (France). Samples were immerged in 30 mL of an MB in DI water solution with an initial concentration of 10 µM. Every time, a beaker was filled with the dye solution and left without sample, to serve as a reference. The beakers were placed under natural solar light with an UV light intensity measured between 2.3 and 2.9 mW/cm^2, according to the weather. The photocatalytic process was monitored by UV–visible spectrophotometry every 30 min for 4 h. To confirm the sample efficiency in conditions closed to the real weather, the same experiments were carried out with a MB and AR14 solution in collected rainwater (pH ~5.2) instead in DI water. The photodegradation efficiency X(%) was calculated using Equation (9) with λ_{max} equal to 664 nm for MB and 515 nm for AR14.

4.2.3. Polluted Runoff Water Photodegradation onto ZnO/Zn Sheets under Solar Light

To study the ZnO/Zn(AG) roof sheet photocatalytic efficiency in reproducible conditions, photodegradation of AR14 was carried out at the laboratory under an artificial solar lamp (Sirius-300 PU, Zolix, Beijing, China, 300 W, 320–2500 nm). The sample was placed under the solar light and a flow of polluted rainwater with AR14 at 10 µM has been imposed to simulate the runoff water comportment. The photocatalytic process was followed after one-pass of 10 mL of polluted rainwater with an imposed flow rate of 0.17–1.03 mL/min. Then, Equation (9) was used to calculate X(%). Between each experiment, called cycle, the sample was rinsed with DI water and replaced under the solar lamp for 15 min. After four cycles, the sample was used for two cycles more of water depollution without solar light irradiation of 15 min and with only DI water rinsing. The results showed a promising potential application of the functionalized Zn roof for water purification. By analogy with our previous work on ZnO NSs based photocatalytic construction materials [14,27], this kind of photocatalytic roof will be also performant for air purification.

5. Conclusions

Some building structures, especially roofs, have the potential to be used as very large photocatalytic reactors, exploiting sunlight to reduce urban pollution. In this context, Zn photocatalytic roof surfaces have been developed by a rapid one-step hydrothermal synthesis. This growth route is particularly attractive and efficient for growing ZnO NSs on a Zn surface, due to its cost effectiveness, ability to treat large areas, and ease of controlling chemical conditions. The resulting nanostructures are homogeneously distributed over the surface and exhibit good crystallinity. The post-annealing improved the ZnO NS crystallinity. The ZnO NSs have excellent photocatalytic properties under UV illumination for the degradation of MB. The photocatalytic activity of the functionalized ZnO/Zn has been demonstrated under solar light for the degradation of AR14 which is known to be less sensitive to photocatalysis under solar light. Finally, a runoff water purification experiment showed a real potential for application that can offer other environmental benefits, such as the removal of atmospheric pollutants, e.g., nitrogen oxides or organic volatile compounds.

Nevertheless, it is important to keep in mind that this study is still a primarily investigation. Therefore, more research work should be performed prior to using these functionalized roofs. The ZnO NSs should be tested for resistance in the acidic pH of water, dust deposition, and other climate factors. The Brunauer–Emmet–Teller (BET) method could be used to determine the specific surface area of the ZnO NSs. The photocatalytic activity of the aged samples should be evaluated in more depth and with real pollutants from rainwater.

Author Contributions: Manipulation realization, M.L.P. and A.P., conceptualization, Y.L.-W. and M.L.P.; methodology, M.L.P. and Y.L.-W.; validation, M.L.P. and Y.L.-W.; formal analysis, M.L.P., A.P., Y.L.-W., S.B. and M.D.; investigation, M.L.P. and Y.L.-W.; writing—original draft preparation, M.L.P. and Y.L.-W.; writing—review and editing, Y.L.-W., M.L.P., S.B. and M.D.; supervision, Y.L.-W. All authors have read and agreed to the published version of the manuscript.

Funding: This research received no external funding.

Acknowledgments: The authors want to thank M. Nicolas Hautiere from Gustave Eiffel University—COSYS Department for his advice to develop zinc-depolluting roof and his support.

Conflicts of Interest: The authors declare no conflict of interest.

References

1. Khayan, K.; Husodo, A.H.; Astuti, I.; Sudarmadji, S.; Djohan, T.S. Rainwater as a source of drinking water: Health impacts and rainwater treatment. *J. Environ. Public Health* **2019**, *2019*, 1760950. [CrossRef] [PubMed]
2. Fernando, L.T.D.; Ray, S.; Simpson, C.M.; Gommans, L.; Morrison, S. Remediation of fouling on painted steel roofing via solar energy assisted photocatalytic self-cleaning technology: Recent developments and future perspectives. *Adv. Eng. Mater.* **2022**, *24*, 2101486. [CrossRef]
3. Motamedi, M.; Yerushalmi, L.; Haghighat, F.; Chen, Z. Recent developments in photocatalysis of industrial effluents: A review and example of phenolic compounds degradation. *Chemosphere* **2022**, *296*, 133688. [CrossRef] [PubMed]
4. Marien, C.B.D.; Le Pivert, M.; Azaïs, A.; M'Bra, I.C.; Drogui, P.; Diarany, A.; Robert, D. Kinetics and mechanism of Paraquat's degradation: UV-C photolysis vs UV-C photocatalysis with TiO_2/SiC foams. *J. Hazard. Mater.* **2019**, *370*, 164–171. [CrossRef] [PubMed]
5. Tang, X.; Ughetta, L.; Shannon, S.K.; Houzé de L'aulnoit, S.; Hen, S.; Gould, R.A.T.; Russel, M.L.; Zhang, J.; Ban-Weiss, G.; Everman, R.L.A.; et al. De-pollution efficacy of photocatalytic roofing granules. *Build. Environ.* **2019**, *160*, 106058. [CrossRef]
6. Singh, V.P.; Sandeep, K.; Kushwaha, H.S.; Powar, S.; Vaish, R. Photocatalytic, hydrophobic and antimicrobial characteristics of ZnO nano needle embedded cement composites. *Construct. Build. Mater.* **2018**, *185*, 285–294. [CrossRef]
7. Zhang, X.; Li, H.; Harvey, J.T.; Liang, X.; Xie, N.; Jia, M. Purification effect on runoff pollution porous concrete with nano-TiO_2 photocatalytic coating. *Transport. Res. Part D* **2021**, *101*, 103101. [CrossRef]
8. Cerro-Prada, E.; Garcia-Salgado, S.; Quijano, M.A.; Varela, F. Controlled synthesis and microstructural properties of sol-gel TiO_2 nanoparticles for photocatalytic cement composites. *Nanomaterials* **2018**, *9*, 26. [CrossRef]
9. Bica, B.O.; Staub, J.V. Concrete blocks nano-modified with zinc oxide (ZnO) for photocatalytic paving: Performance comparison with tintanium dioxide (TiO_2). *Construct. Build. Mater.* **2020**, *252*, 119120. [CrossRef]
10. Liu, Y.; Kang, Z.H.; Shafiq, I.; Zapien, J.A.; Bello, I.; Zhang, W.J.; Lee, S.T. Synthesis, characterization, and photocatalytic application of different ZnO nanostructures in array configurations. *Cryst. Growth Des.* **2009**, *9*, 3222–3227. [CrossRef]
11. Asadi, S.; Hassan, M.M.; Kevern, J.T.; Rupnow, T.D. Development of photocatalytic pervious concrete pavement for air and storm water improvements. *J. Transport. Res. Board* **2012**, *2290*, 161–167. [CrossRef]
12. Guo, M.Z.; Ling, T.C.; Poon, C.S. Photocatalytic NO_x degradation of concrete surface layers intermixed and spray-coated with nano-TiO_2: Influence of experimental factors. *Cem. Concr. Compos.* **2017**, *83*, 279–289. [CrossRef]
13. Le Pivert, M.; Martin, N.; Leprince-Wang, Y. Hydrothermally grown ZnO nanostructures for water purification by photocatalysis. *Crystals* **2022**, *12*, 308. [CrossRef]
14. Le Pivert, M.; Zerelli, B.; Martin, N.; Capochihi-Gnambodoe, M.; Leprince-Wang, Y. Smart ZnO decorated optimized engineering materials for water purification under natural sunlight. *Const. Build. Mater.* **2020**, *257*, 119592. [CrossRef]
15. Baruah, S.; Dutta, J. Hydrothermal growth of ZnO nanostructures. *Sci. Technol. Adv. Mater.* **2009**, *10*, 013001. [CrossRef]
16. Demes, T.; Ternon, C.; Riassetto, D.; Stambouli, V.; Langlet, M. Comprehensive study of hydrothermally grown ZnO nanowires. *J. Mater. Sci.* **2016**, *51*, 10652–10661. [CrossRef]
17. Yu, Z.; Moussa, H.; Chouchene, B.; Schneider, R.; Wang, W.; Moliere, M.; Liao, H. Tunable morphologies of ZnO films via the solution precursor plasma spray process for improved photocatalytic degradation performance. *App. Surf. Sci.* **2018**, *455*, 970–979. [CrossRef]
18. Joshi, S.; Jones, L.A.; Sabri, Y.M.; Bhargava, S.K.; Sunkara, M.V.; Ippolito, S.J. Facile conversion of zinc hydroxide carbonate to Cao-ZnO for selective CO_2 detection. *J. Colloid Interface Sci.* **2020**, *558*, 310–322. [CrossRef] [PubMed]
19. Kumar, V.; Prakash, J.; Singh, J.P.; Chae, K.H.; Swart, C.; Ntwaeaborwa, O.M.; Swart, H.C.; Dutta, V. Role of silver doping on the defects related photoluminescence and antibacterial behaviour of zinc oxide nanoparticles. *Colloids Surf. B Biointerfaces* **2017**, *159*, 191–199. [CrossRef]
20. Chevalier-César, C.; Capochichi-Gnambodoe, M.; Lin, F.; Yu, D.; Leprince-Wang, Y. Effect of growth time and annealing on the structural defect concentration of hydrothermally grown ZnO nanowires. *AINS Mater. Sci.* **2016**, *3*, 562–572. [CrossRef]
21. Sun, Y.; Fuge, G.M.; Fox, N.A.; Riley, D.J.; Ashfold, M.N.R. Synthesis of Aligned Arrays of Ultrathin ZnO Nanotubes on a Si wafer coated with a thin ZnO film. *Adv. Mater.* **2005**, *85*, 2477–2481. [CrossRef]
22. Sun, Y.; Ndifor-Angwafor, G.N.; Riley, D.J.; Ashfold, M.N.R. Synthesis and photoluminescence of ultra-thin ZnO nanowire/nanotube arrays formed by hydrothermal growth. *Chem. Phys. Lett.* **2006**, *431*, 352–357. [CrossRef]
23. Sheng, Y.; Jiang, Y.; Lan, X.; Wang, C.; Li, S.; Liu, X.; Zhong, H. Mechanism and growth of flexible ZnO nanostructure arrays in a facile controlled way. *J. Nanomater.* **2011**, *2011*, 473629. [CrossRef]
24. Liu, B.; Zeng, H.C. Fabrication of ZnO "dandelions" via a modified kirkendall process. *J. AM. Chem. Soc.* **2004**, *126*, 16744–16746. [CrossRef]
25. Tak, Y.; Yong, K. Controlled growth of well-aligned ZnO nanorod array using a novel solution method. *J. Phys. Chem. B* **2005**, *109*, 19263–19269. [CrossRef]
26. Jung, S.H.; Oh, E.; Lee, K.-H.; Park, W.; Jeong, S.H. A sonochemical method for fabrication aligned ZnO nanorods. *Adv. Mater.* **2007**, *19*, 749–753. [CrossRef]
27. Le Pivert, M.; Kerivel, O.; Zerelli, B.; Leprince-Wang, Y. ZnO nanostructures based innovative photocatalytic road for air purification. *J. Clean. Prod.* **2021**, *318*, 128447. [CrossRef]

28. Cho, S.; Kim, S.; Jang, J.W.; Jung, S.H.; Oh, E.; Lee, B.R.; Lee, K.H. Large-scale fabrication of sub-20-nm-diameter ZnO nanorod arrays at room temperature and their photocatalytic activity. *J. Phys. Chem.* **2009**, *113*, 10452–10458. [CrossRef]
29. Bora, T.; Sathe, P.; Laxman, K.; Dobrestov, S.; Dutta, J. Defect engineered visible light active ZnO nanorods for photocatalytic treatment of water. *Catal. Today* **2017**, *284*, 11–18. [CrossRef]
30. Khoa, N.T.; Kim, S.W.; Thuan, D.V.; Yoo, D.H.; Kim, E.J.; Han, S.H. Hydrothermally controlled ZnO nanosheet self-assembled hollow spheres/hierarchical aggreagates and their photocatalytic activities. *CrystEngComm.* **2014**, *16*, 1344–1350. [CrossRef]
31. Akbal, F. Photocatalytic degradation of organic dyes in the presence of titanium dioxide under UV and solar light: Effect of operational parameters. *Environ. Prog.* **2005**, *24*, 317–322. [CrossRef]
32. Zinatloo-Ajabshir, S.; Heidari-Asil, S.A.; Salavati-Niasari, M. Recyclable magnetic ZnC2O4-based ceramic nanostructure materials fabricated by simple sonochemical route for effective sunlight-driven photocatalytic degradation of organic pollution. *Ceram. Intern.* **2021**, *47*, 8959–8972. [CrossRef]
33. Miao, J.; Lu, H.B.; Khiadani, D.; Kiadani, M.H.; Zhang, L.C. Photocatalytic degradatation of the Azo Dye Acid Red 14 in nanosized TiO2 suspension under simulated solar light. *Clean Soil Air Water* **2015**, *43*, 1037–1043. [CrossRef]

Article

The Effect of Arsenic on the Photocatalytic Removal of Methyl Tet Butyl Ether (MTBE) Using Fe_2O_3/MgO Catalyst, Modeling, and Process Optimization

Akbar Mehdizadeh [1], Zahra Derakhshan [2], Fariba Abbasi [1], Mohammad Reza Samaei [1,*], Mohammad Ali Baghapour [1], Mohammad Hoseini [1], Eder Claudio Lima [3] and Muhammad Bilal [4,*]

[1] Department of Environmental Health Engineering, School of Health, Shiraz University of Medical Sciences, Shiraz 7153675541, Iran
[2] Research Center for Health Sciences, Department of Environmental Health, School of Health, Shiraz University of Medical Sciences, Shiraz 7153675541, Iran
[3] Institute of Chemistry, Federal University of Rio Grande do Sul (UFRGS), Porto Alegre 91501-970, Brazil
[4] School of Life Science and Food Engineering, Huaiyin Institute of Technology, Huai'an 223003, China
* Correspondence: mrsamaei@sums.ac.ir (M.R.S.); bilaluaf@hotmail.com (M.B.)

Citation: Mehdizadeh, A.; Derakhshan, Z.; Abbasi, F.; Samaei, M.R.; Baghapour, M.A.; Hoseini, M.; Lima, E.C.; Bilal, M. The Effect of Arsenic on the Photocatalytic Removal of Methyl Tet Butyl Ether (MTBE) Using Fe_2O_3/MgO Catalyst, Modeling, and Process Optimization. *Catalysts* 2022, 12, 927. https://doi.org/10.3390/catal12080927

Academic Editor: John Vakros

Received: 29 June 2022
Accepted: 18 August 2022
Published: 22 August 2022

Publisher's Note: MDPI stays neutral with regard to jurisdictional claims in published maps and institutional affiliations.

Copyright: © 2022 by the authors. Licensee MDPI, Basel, Switzerland. This article is an open access article distributed under the terms and conditions of the Creative Commons Attribution (CC BY) license (https://creativecommons.org/licenses/by/4.0/).

Abstract: MTBE is an aliphatic matter successfully removed from contaminated water by an advanced oxidation process. Additionally, arsenic is a toxic metalloid that is detected in some water supplies, such as in Iran. Concerning the oxidation potential of arsenic in an aqueous solution, it is expected that its interference in the photocatalytic removal of organic matter includes MTBE. Nevertheless, there is a lack of observation of this effect. In this study, the effect of arsenic on the photocatalytic removal of MTBE using an Fe_2O_3/MgO catalyst under UV radiation was investigated. Using an experimental design, modeling, and optimizing operational parameters, such as the arsenic and MTBE concentrations, catalyst dosage, pH, and reaction time, were studied. The synthesized nanocatalyst had a uniform and spherical morphological structure and contained 33.06% Fe_2O_3 and 45.06% MgO. The results indicate that the best model is related to the quadratic (*p*-value < 0.0001, R^2 = 0.97) and that the effect of the MTBE concentration is greater than the others. The highest removal efficiency was taken in an initial concentration of 37.5 mg/L MTBE, 1.58 mg/L Fe_2O_3/MgO, pH 5, and a reaction time of 21.41 min without any As. The removal efficiency was negatively correlated with the initial MTBE concentration and pH, but it was positively associated with the Fe_2O_3/MgO dosage and reaction time. Finally, the presence of arsenic decreased the removal efficiency remarkably (90.90% As = 0.25 μg/L and 61% As = 500 μg/L). Consequently, MTBE was removed by the photocatalytic process caused by Fe_2O_3/MgO, but the presence of arsenic was introduced as a limiting factor. Therefore, pretreatment for the removal of arsenic and more details of this interference effect are suggested.

Keywords: MTBE; arsenic; Fe_2O_3/MgO catalyst; AOP; modeling; optimization

1. Introduction

Water pollution due to petroleum compounds such as MTBE is one of the main problems in water resources [1,2]. MTBE is classified as an oxygenated compound added to gasoline to restore combustion [3]. MTBE has expanded as an octane booster and anti-knocking agent in the past two decades due to its low cost and easy usage, transition, and distribution. MTBE, $(CH_3)_3COCH_3$, is a volatile, colorless, and flammable liquid, and it is soluble in water (48 g/L at 25 °C). However, it causes environmental and health effects (IARC), such as nervous system reactions, dizziness, distraction, nausea, and forgetfulness [4,5]. Moreover, it has a low-threshold taste and odor (40 and 15 μg/L, respectively) [6]. The maximum contaminant level of MTBE in drinking water has been determined to be

20–40 µg/L by the USEPA. MTBE enters groundwater in various ways, such as leakage from road accidents, underground storage tanks (USTs), and pipelines [7,8].

The effective removal of MTBE from water is essential. Many attempts have been made over the past decades to eliminate MTBE from groundwater. However, MTBE removal using conventional water treatment processes is difficult due to its high solubility, low tendency to the soil, resistance to degradation, the presence of ether bands, and its resistant carbon chain [9–13]. The most commonly used processes for MTBE removal are adsorption using activated carbon, air stripping, and advanced oxidation processes (AOPs). According to previous studies, AOPs, especially photocatalytic oxidation, are the most effective for MTBE decomposition [14–18]. AOPs can remove MTBE effectively, but their efficiency depends on the water-quality characteristics and the level of other contaminants, such as metals [19–21], because the oxidation of metals during AOPs might decrease the efficiency of MTBE removal.

Arsenic is a highly toxic metalloid, and its presence in water is a global problem. Elevated concentrations of arsenic have been detected in some parts of the world, including Bangladesh; India; some South American countries, such as Argentina; and some areas of Kurdistan and Hashtrood in Iran [22]. Arsenic leaches into water from natural and anthropogenic sources, including biological activities, volcanic releases, and human activities, such as the use of pesticides and minerals. Therefore, investigating the removal efficiency of organic matter in arsenic-contaminated water is essential. As(III), oxidized to As(V) during photocatalytic oxidation [23], can be considered a competitor for the removal of organic matter (such as MTBE). One suggested solution is a combination process with a high oxidation capacity, such as the photocatalytic process. The initial energy obtained from the photo resource and the activation energy is decreased using catalysts in the photocatalytic process.

In this study, Fe_2O_3/MgO was used as a catalyst because it is a porous structure with a high capacity for adsorbing organic matter [24,25] and generating free radicals [26]. Moreover, its band gap is 5.4–5.45 ev, so the electron in the band layer excites the current layer [24]. Therefore, its photocatalytic effect is performed when it is exposed to UV radiation. In other words, MgO doped with a metal oxide improves photocatalytic activity [27,28]. So, the removal efficiency of organic dye using Fe_2O_3-doped MgO is estimated to be above 92% after five regenerations [29] therefore, the combination of MgO and Fe_2O_3 can improve the removal efficiency of organic matter, such as MTBE, which is used for the first time in this study. In an earlier study, the photocatalytic process was used for MTBE removal [30], but there was less attention paid to the effect of metal, especially arsenic, on the MTBE's removal efficiency. For example, Zhang et al. (2018) reported that nickel decreased the adsorption capacity of MTBE using ZSM-5 zeolite [6]. However, for the first time, this study focuses on the decreasing trend in the photocatalytic oxidation of MTBE using Fe_2O_3/MgO due to arsenic.

Arsenic, as a semimetal, can participate in electron-hole processes in aqueous solutions so that it turns into arsenic with an oxidation number of five during the redox reaction. Since the photocatalytic process is a suitable method for mineralizing organic compounds such as MTBE, the released radicals ultimately cause water treatment. Therefore, the paths for both arsenic deposition and photocatalytic reactions are the same, which greatly increases the possibility of interference. Numerous studies have examined the effect of some metals on interference in the photocatalytic process, including the formation, recombination, and formation rate of radicals. However, these effects have been less investigated when considering semimetals, especially arsenic. Due to the existence of a variety of arsenic concentrations in the water resources of Iran and other regions of the world; the high potential of semiconductors for active participation in electron transfer processes; and the strong tendency for arsenic to change its oxidation number and phase from soluble to insoluble forms in aqueous solutions exposed to the photocatalytic process, the present study aims to model and determine the interference effect on a laboratory scale for the first time in order to consider the type and amount of interference effect in the treatment of water resources using the photocatalytic process on a real scale and optimize the parameters accordingly.

Arsenic is a toxic metalloid detected in some water supplies, such as in Iran. Concerning the oxidation potential of arsenic in an aqueous solution, it is expected that its interference in the photocatalytic removal of organic matter includes MTBE. Regarding the significant concern about MTBE, the high concentrations of arsenic that have been reported in some areas of Iran, and its interference effect on the mineralization and degradation of MTBE, as well as the lack of documents, imply the effect of metals on the removal efficiency of organic matter. Therefore, this study aims to investigate the effect of arsenic on the photocatalytic oxidation process of MTBE using Fe_2O_3/MgO as a catalyst. Moreover, the interactions between the effective factors and the optimization of the process were determined by the response surface methodology (RSM) method.

2. Results
2.1. Specification of the Catalyst
2.1.1. SEM–EDX Analysis

SEM determined the microstructure and morphology of the catalyst, and its images are shown in Figure 1.

The synthesized catalyst has a uniform and aggregated structure, making separation from the liquid effluent easier after the photocatalytic process [31]. In addition, it has good dispersion, and the sample has a diameter of approximately 2 µm. In Figure 1b, the elements' quantitative levels in the nanocatalyst were determined using EDX, confirming the presence of the elements MgO (45.06%) and Fe_2O_3 (33.06%) in the synthesized catalyst.

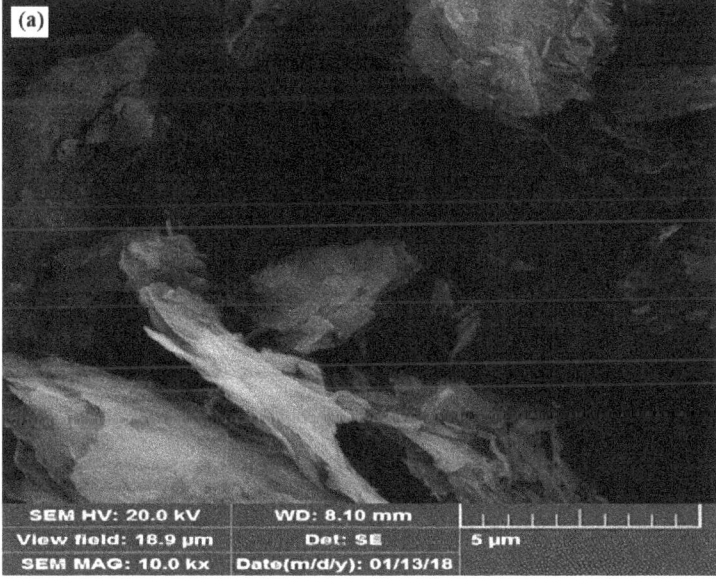

Figure 1. *Cont.*

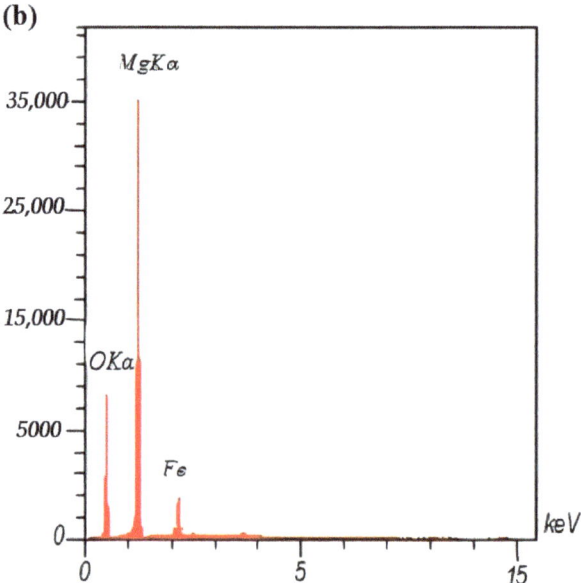

Figure 1. SEM (**a**) and EDX (**b**) images for the nanocatalyst.

2.1.2. XRD Analysis

The XRD analysis (10–80° 2θ) of the nanocatalyst is shown in Figure A1. The average crystal size of the sample was calculated using the Scherrer equation: $(d = K\lambda/(\beta \cos \theta))$ The strong and sharp peaks in Figure A1 suggest that the nanocatalyst has a crystalline structure. The values for Fe_2O_3 and MgO were 91.1 and 27.9 nm, respectively. As shown in the figure, the diffraction peaks of both Fe_2O_3 and MgO could be observed. The blue peaks are associated with Fe_2O_3, and the red peaks are related to MgO. This nanocatalyst had two peaks at 33.06° and 36.5° that can be associated with Fe_2O_3 and MgO, respectively. According to Figure A1, the peaks related to Fe_2O_3 are wholly covered over by MgO's peaks, indicating the high precision of the sol–gel process.

2.2. Photocatalytic Removal of MTBE

The interactions between independent and dependent variables while removing MTBE were investigated in DE7.

The effect of the Fe_2O_3/MgO nanocatalyst dosage, pH, and initial MTBE concentrations on the photocatalytic removal efficiency and optimization of these variables was investigated by Factorial Experimental Design. The MTBE's lowest, highest, and average removal efficiencies were 61%, 90.90% and 73%, respectively. The ANOVA results for the proposed model for the photocatalytic removal of MTBE based on the experiment's results and the effect of the parameters are presented in Table 1. The F-value and p-value for MTBE were 835.92 and >0.0001, respectively. Thus, the p-value of the investigated parameters ($p \leq 0.05$) shows that the concentration of MTBE and the concentration of catalyst are effective parameters for the photocatalytic oxidation of MTBE (Table 1).

In the model, the F-value also shows that the defined parameters in the models have significant effects, and the effect of the initial MTBE concentration was more pronounced than other parameters. However, only a 0.01% chance of the occurrence of the F-value was due to noise. The effects of the significant parameters based on the suggested model are presented in Equation (1) and Figure A2, which are used for predicting the photocatalytic oxidation values of MTBE.

$$\text{MTBE} = 18.70 + 11.32^a - 0.89C + 1.03BC - 1.43DE - 1.38B^2 \tag{1}$$

The MTBE residual was increased based on the model equation, increasing the MTBE (A) concentration and decreasing the catalyst (C). Furthermore, the factors for interactions, such as concentration of arsenic × concentration of catalyst, which had a positive effect and the effect of pH × time, which had a negative effect on the response, and the response affected the quadratic of the arsenic concentration. Based on any of the coefficients, the effect of the catalyst was higher than other parameters.

Table 1. ANOVA results for the Response Surface Quadratic Model (MTBE).

Source	Sum of Squares	df	Mean Squares	F-Value	p-Value
Model	5624.73				
A—MTBE	5122.75	20	281.24	45.84	
B—Arsenic	15.89	1	5122.75	834.92	<0.0001
C—Catalyst	31.60	1	15.89	2.59	<0.0001
D—pH	5.23	1	31.60	5.15	0.1184
E—Time	9.19	1	5.23	0.85	0.0309
AB	6.53	1	9.19	1.50	0.3633
AC	9.26	1	6.53	1.06	0.2309
AD	0.043	1	9.26	1.51	0.3109
AE	8.09	1	0.043	6.972	0.2292
BC	33.97	1	8.09	1.32	0.9340
BD	4.34	1	33.97	5.54	0.2602
BE	0.10	1	4.34	0.71	0.0256
CD	23.27	1	0.10	0.016	0.4070
CE	8.85	1	23.27	3.79	0.8992
DE	65.24	1	8.85	1.44	0.0612
A2	14.39	1	65.24	10.63	0.2394
B2	60.78	1	14.39	2.35	0.0028
C2	183.26	1	60.78	9.91	0.1365
D2	21.88	1	183.26	29.87	0.0038
E2	0.078	1	21.88	3.57	<0.0001
Residual	177.93	1	0.078	0.013	0.0690
Lack of Fit	159.44	29	6.14		0.9111
Pure Error	18.49	22	7.25	2.74	
Cor Total	5802.67	7	2.64		0.0868
		49			

Moreover, the perturbation plot of independent variables on the photocatalytic oxidation of MTBE is shown in Figure A2.

According to Figure A2, the most influential factor was MTBE concentration. Moreover, the effect of arsenic concentration on the MTBE photocatalytic process was insignificant.

The proportion of predicted values of the photocatalytic oxidation efficiency of As(III) in the model and the actual values obtained from the experiments are shown in Figure 2.

The determination coefficient for the actual and predicted values of residual MTBE was acceptable (R^2 = 0.9693). The difference between the adjusted determination coefficient (Adj R^2) and the predicted determination coefficient (Pred R^2) should not be more than 0.2 [32]. Based on the results, this difference was 0.0498, indicating the model's adequacy. Furthermore, R^2 0.97 for MTBE shows that the model can explain about 97 data variations. The calculated lack of fit in these models is 2.74, which is not significant; therefore, the disordered data do not affect the model.

The Simultaneous Effect of the Concentration of Fe_2O_3/MgO and Initial MTBE

The efficiency of MTBE removal at various initial concentrations of MTBE and the Fe_2O_3/MgO catalyst is shown in Figure 3.

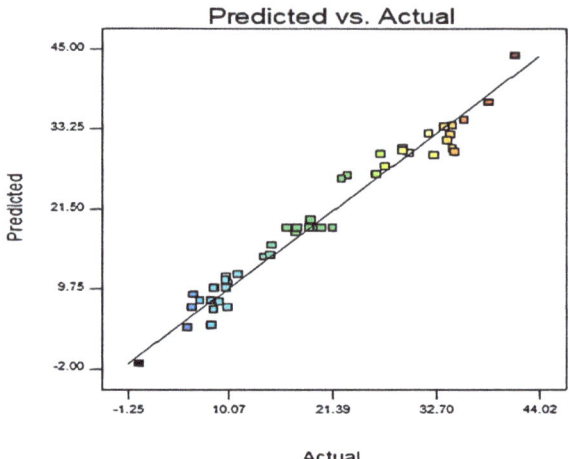

Figure 2. Predicted versus actual experimental results for residual MTBE.

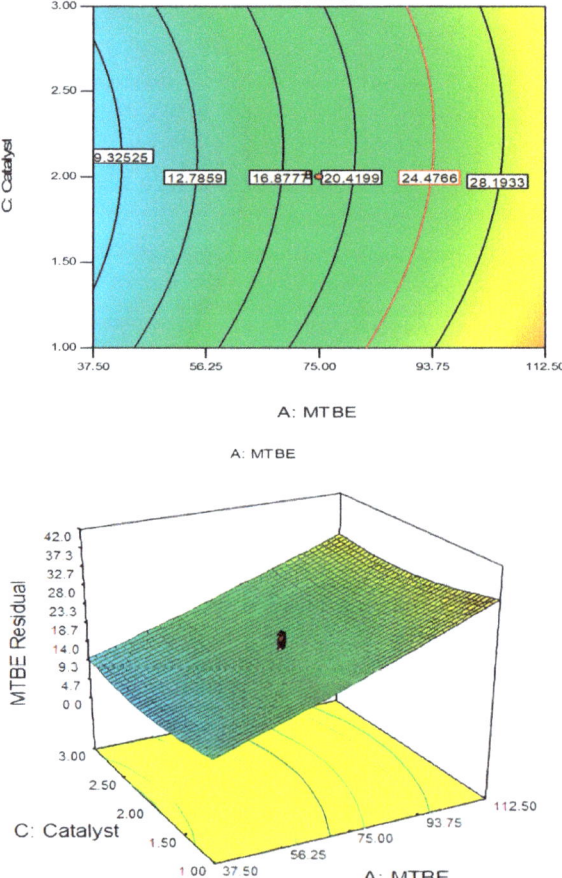

Figure 3. Contour and 3D Plots of the interactive effect of catalyst dosage (g/L) and MTBE (mg/L).

According to Figure 3, 1–2.5 g/L of catalyst can slightly increase MTBE's removal efficiency, while a catalyst concentration of more than 4 g/L had a negative effect. This is because the number of photons and adsorbed molecules in a high catalyst level also increases. This reaction increased the particle density in the brightness range and the amount of free radicals that improved the degradation of MTBE. However, a further increase in the catalyst dosage caused the reduction in active sites on the surface of Fe_2O_3/MgO. Therefore, the removal efficiency of MTBE was decreased [33,34]. Concerning this result, the best concentration of nanoparticles was 1.54 g/L, but the higher concentrations increased costs.

3. Materials and Methods

3.1. Materials

All the analytical-grade chemical materials, including MTBE, sodium arsenite ($NaAsO_2$), magnesium oxide (MgO), hematite (Fe_2O_3), sodium hydroxide (NaOH), hydrochloric acid (HCL), citric acid ($C_6H_8O_7$), acetone (C_3H_6O), hexane (C_6H_{14}), and ethanol (C_2H_6O), were purchased from Sigma Aldrich Company (St. Louis, MI, USA). NaOH and HCl were used for adjusting the pH. The double-distilled water (18 MΩ cm) was applied to prepare the solutions. A UV lamp (11W) was used for the light supply. A powder X-ray diffractometer (XRD) (Philips PW 1800, Round Rock, TX, USA) was used to determine the gel structure. Scanning electron microscopy (SEM) (Carl-Zeiss-Promenade, Jena, Germany) measurement (HV: 20.0 kV) was used to envisage the nanoparticles' surface morphology.

3.2. Preparation of the Catalyst

In a typical process, highly active MgO nanoparticles were prepared via a simple sol–gel method. First, the MgO nanoparticles were dissolved in ethanol at 60 °C for 5 h, and citric acid was added slowly to form the gelling solution by stirring. When the solution was completely dispersed, the compound was concentrated at 65 °C to form a wet gel. The wet gel was then dried at 110 °C for 12 h. Finally, the above-obtained product was calcinated at 550 °C with a heating rate of 4 °C min^{-1} for 4 h. The Fe_2O_3 solution was synthesized by the sol–gel method. This solution was joined to make the final coating with the MgO gel to obtain the final product of Fe_2O_3/MgO [30].

3.3. Stability and Reusability of the Catalyst

To examine the stability of the catalyst, 1 g/L of the catalyst was placed in the simulated conditions of the experiments, and then, the concentrations of iron and magnesium remaining in the solution were measured after 0–50 min as described in Table 2.

Table 2. Concentration of iron and magnesium remaining in the solution after 0–50 min.

Time (min)	0	10	20	30	40	50
Mg (mg/L)	0	2.02	3	3.9	4.3	6.1
Fe (mg/L)	0	5.7	12.6	17.2	21.4	26.3

The leakage rate of both magnesium and iron in the aqueous solution was less than 5% during the laboratory conditions, which was statistically confirmed.

After performing the photocatalytic process, the studied reactor was placed in the magnetic field to investigate the secondary application of the catalyst in laboratory conditions during the study period, the efficiency of which was favorable. After drying the catalyst at 70 °C, the collected catalyst was weighed on a scale, and the weight difference was estimated to be less than 15%. Moreover, the catalyst was photographed after the photocatalytic process, which did not differ significantly from the initial images. Thus, it can be said that the MTBE mineralization and removal operations were carried out under suitable conditions.

3.4. Sample Preparation

This study used a high-purity MTBE solution (99.9%). An arsenite stock solution (1000 mg/L) was prepared by dissolving potassium arsenite ($KAsO_2$) in distilled water (acid-distilled twice) with HCl (with a final concentration of 2.5%). The samples were stored in a refrigerator at 4 °C after preparation. NaOH and HCL were used to adjust the reaction mixture's pH.

3.5. Chemical Reactor and Optimization

The photocatalytic removal of MTBE was carried out in a laboratory-scale batch slurry reactor. The reactor comprised transparent Plexiglas with dimensions (length, width, and height) of 20 × 10 × 14 cm and 1 L effective volume. A UV lamp (11 W) was submerged in the solution to provide UV radiation for the reactor. First, the reactor was tightly closed and stirred completely. Then, the reactor was fixed and made in the two-seated form. The temperature of the reactor was controlled at 25–30 °C. Then, the solution was centrifuged (at 4000 rpm for 30 min) to separate the nanocatalyst.

The design of the experiments was performed based on a partial factorial method. Table 3 lists the values and variables used in the design of the experiments.

Table 3. Natural and coded levels of independent variables based on the central composite design.

Independent Variable	Symbol	Coded Level				
		−2	−1	0	1	2
				Natural Level		
MTBE (mg/L)	A	0	37.5	75	112.5	150
Arsenic (mg/L)	B	0	0.25	0.5	0.75	1
Catalyst (g/L)	C	0	1	2	3	4
pH	D	3	5	7	9	11
Time (min)	E	10	20	30	40	50

The values of the natural and coded independent variables were designed based on the central composite design (CCD). The order of each experiment was selected randomly by Design Expert 7 (DE7) software based on past studies and observed values in Iran's groundwaters. To more closely look at the effect of each variable on the process, one of the values of MTBE, the catalyst or arsenic, is considered zero in the DE7. Finally, 50 experiments were designed as Factorial Designs.

3.6. MTBE Extraction Method

MTBE was extracted using the dispersive liquid–liquid microextraction (DLLME) method. DLLME is a simple, fast, and sensitive method for extracting organic compounds from aqueous solutions [35]. First, 10 mL of the sample was added to a glass tube with a conical bottom centrifuge tube. The acetonitrile (1 mL) and hexane (0.2 mL) were used as dispersing and extraction solutions, respectively. Next, the mixture solution was centrifuged at 5000 rpm for 2 min. After this stage, hexane droplets were separated, removed with a 10 µL Hamilton syringe, and injected into the GC.

3.7. Data Analysis

Data analysis was performed using the RSREG method in DE7. The regression coefficients of the empirical data were generalized as a quadratic polynomial model [36]. This model is as follows in Equation (2):

$$Y = \beta_0 + \sum_{i=1}^{3} \beta_i X_i + \sum_{i=1}^{3} \beta_{ii} X_i^2 + \sum_{i<1=1}^{3} \beta_{ij} X_i X_j \tag{2}$$

Y = the amount of MTBE, β_0 = interactive regression coefficients, β_i = linear regression coefficients, β_{ii} = quadratic regression coefficients, β_{ij} = interactive regression coefficients, and X_i and X_j = dependent variables

ANOVA was used to determine the significance of the model and independent variables and determine the values of R^2, $R^2_{adjusted}$, and $R^2_{predicted}$. The p-value and lack of fit in the variance analysis table, the normal curves, and the predicted values versus the real values were used to check the model's desirability and data. The statistically significant level of the p-value was 0.05 [37]. The lack of fit value should not be meaningful (meaning any distorted data in the model). The contours and three-dimensional charts were plotted to analyze the removal rate of MTBE and the interactions of independent variables. Finally, the RSM-obtained optimal conditions for removing MTBE and arsenic were obtained using predicted equations.

4. Conclusions

Although the photocatalytic process is a suitable method for treating MTBE in an aqueous solution, the presence of some contaminants can be a limiting factor for efficient removal. Arsenic is a metalloid that oxides during the photocatalytic process and competes with MTBE for oxidation. In general, it can be concluded that the method used in this study is more suitable for lower concentrations of MTBE. Although less attention was paid to the effect of As on water treatment, this study showed that the presence of arsenic in water could limit the efficiency of removal of MTBE. Since there is arsenic in some water resources and MTBE leaching, pretreatment for the removal of arsenic is essential. The photocatalytic process is more cost-effective than other methods used to remove organic compounds, such as air stripping, GAC, incineration, and ozonation. This process generally destroys organic pollutants at an ambient temperature and pressure without any need for direct oxygen injection. However, this treatment method is applicable in conditions that are optimal in economic, environmental, and operational terms. The presence of some inhibitory and disruptive factors, such as arsenic, in the water environment could reduce the process efficiency. Therefore, it could be argued that arsenic separation and removal before the photocatalytic process is one of the best options for decreasing this inhibitory effect, which reduces downstream treatment costs.

In this study, the inhibitory effect of arsenic on the photocatalytic oxidation process of MTBE using Fe_2O_3/MgO was investigated. The Fe_2O_3/MgO nanocatalyst was successfully synthesized based on the SEM, EDX, and XRD analyses. The results showed that MTBE was removed successfully by photocatalytic oxidation and the best model was the quadratic model ($p < 0.0001$, $F = 45.84$, $R^2 = 0.9693$, R^2 adjusted = 0.9482, and R^2 predicted = 0.8984). According to the quadratic model and ANOVA analysis, the most effective factors were the initial concentration of MTBE and the catalyst dosage. The process optimization expressed that the efficiency of the MTBE oxidation process was improved by increasing both the catalyst dosage and time and decreasing MTBE and the pH. The predicted and experimental removal efficiencies were 90.90% and 88.73%, respectively, under the optimal conditions (initial concentration of MTBE = 37.5 mg/L, pH = 5, Catalyst = 1.54 g/L, and Time = 21.41 min). Although the process successfully removed MTBE, arsenic was introduced as an inhibitor factor. So, the highest inhibitory effect related to the initial arsenic is 0.5 mg/L. Concerning these results, the pretreatment of arsenic from water resources and more details about the inhibitory effect of arsenic on the photocatalytic removal of MTBE, such as the formation of hole-electrons, is essential. Moreover, further studies are suggested to determine the effect of other heavy metals and metalloids on organic matter removal efficiency.

Author Contributions: Conceptualization, A.M. and M.R.S.; methodology, F.A.; software, M.A.B. and A.M.; validation, M.H.; investigation, Z.D.; resources, E.C.L.; writing—review and editing, M.B. and E.C.L. All authors have read and agreed to the published version of the manuscript.

Funding: This research received no external funding.

Data Availability Statement: The data presented in this study are available on request from the corresponding author.

Acknowledgments: This article was extracted from a thesis written by Akbar Mehdizadeh (proposal No. 14140), which Shiraz University of Medical Sciences financially supported.

Conflicts of Interest: The authors declare no conflict of interest.

Appendix A

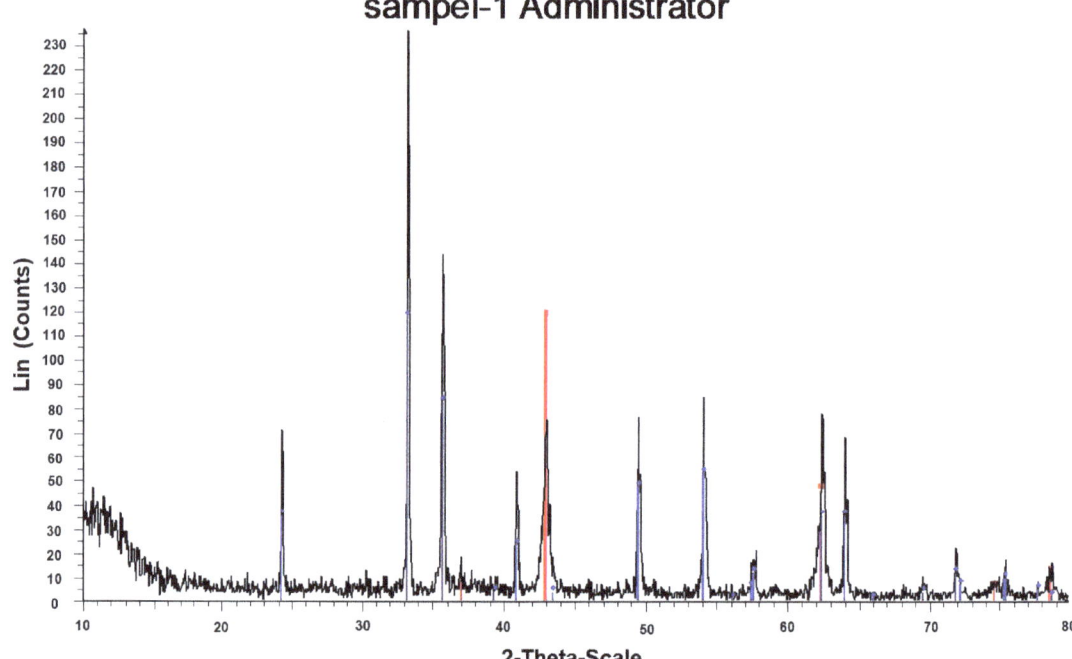

Figure A1. XRD patterns of Fe_2O_3 and MgO.

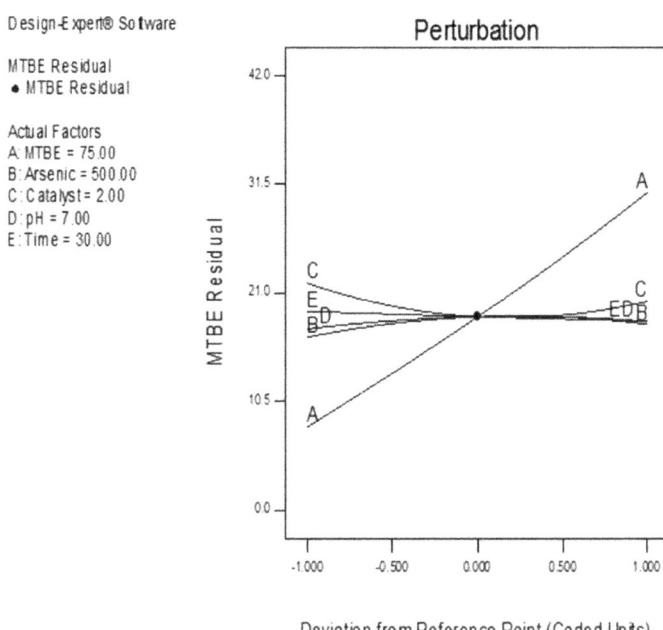

Figure A2. The perturbation Plot of independent variables in the photocatalytic oxidation of MTBE.

References

1. Jorfi, S.; Samaei, M.R.; Soltani, R.D.C.; Talaiekhozani, A.; Ahmadi, M.; Barzegar, G.; Reshadatian, N.; Mehrabi, N. Enhancement of the bioremediation of pyrene-contaminated soils using a hematite nanoparticle-based modified Fenton oxidation in a sequenced approach. *Soil Sediment Contam. Int. J.* **2017**, *26*, 141–156. [CrossRef]
2. Ekinci, E.K. Mesoporous magnesia sorbent for removal of organic contaminant methyl tert -butyl ether (MTBE) from water. *Sep. Sci. Technol.* **2022**, *57*, 843–853. [CrossRef]
3. Shamim, M.; Aalam, C.S.; Manivannan, D.; Kumar, S. Characterization of Gasoline Engine Using MTBE and DIE Additives. *Int. Res. J. Eng. Technol.* **2017**, *4*, 191–199.
4. International Agency for Research on Cancer. *IARC Monographs on the Evaluation of Carcinogenic Risks to Humans*; IARC: Lyon, France, 1994; pp. 389–433.
5. Mahmoodsaleh, F.; Ardakani, M.R. Methyl tertiary butyl ether biodegradation by the bacterial consortium isolated from petrochemical wastewater and contaminated soils of Imam Khomeini Port Petrochemical Company (Iran). *Bioremed. J.* **2021**, *26*, 127–137. [CrossRef]
6. Zhang, Y.; Jin, F.; Shen, Z.; Lynch, R.; Al-Tabbaa, A. Kinetic and equilibrium modelling of MTBE (methyl tert-butyl ether) adsorption on ZSM-5 zeolite: Batch and column studies. *J. Hazard. Mater.* **2018**, *347*, 461–469. [CrossRef]
7. Samaei, M.R.; Maleknia, H.; Azhdarpoor, A. A comparative study of removal of methyl tertiary-butyl ether (MTBE) from aquatic environments through advanced oxidation methods of $H_2O_2/nZVI$, $H_2O_2/nZVI$/ultrasound, and $H2O2/nZVI$/UV. *Desalination Water Treat.* **2016**, *57*, 21417–21427. [CrossRef]
8. Beryani, A.; Pardakhti, A.; Ardestani, M.; Zahed, M.A. Benzene and MTBE removal by Fenton's process using stabilized Nano Zero-Valent Iron particles. *J. Appl. Res. Water Wastewater* **2017**, *4*, 343–348.
9. Lindsey, D.; Ayotte, J.D.; Jurgens, B.C.; Desimone, L.A. Using groundwater age distributions to understand changes in methyl tert-butyl ether (MtBE) concentrations in ambient groundwater, northeastern United States. *Sci. Total Environ.* **2017**, *579*, 579–587. [CrossRef]
10. Khademi, S.M.S.; Tabrizchi, M.; Telgheder, U.; Valadbeigi, Y.; Ilbeigi, V. Determination of MTBE in drinking water using corona discharge ion mobility spectrometry. *Int. J. Ion Mobil. Spectrom.* **2017**, *20*, 15–21. [CrossRef]
11. Abbas, A.; Sallam, A.S.; Usman, A.R.A.; Al-Wabel, M.I. Organoclay-based nanoparticles from montmorillonite and natural clay deposits: Synthesis, characteristics, and application for MTBE removal. *Appl. Clay Sci.* **2017**, *142*, 21–29. [CrossRef]
12. Pirsaheb, M.; Dargahi, A.; Khamutian, R.; Asadi, F.; Atafar, Z. A survey of methyl tertiary butyl ether concentration in water resources and its control procedures. *J. Maz. Univ. Med. Sci.* **2014**, *24*, 119–128.
13. Iraji, G.; Givianrad, M.H.; Tehrani, M.S. Highly efficient degradation of MTBE by $\gamma\text{-}Al_2O_3/NiO/TiO_2$ core-shell nanocomposite under visible light irradiation. *Int. J. New Chem.* **2021**, *8*, 222–228. [CrossRef]

14. Andreozzi, R.; Caprio, V.; Insola, A.; Marotta, R. Advanced oxidation processes (AOP) for water purification and recovery. *Catal. Today* **1999**, *53*, 51–59. [CrossRef]
15. Tsimas, E.S.; Tyrovola, K.; Xekoukoulotakis, N.P.; Nikolaidis, N.P.; Diamadopoulos, E.; Mantzavinos, D. Simultaneous photocatalytic oxidation of As (III) and humic acid in aqueous TiO_2 suspensions. *J. Hazard. Mater.* **2009**, *169*, 376–385. [CrossRef]
16. López-Muñoz, M.J.; Arencibia, A.; Segura, Y.; Raez, J.M. Removal of As (III) from aqueous solutions through simultaneous photocatalytic oxidation and adsorption by TiO_2 and zero-valent iron. *Catal. Today* **2017**, *280*, 149–154. [CrossRef]
17. Barkoula, N.-M.; Alcock, B.; Cabrera, N.O.; Peijs, T. Fatigue properties of highly oriented polypropylene tapes and all-polypropylene composites. *Polym. Polym. Compos.* **2008**, *16*, 101–113. [CrossRef]
18. Pal, D.; Lavania, R.; Srivastava, P.; Singh, P.; Srivastava, K.; Madhav, S.; Mishra, P. Photocatalytic degradation of methyl tertiary butyl ether from wastewater using CuO/CeO_2 composite nanofiber catalyst. *J. Environ. Chem. Eng.* **2018**, *6*, 2577–2587. [CrossRef]
19. Tawabini, B.; Makkawi, M. Remediation of MTBE-contaminated groundwater by integrated circulation wells and advanced oxidation technologies. *Water Sci. Technol. Water Supply* **2018**, *18*, 399–407. [CrossRef]
20. Smedley, P.L.; Kinniburgh, D.G. Source and behaviour of arsenic in natural waters. In *United Nations Synthesis Report on Arsenic in Drinking Water*; World Health Organization: Geneva, Switzerland, 2001; pp. 1–61.
21. Ghayurdoost, F.; Assadi, A.; Mehrasbi, M.R. Removal of MTBE From Groundwater Using A PRB of ZVI/Sand Mixtures: Role of Nitrate And Hardness. *Res. Square* **2021**. [CrossRef]
22. Azhdarpoor; Nikmanesh, R.; Samaei, M.R. Removal of arsenic from aqueous solutions using waste iron columns inoculated with iron bacteria. *Environ. Technol* **2015**, *36*, 2525–2531. [CrossRef]
23. Eslami, H.; Ehrampoush, M.H.; Esmaeili, A.; Ebrahimi, A.A.; Salmani, M.H.; Ghaneian, M.T.; Falahzadeh, H. Efficient photocatalytic oxidation of arsenite from contaminated water by Fe_2O_3-Mn_2O_3 nanocomposite under UVA radiation and process optimization with experimental design. *Chemosphere* **2018**, *207*, 303–312. [CrossRef]
24. Balakrishnan, J.; Sreeshma, D.; Siddesh, B.M.; Jagtap, A.; Abhale, A.; Rao, K.K. Ternary alloyed HgCdTe nanocrystals for short-wave and mid-wave infrared region optoelectronic applications. *Nano Express* **2020**, *1*, 020015. [CrossRef]
25. Beltrán, D.E.; Uddin, A.; Xu, X.; Dunsmore, L.; Ding, S.; Xu, H.; Zhang, H.; Liu, S.; Wu, G.; Litster, S. Elucidation of Performance Recovery for Fe-Based Catalyst Cathodes in Fuel Cells. *Adv. Energy Sustain. Res.* **2021**, *2*, 2100123. [CrossRef]
26. Chan, S.H.S.; Wu, T.Y.; Juan, J.C.; Teh, C.Y. Recent developments of metal oxide semiconductors as photocatalysts in advanced oxidation processes (AOPs) for treatment of dye waste-water. *J. Chem. Technol. Biotechnol.* **2011**, *86*, 1130–1158. [CrossRef]
27. Sierra-Fernandez, A.; De la Rosa-García, S.C.; Yañez-Macías, R.; Guerrero-Sanchez, C.; Gomez-Villalba, L.S.; Gómez-Cornelio, S.; Rabanal, M.E.; Schubert, U.S.; Fort, R.; Quintana, P. Sol-gel synthesis of $Mg(OH)_2$ and $Ca(OH)_2$ nanoparticles: A comparative study of their antifungal activity in partially quaternized p(DMAEMA) nanocomposite films. *J. Sol-Gel Sci. Technol.* **2019**, *89*, 310–321. [CrossRef]
28. Popov, A.; Shirmane, L.; Pankratov, V.; Lushchik, A.; Kotlov, A.; Serga, V.; Kulikova, L.; Chikvaidze, G.; Zimmermann, J. Comparative study of the luminescence properties of macro- and nanocrystalline MgO using synchrotron radiation. *Nucl. Instrum. Methods Phys. Res. Sect. B Beam Interact. Mater. Atoms* **2013**, *310*, 23–26. [CrossRef]
29. Zheng, S.; Zhou, Q.; Chen, C.; Yang, F.; Cai, Z.; Li, D.; Geng, Q.; Feng, Y.; Wang, H. Role of extracellular polymeric substances on the behavior and toxicity of silver nanoparticles and ions to green algae Chlorella vulgaris. *Sci. Total Environ.* **2019**, *660*, 1182–1190. [CrossRef] [PubMed]
30. Aghamohammadi, S.; Haghighi, M.; Karimipour, S. A comparative synthesis and physicochemical characterizations of Ni/Al_2O_3–MgO nanocatalyst via sequential impregnation and sol-gel methods used for CO_2 reforming of methane. *J. Nanosci. Nanotechnol.* **2013**, *13*, 4872–4882. [CrossRef]
31. Berijani, S.; Assadi, Y.; Anbia, M.; Hosseini, M.-R.M.; Aghaee, E. Dispersive liquid-liquid microextraction combined with gas chromatography-flame photometric detection: A very simple, rapid and sensitive method for the determination of organophosphorus pesticides in water. *J. Chromatogr. A* **2006**, *1123*, 1–9. [CrossRef]
32. Khuri, I. *Response Surface Methodology and Related Topics*; World Scientific: Singapore, 2006.
33. Salmani, M.H.; Mokhtari, M.; Raeisi, Z.; Ehrampoush, M.H.; Sadeghian, H.A. Evaluation of removal efficiency of residual diclofenac in aqueous solution by nanocomposite tungsten-carbon using design of experiment. *Water Sci. Technol.* **2017**, *76*, 1466–1473. [CrossRef]
34. Habibi, M.H.; Mosavi, V. Synthesis and characterization of $Fe_2O_3/Mn_2O_3/FeMn_2O_4$ nanocomposite alloy coated glass for photocatalytic degradation of Reactive Blue 222. *J. Mater. Sci. Mater. Electron.* **2017**, *28*, 11078–11083. [CrossRef]
35. Akyol, A.; Can, O.T.; Demirbas, E.; Kobya, M. A comparative study of electrocoagulation and electro-Fenton for treatment of wastewater from the liquid organic fertilizer plant. *Sep. Purif. Technol.* **2013**, *112*, 11–19. [CrossRef]
36. Garcia, J.C.; Takashima, K. Photocatalytic degradation of imazaquin in an aqueous suspension of titanium dioxide. *J. Photochem. Photobiol. A Chem.* **2003**, *155*, 215–222. [CrossRef]
37. Nikazar, M.; Gholivand, K.; Mahanpoor, K. Photocatalytic degradation of azo dye Acid Red 114 in water with TiO_2 supported on clinoptilolite as a catalyst. *Desalination* **2008**, *219*, 293–300. [CrossRef]

Article

Photocatalytic Degradation of Pharmaceutical Amisulpride Using g-C₃N₄ Catalyst and UV-A Irradiation

Maria Antonopoulou [1,*], Maria Papadaki [2], Ilaeira Rapti [3] and Ioannis Konstantinou [3,4]

[1] Department of Sustainable Agriculture, University of Patras, GR-30100 Agrinio, Greece
[2] Department of Agriculture, University of Patras, Nea Ktiria, GR-30200 Messolonghi, Greece
[3] Department of Chemistry, University of Ioannina, GR-45110 Ioannina, Greece
[4] Institute of Environment and Sustainable Development, University Research Center (URCI), GR-45110 Ioannina, Greece
* Correspondence: mantonop@upatras.gr; Tel.: +30-264-107-4114

Abstract: In the present study, the photocatalytic degradation of amisulpride using g-C₃N₄ catalyst under UV-A irradiation was investigated. The photocatalytic process was evaluated in terms of its effectiveness to remove amisulpride from ultrapure and real municipal wastewater. High removal percentages were achieved in both aqueous matrices. However, a slower degradation rate was observed using wastewater as matrix that could be attributed to its complex chemical composition. The transformation products (TPs) were identified with liquid chromatography–mass spectrometry (LC–MS) in both ultrapure and real municipal wastewater. Based on the identified TPs, the photocatalytic degradation pathways of amisulpride are proposed which include mainly oxidation, dealkylation, and cleavage of the methoxy group. Moreover, the contribution of reactive species to the degradation mechanism was studied using well-documented scavengers, and the significant role of h^+ and $O_2^{\bullet-}$ in the reaction mechanism was proved. The evolution of ecotoxicity was also estimated using microalgae *Chlorococcum* sp. and *Dunaliella tertiolecta*. Low toxicity was observed during the overall process without the formation of toxic TPs when ultrapure water was used as matrix. In the case of real municipal wastewater, an increased toxicity was observed at the beginning of the process which is attributed to the composition of the matrix. The application of heterogeneous photocatalysis reduced the toxicity, and almost complete detoxification was achieved at the end of the process. Our results are in accordance with literature data that reported that heterogeneous photocatalysis is effective for the removal of amisulpride from aqueous matrices.

Keywords: photocatalysis; g-C₃N₄; pharmaceuticals; amisulpride; mechanism; reactive species; transformation products; ecotoxicity

1. Introduction

Aquatic pollution derived from chemicals that are generated during various anthropogenic activities is an environmental issue of considerable importance. The increasing use of chemicals due to the current model of life and the continued growth of the population is expected to cause greater pressure on natural ecosystems and humans in the near future. Pollution caused by a multitude of pharmaceutical compounds used by humans is an environmental problem that has attracted scientific interest. Pharmaceutical substances are a large and diverse group, usually of organic compounds, which are used in high quantities throughout the world. They are also considered to be a unique class of pollutants because of their characteristics. In many cases, their behavior and fate cannot be simulated with other categories of organic pollutants [1,2].

The main source of pharmaceutical residues in the aquatic environment is wastewater treatment plants (WWTPs), which exhibit frequently limited capacity to remove these pollutants, since most of them cannot be metabolized by microorganisms [3,4]. As a result, pharmaceutical compounds reach aquatic systems at concentration levels from ng L^{-1} to µg L^{-1}, but even at these very low concentrations, they can cause toxicological risks on living organisms [5,6].

Psychiatric drugs are a class of pharmaceuticals often prescribed and used in a wide range of mental health problems, and their use is increasing worldwide. Up to now, many studies have been carried out to investigate the presence of psychiatric compounds in the environment. Amisulpride is a typical antipsychotic drug [7], and its occurrence in the environment has been highlighted in recent monitoring studies. For instance, in hospital WWTP effluent of Ioannina (northwestern Greece), amisulpride has been detected at concentration ranges from 102.0 ng L^{-1} to 929.4 ng L^{-1} [8]. In seawater of the Eastern Mediterranean Sea, amisulpride was detected in the range of <0.2–5.5 ng L^{-1} [9]. Moreover, Gago–Ferrero at al. reported amisulpride detection in the effluent of WWTP of Athens at a concentration of 0.07 ng L^{-1} [10]. During the first wave of COVID-19, amisulpride was quantified at 16.8 ng L^{-1} in wastewater samples of Milano and Monza (Italy) WWTPs [11]. Amisulpride has also been identified as a persistent contaminant that can be introduced into the aquatic environment and able to reach groundwaters. Moreover, the formation of its characteristic non-biodegradable N-oxide product has been verified during its treatment by conventional methods, revealing the need for advanced treatment [12,13]. Incomplete removal of various pollutants including pharmaceuticals from conventional WWTPs clearly shows the need for applying innovative technologies [3,4]. Advanced oxidation processes (AOPs) have been proposed as a tertiary treatment for wastewater [14,15]. Among different AOPs, heterogeneous photocatalysis is a promising method for the removal of organic compounds [13]. Graphitic carbon nitride (g-C$_3$N$_4$), an organic semiconductor, has drawn widespread attention and become a hotspot for a wide variety of photocatalytic applications (e.g., degradation of organic pollutants, CO$_2$ and NOx reduction, hydrogen production, organic selective synthesis, and water splitting). This visible light photocatalyst is an outstanding option due to its properties such as graphene-like structure, chemical and thermal stability, good visible light absorption, and low-cost [16–19]. Previously, g-C$_3$N$_4$ has been used in photocatalytic experiments in various aqueous matrices including real hospital wastewaters [8]. However, the application of g-C$_3$N$_4$ catalysts for the degradation of amisulpride under various matrix parameters, the identification of transformation products (TPs), and the ecotoxicological assessment of the photocatalytic treatment have not been studied so far. For this purpose, the main goal of this study was to investigate the photocatalytic degradation of amisulpride under UV-A irradiation using g-C$_3$N$_4$ catalyst and the identification of the TPs formed in the photocatalytic process. Furthermore, the toxicity by means of microalgae *Chlorococcum* sp. and *Dunaliella tertiolecta* was evaluated. Finally, the contribution of reactive species to the degradation mechanism was estimated.

2. Results and Discussion

2.1. Photocatalytic Degradation Kinetics

Before the study of amisulpride photocatalytic degradation, preliminary control experiments (adsorption and UV-A photolysis) were conducted in ultrapure water (UW), and the results are depicted in Figure 1. Adsorption experiments showed that the equilibrium succeeded at 30 min, and about 13% removal of amisulpride was observed. Photolysis under UV-A light resulted in a negligible decrease (~6.0%) of amisulpride and can be explained by the UV-Vis spectra of the pharmaceutical (Figure S1). Direct photolysis under UV-A light and the employed conditions was not expected since amisulpride does not absorb at 365 nm (maximum emission wavelength of the employed irradiation source), justifying the observed trend. Negligible photodegradation of the studied pharmaceutical under UV-A irradiation [13] and simulated solar irradiation [8] was also reported in previous studies.

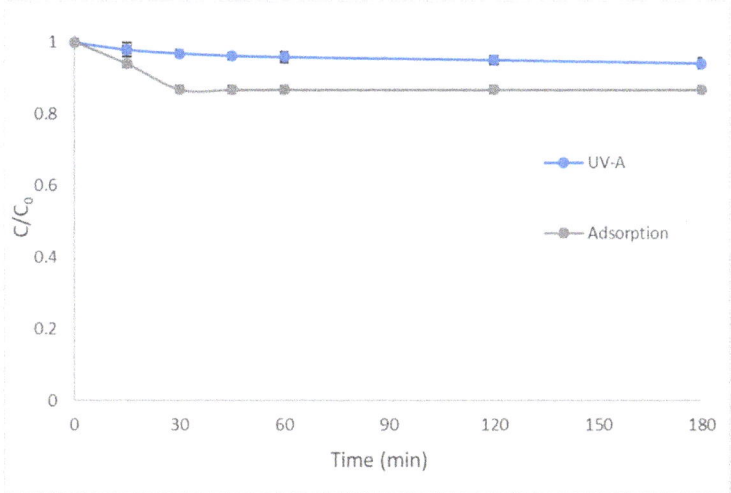

Figure 1. Adsorption and UV-A photolysis of amisulpride in UW ([amisulpride]$_0$ = 1 mg L^{-1}; [g-C$_3$N$_4$] = 300 mg L^{-1}).

The photocatalytic performance of g-C$_3$N$_4$ was evaluated under UV-A light in both UW and treated municipal wastewater (WW), and the results are presented in Figure 2. The photocatalytic study reveals that g-C$_3$N$_4$ can be activated with the adsorption of photon energy larger than its bandgap (~2.82 eV) and subsequently generates a photo-induced electron–hole pair (e− - h+) as well as other reactive species responsible for amisulpride degradation. The reaction kinetics of amisulpride degradation were studied in both matrices, the experimental data were fitted in pseudo-first order reaction kinetics, and the corresponding kinetics parameters (degradation rate constants, half-lives, and correlation coefficients) are reported in Table 1. High removals were achieved in both matrices (Figure 2). The degradation percentages were about 86.0% within 60 min in UW and about 67.0% within 180 min in WW. Our results are in agreement with previous photocatalytic studies in which high removal percentages (up to 98%) of amisulpride were achieved in aqueous matrices using various photocatalysts, i.e., TiO$_2$, C-doped TiO$_2$, g-C$_3$N$_4$, and SrTiO$_3$/g-C$_3$N$_4$ [8,13].

A higher degradation rate was observed in UW than WW. In particular, a rate constant of k = 3.04 × 10^{-2} min^{-1} and a half-life of t$_{1/2}$ = 22.80 min were calculated for amisulpride in UW. The amisulpride degradation was significantly lower in WW (k = 0.7 × 10^{-2} min^{-1} and t$_{1/2}$ = 99.02 min).

This trend indicates the significant effect of the constitution of the water matrices on the photocatalytic performance. It is well-documented in the literature that the organic as well as inorganic content (e.g., Cl$^-$, HCO$_3$$^-$, NO$_3$$^-$, SO$_4$$^{2-}$) that co-exist in WW can affect the photocatalytic performance, acting mainly as scavenger of the reactive species [20,21]. Moreover, the constituents of WW can be adsorbed onto a catalyst's surface leading to the change of the surface charges as well as to the reduction of the available active sites [8]. Metal ions, also present in WW, can significantly affect the photocatalytic performance. The effects of Cu^{2+}, Zn^{2+}, Fe^{3+}, and Al^{3+} were studied, and an inhibition influence on the photocatalytic performance was observed in all cases under the studied experimental conditions (Figure S2). The addition of metal ions can affect the photocatalytic degradation through (i) the decrease of O$_2$ reduction by photogenerated conduction electrons and subsequent suppression of reactive oxygen species formation and (ii) alteration of the pollutant's adsorption [22].

Table 1. Kinetic parameters (rate constants, correlation coefficients (R^2), half-lives ($t_{1/2}$)) of amisulpride photocatalytic degradation in UW and WW.

Matrix	$k \times 10^{-2}$ (min^{-1})	$t_{1/2}$ (min)	R^2
UW	3.04	22.80	0.9420
WW	0.7	99.02	0.9809

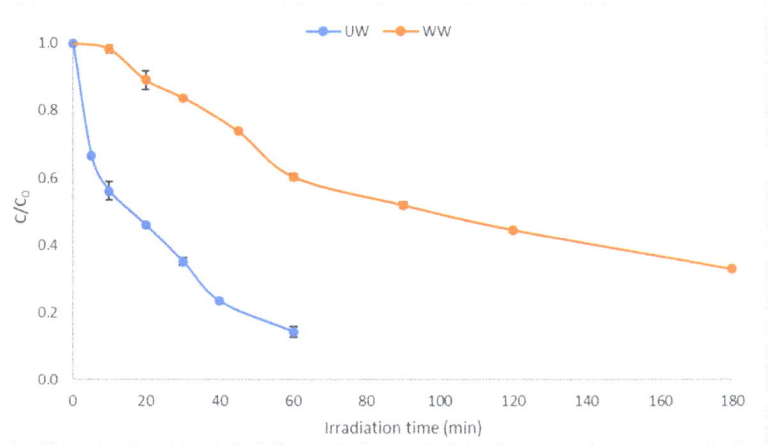

Figure 2. Photocatalytic degradation of amisulpride under UVA irradiation in UW and WW ([amisulpride]$_0$ = 1 mg L^{-1}; [g-C$_3$N$_4$] = 300 mg L^{-1}).

The effect of initial pH was also evaluated for the degradation of amisulpride under similar conditions in WW as matrix (Figure S3). The highest removal of amisulpride was observed at natural pH (about 7.6). The pH$_{PZC}$ of the used catalyst was determined to be ~4.9. Below or above this value, the charge of the catalyst surface is positive and negative, respectively. Similarly, amisulpride (pKa 9.37) molecules are mainly protonated at pH ~ 7.6. Subsequently, under the experimental conditions, electrostatic attractions occur between amisulpride and negatively charged catalyst's surface also leading to the highest removal. On the other hand, under pH 4, electrostatic repulsions of positive charged catalyst, and amisulpride molecules significantly decrease the degradation. Under pH 10, some electrostatic repulsion of negative charged catalysts and partially negatively charged amisulpride molecules could also be considered that verified the slight decrease of the observed degradation.

2.2. Role of Reactive Species to the Degradation Mechanism

The contribution of HO$^{\bullet}$, h^{+}, and O$_2^{\bullet-}$ was evaluated using isopropanol (i-PrOH), methanol (MeOH), and p-benzoquinone (p-BQ), respectively. The scavengers were selected due to their high-rate constant reaction with the corresponding species. More specifically, i-PrOH is a well-documented HO$^{\bullet}$ scavenger presenting a high-rate constant reaction with the radical equal to 1.9×10^9 L mol^{-1} s^{-1} [23,24]. MeOH can quench both HO$^{\bullet}$ and h$^+$ and react with HO$^{\bullet}$ with a rate constant of 1×10^9 L mol^{-1} s^{-1} [20,25]. Para-benzoquinone (p-BQ) can quench O$_2^{\bullet-}$ with a rate constant of 1.9×10^9 L mol^{-1} s^{-1} [20]. The pseudo-first order rate constants as well as the degradation profile in the presence of scavengers are presented in Table 2 and Figure 3, respectively.

Table 2. Kinetic parameters (rate constants, correlation coefficients (R^2)), and percentages of inhibition (% Δk) of amisulpride photocatalytic degradation in the presence of scavengers.

Scavenger	$k \times 10^{-2}$ (min^{-1})	R^2	Δk (%)
Without Scavenger	3.04	0.9420	-
i-PrOH	1.76	0.9800	42.1
MeOH	0.21	0.9292	93.1
p-BQ	0.80	0.9720	73.7

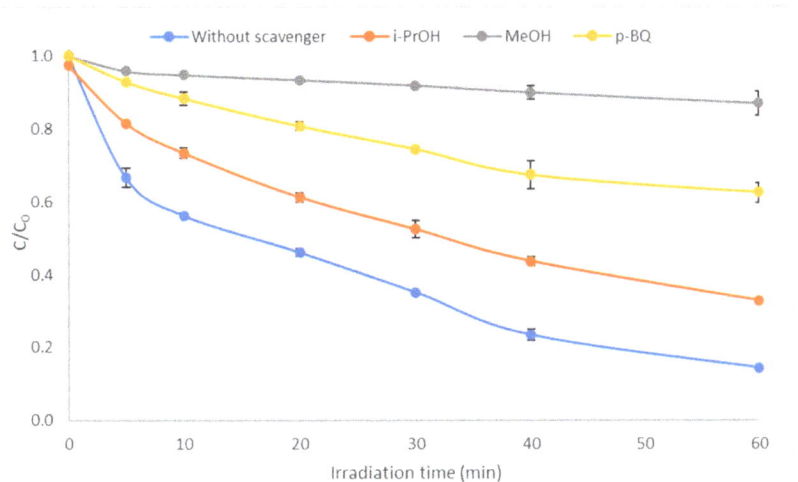

Figure 3. Photocatalytic degradation of amisulpride in the presence of scavengers ([amisulpride]$_0$ = 1 mg L^{-1}; [g-C$_3$N$_4$] = 300 mg L^{-1}; [i-PrOH] = 0.1 M; [MeOH]= 0.1 M; [p-BQ] = 0.2 mM, matrix: UW).

Among the employed scavengers, MeOH caused the highest inhibition (93.1%), as it can quench both HO$^\bullet$ and h$^+$. In contrast, the addition of i-PrOH provoked only 42.1% inhibition, highlighting the significant role of h$^+$ and the low contribution of HO$^\bullet$. This is in agreement with the literature data, as during the activation of g-C$_3$N$_4$ photocatalysts, the formation of HO$^\bullet$ through direct water or OH$^-$ oxidation by holes is not promoted, since VB edges of g-C$_3$N$_4$ catalysts are still less positive than the redox potential of OH$^-$/HO$^\bullet$ (1.99 eV vs. NHE) [26,27]. The significant role of h$^+$ was further verified with experiments using acetonitrile as reaction media (data not shown). The addition of p-BQ led to an inhibition of 73.7%. Superoxide anions can directly degrade organic pollutants through oxidative mechanisms. In addition, $O_2^{\bullet-}$ can lead to the formation of HO$^\bullet$ through the following reactions [27]:

$$O_2^{\bullet-} + H^+ \rightarrow HO_2^\bullet \tag{1}$$

$$2\,HO_2^\bullet \rightarrow O_2 + H_2O_2 \tag{2}$$

$$H_2O_2 + O_2^\bullet \rightarrow HO^\bullet + OH^- + O_2 \tag{3}$$

Our results are consistent with a previous study [28] that reported that h$^+$ and $O_2^{\bullet-}$ are the main species in the photocatalytic degradation of refractory contaminants using g-C$_3$N$_4$- based catalysts. According to the above analysis, the proposed photocatalytic mechanism involves the formation of electron–hole pairs in the conduction band (CB) and valence band (VB) of g-C$_3$N$_4$. The holes (h$^+$) in the VB of g-C$_3$N$_4$ participate directly in the degradation process, whereas the e- on the CB can reduce O$_2$ to $O_2^{\bullet-}$.

2.3. Photocatalytic Degradation Mechanism

The TPs that are produced at the first stages of the photocatalytic processes were identified in both UW and WW using mass spectrometry techniques and interpretation of their mass spectra obtained in positive ionization mode. In total, three TPs were detected in both matrices, and their structural identification was based on their molecular and fragment ions (Table 3) as well as on previous literature data [13].

Table 3. Mass measurements ([M + H]$^+$) and molecular formula of the identified TPs.

TPs	[M + H]$^+$	Molecular Formula
AMI	370.30	$C_{17}H_{27}N_3O_4S$
AMI1	340.37	$C_{16}H_{25}N_3O_3S$
	342.39	
AMI2	242.05	$C_{15}H_{23}N_3O_4S$
	196.01	
	386.26	
AMI3	196.01	$C_{17}H_{27}N_3O_5S$

Three pathways are proposed and depicted in Figure 4. The first pathway includes the scission of the methoxy group of the amisulpride molecule and the formation of AMI1. This TP was also identified by Skibiński et al., 2011 [29] during the photodegradation of the studied pharmaceutical. Similarly, it was identified during the photocatalytic treatment of pharmaceuticals mixture also containing amisulpride [13]. The other pathway proceeds through dealkylation and more specifically through the cleavage of the ethyl group linked to the N-atom (AMI 2). The last route leads to the formation of a characteristic N-oxide TP viaan oxygen transfer mechanism [24]. N-oxide amisulpride is a characteristic TP that has been identified during the treatment of amisulpride by photolysis [29] and heterogeneous photocatalysis [13]. The absence of hydroxylated TPs verifies the low contribution of HO$^\bullet$ to the degradation mechanism, reported in the previous section. The proposed degradation routes are consistent with previous work focused on the photocatalytic degradation of pharmaceuticals mixture also containing amisulpride [13].

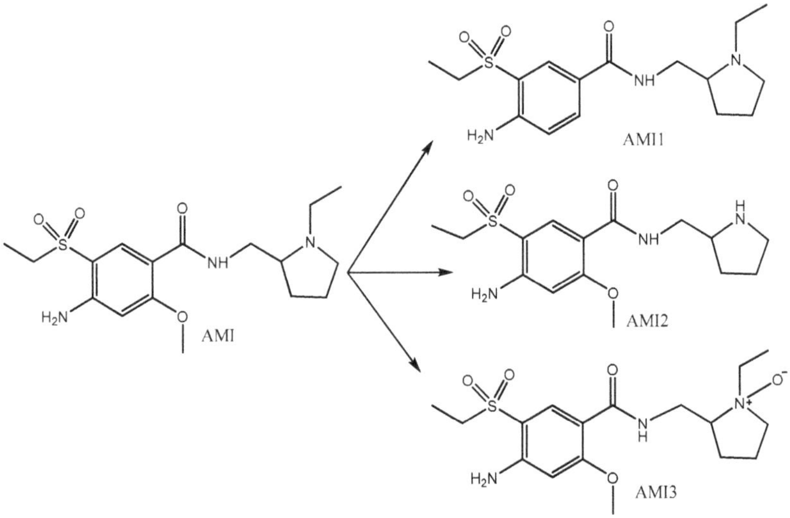

Figure 4. Photocatalytic degradation pathways of amisulpride.

The formation of TPs was also followed during the photocatalytic treatment. As presented in Figure 5, the TPs in UW and WW attained their maximum area in 20 min and

60 min, respectively. Thereafter, the area of the three TPs in both matrices were readily reduced. Subsequently, the cleavage of the rings can take place, and low molecular weight TPs can be produced before complete mineralization. After prolonged application of heterogeneous photocatalysis, two low-molecular-weight carboxylic acids, i.e., formic and acetic acid, were identified.

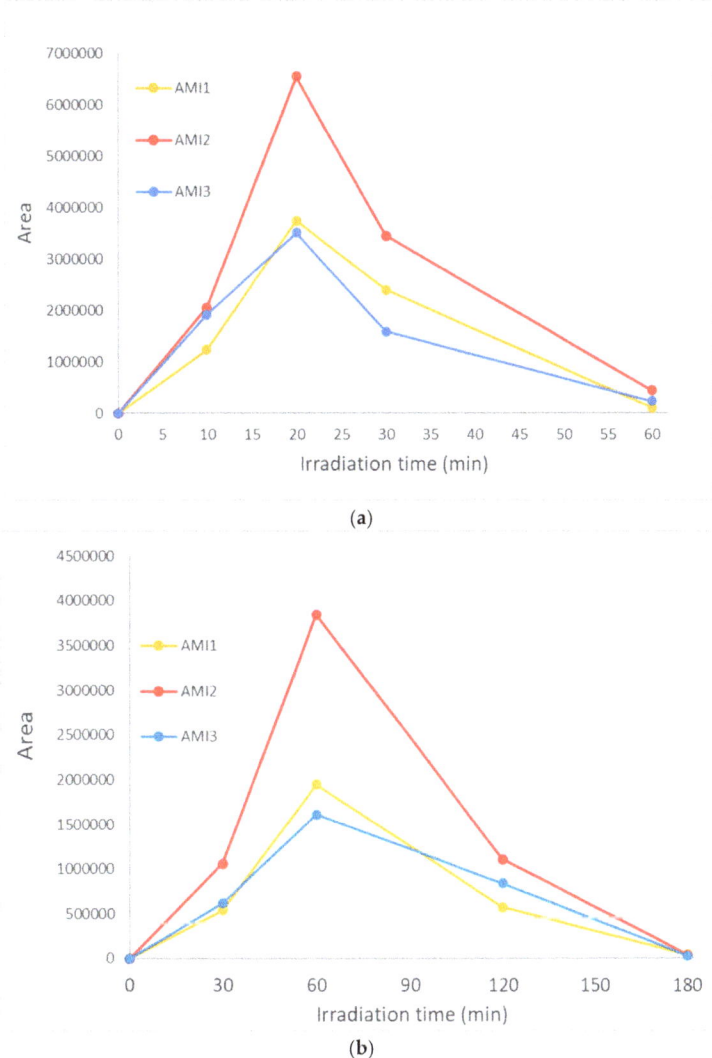

Figure 5. Evolution profiles of TPs identified during the photocatalytic degradation of amisulpride in: (**a**) UW; and (**b**) WW ([amisulpride]$_0$ = 1 mg L^{-1}; [g-C$_3$N$_4$] = 300 mg L^{-1}).

2.4. Ecotoxicity Evolution

The evolution of toxicity against freshwater and marine microalgae was evaluated in both matrices (Figures 6 and 7). According to the results, low toxicity was observed at 0 min of the process in UW (Figures 6a and 7a). In contrast, the untreated solution of amisulpride in WW (0 min) led to over 25% inhibition of growth rate for both microalgae at contact time of 72 h (Figures 6b and 7b). The high growth inhibition rates observed in

WW can be correlated with the composition of this matrix that can contain compounds that exert toxic effects to the exposed microalgae. This was verified by the initial toxicity of the wastewater sample which showed an inhibition of 18.1%. Similar trends were also observed by Gomes et al., 2019 [30] and Antonopoulou et al., 2016 [20]. Only a slight increase in toxicity was observed at the first stages of the photocatalytic treatment in both matrices. This increase can be correlated with the formation of the identified TPs. It is worth mentioning that the highest % growth inhibition rates were observed at 20 min and 60 min in UW and WW, respectively, where the TPs showed their maximum formation. Similar to our results, enhanced toxicity at the first stages was also noticed during the photocatalytic degradation of various contaminants in aqueous matrices by different AOPs [20,31]. However, the contribution of the TPs in the overall toxicity can be characterized as low. With the application of photocatalytic treatment, the % inhibition was significantly reduced in all cases. This clearly indicates that heterogeneous photocatalysis using g-C_3N_4 can proceed without the formation of toxic TPs. It is also able to significantly reduce the toxicity derived from municipal wastewater.

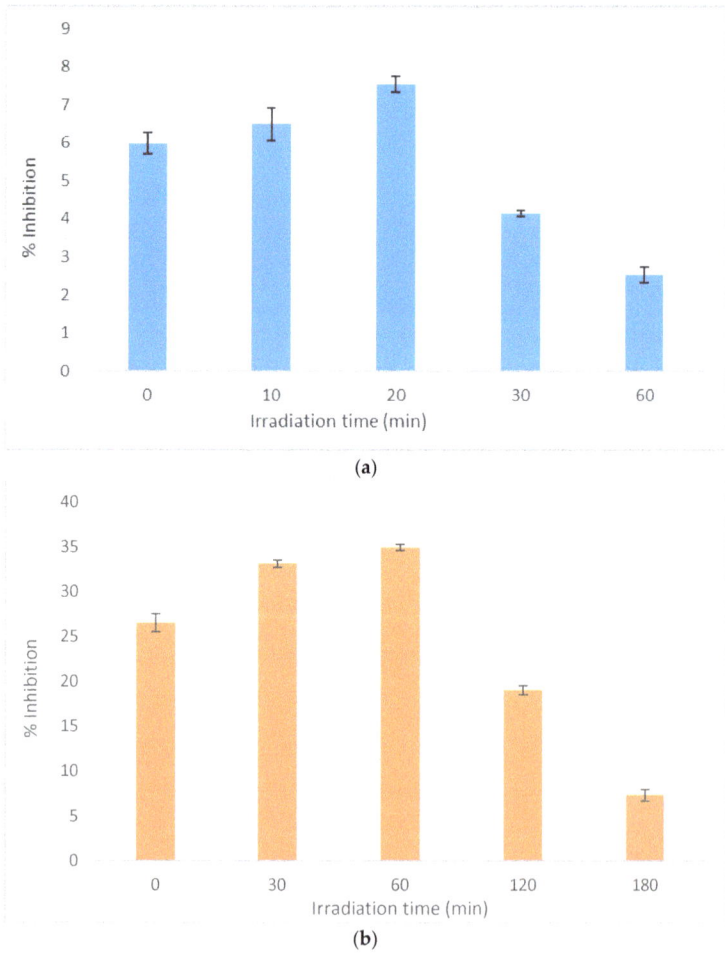

Figure 6. Evolution of toxicity during the photocatalytic process in: (**a**) UW; and (**b**) WW using *Chlorococcum* sp. ([amisulpride]$_0$ = 1 mg L^{-1}; [g-C3N4] = 300 mg L^{-1}).

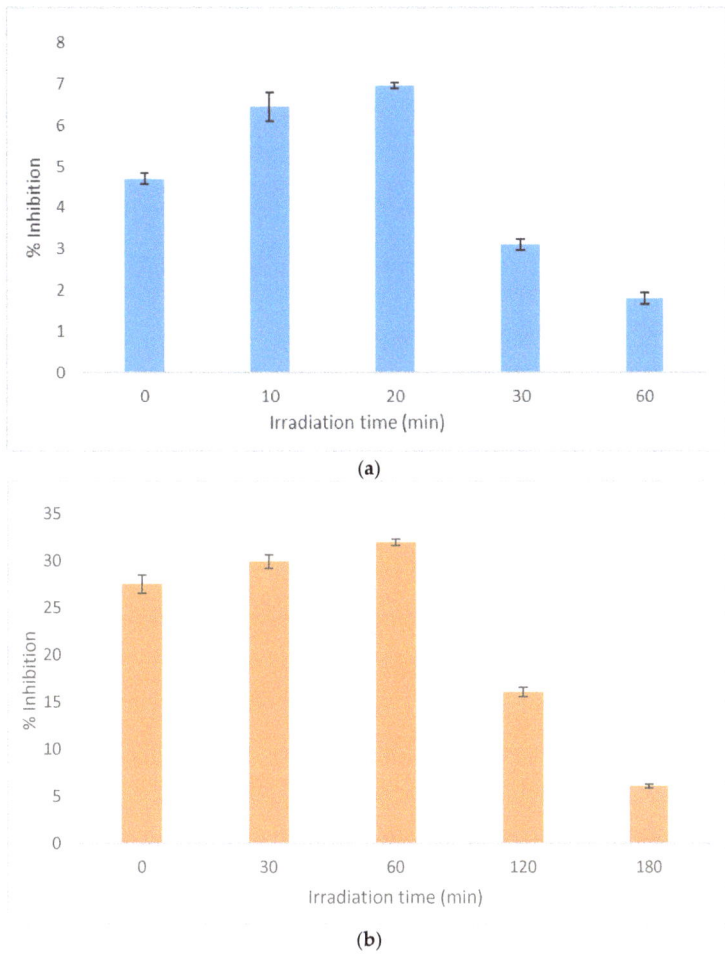

Figure 7. Evolution of toxicity during the photocatalytic process in: (**a**) UW; and (**b**) WW using *Dunaliella tertiolecta* ([amisulpride]$_0$ = 1 mg L^{-1}; [g-C$_3$N$_4$] = 300 mg L^{-1}).

3. Materials and Methods

3.1. Chemicals and Reagents

Amisulpride (98%) was obtained from Tokyo chemical industry Co., Ltd. (Tokyo, Japan). Graphitic carbon nitride (g-C$_3$N$_4$) (BET SSA 35 m^{2}g^{-1}, particle size of 25 nm, E$_g$ = ~2.82 eV) was used as the photocatalyst [8,32]. Acetonitrile, isopropanol, methanol, and water of HPLC-grade solvents were supplied by Fisher Chemical (Waltham, MA, USA). *p*-benzoquinone (≥98%), formic acid, BG-11 medium (Cyanobacteria BG-11 Freshwater Solution), and F/2 medium (Guillard's (F/2) Marine Water Enrichment Solution) were purchased from Sigma-Aldrich (St. Louis, MO, USA). Ultrapure water and treated wastewater (pH = 7.45 ± 0.02; conductivity = 312.81 ± 20.3 µS cm^{-1}; total suspended solids = 1.97 ± 0.022 mg L^{-1}; chemical oxygen demand = 17.6 ± 1.92 mg L^{-1}; PO$_4$$^{3-}$ = 3.85 ± 0.024 mg L^{-1}; SO$_4$$^{2-}$ = 32.1 ± 1.56 mg L^{-1}; NO$_3$$^-$ = 24.2 ± 0.78 mg L^{-1}) were used as matrices. The microalgae species *Chlorococcum* sp. (strain SAG 22.83) and *Dunaliella tertiolecta* (strain CCAP 19/6B) were purchased from the bank SAG collection of the Göttingen University (Germany) and Scottish Marine Institute, Oban, Argyll, Scotland, respectively.

3.2. Photocatalytic Experiments

Photocatalytic experiments were carried out on a system equipped with 4 black light tubes from Sylvania (maximum emission at 365 nm) and electric fans to avoid overheating. A cylindrical pyrex glass cell with a maximum capacity of 200 mL was used as a photoreactor. In a typical experiment, the following experimental conditions [amisulpride]$_0$ = 1 mg L^{-1}; [g-C$_3$N$_4$] = 300 mg L^{-1} were adopted. The suspensions were kept in the dark for 30 min under continuous stirring to achieve adsorption equilibrium before starting the radiation. The photon flux (I_0) entering the solution was calculated to be 1.1 ± 0.02 µEinstein s^{-1} under the applied conditions, periodically using potassium ferrioxalate as actinometer [33,34]. PVDF 0.22 µm by Millex-GV was used to remove the catalyst particles.

3.3. Scavenging Experiments

The contribution of hydroxyl radicals (HO$^\bullet$), superoxide anion radicals (O$_2^{\bullet-}$), and positive holes (h$^+$) in the degradation mechanism was evaluated using well-known scavengers, i.e., isopropanol (0.1M), p-benzoquinone (0.2mM), and methanol (0.1M), respectively [13].

3.4. Analytical Methods

The concentration of amisulpride was quantified by a Dionex (Thermo Scientific) Ultimate 3000 UHPLC using an Acclaim™ RSLC 120 C18 (2.2 µm, 2.1 × 100 mm) column (Thermo Scientific) and an Acquity UPLC BEH C18 VanGuard™ pre-column (1.7 µm, 2.1 × 5 mm) from Waters. The mobile phase was a mixture of ultrapure water with 0.1% formic acid (80%) and acetonitrile (20%) with a flow rate of 0.15 mL min^{-1}. The detection was performed at pollutant's λ_{max}.

3.5. UHPLC/MS Analysis

TPs were identified by a UHPLC/MS system (Ultimate 3000 RSLC System (Thermo Scientific)/amaZon SL ion trap mass spectrometer from Bruker with an ESI source). A full description of the analysis is reported in our previous work [13]. For the determination of small-molecules TPs (carboxylic acids) that can be generated after the decay of the first-stage TPs in heterogeneous photocatalysis, a Dionex P680 HPLC equipped with a Dionex PDA-100 Photodiode Array Detector and a Themo Scientific AQUASIL C18 (250 mm length × 4.6 mm ID × 5 µm particle size) analytical column with a flow rate of 1 mL min^{-1} was used. The mobile phase consisted of 1% acetonitrile and 99% 0.05 M KH$_2$PO$_4$, pH 2.8. The detection was performed at 210 nm.

3.6. Algal Biotest

Algal bioassays were conducted using *Chlorococcum* sp. (strain SAG 22.83) and *Dunaliella tertiolecta* (CCAP19/6B) according to OECD 201 protocol [35], under sterile conditions and continuous illumination (4300 lux). BG-11 and F/2 were used as culture mediums for fresh and saltwater algal strains, respectively. The experiments were initiated by appropriate transfers of stock algal cultures to conical flasks containing the appropriate medium to maintain a supply of cells (1×10^4 cells mL^{-1}) in the logarithmic growth phase (final volume 100 mL). The samples collected at different stages were tested in duplicate cultures for 72 h with continuous stirring under the abovementioned conditions. The cell numbers were determined by using a Neubauer hemocytometer. Thereafter, the growth rate (µ) and the % inhibition of growth rate were calculated. The results are expressed as the mean ± SD.

4. Conclusions

The photocatalytic degradation of the pharmaceutical amisulpride was studied in UW and treated WW using UV-A irradiation and g-C$_3$N$_4$ as the photocatalyst. High removal percentages were observed in both matrices. However, a slower degradation rate was observed in WW that could be attributed to its complex composition containing both inorganic and organic substances. A scavenging study proved the significant contribution of h^+ and $O_2^{\bullet-}$ in the reaction mechanism. Oxidation, dealkylation, and cleavage of methoxy group were found to take place during the photocatalytic degradation of the studied pharmaceutical. Low inhibition in the growth rates and consequently low toxicity were observed at the beginning and during the photocatalytic process when UW was used as matrix, using microalgae *Chlorococcum* sp. and *Dunaliella tertiolecta*. In contrast, higher adverse effects were observed when WW was used as matrix. However, an overall abatement of the effects was noticed at 180 min. Based on the results, heterogeneous photocatalysis using g-C_3N_4 showed good performance for the removal of amisulpride without the formation of harmful TPs. Considering that g-C_3N_4 has a response to visible light and consequently solar light can be used for its activation, heterogeneous photocatalysis using g-C_3N_4 is considered to be a promising method for removal of pharmaceuticals after the efficient separation of the photocatalyst or its immobilization on appropriate supports.

Supplementary Materials: The following supporting information can be downloaded at: https://www.mdpi.com/article/10.3390/catal13020226/s1, Figure S1: UV-Vis spectrum of amisulpride ([amisulpride]$_0$ = 1 mg L^{-1}); Figure S2: Effect of metal ions on the photocatalytic degradation of amisulpride in UW ([amisulpride]$_0$ = 1 mg L^{-1}, [metal ion]$_0$ = 10 mg L^{-1}, [g-C_3N_4] = 300 mg L^{-1}); Figure S3: Effect of pH on the photocatalytic degradation of amisulpride in WW ([amisulpride]$_0$ = 1 mg L^{-1}, [g-C_3N_4] = 300 mg L^{-1}).

Author Contributions: Conceptualization, M.A.; methodology M.A., I.K. and I.R.; validation M.A.; formal analysis M.A.; investigation M.A., M.P., I.K. and I.R.; resources M.A.; data curation, M.A.; writing—original draft M.A. and I.R.; writing—review and editing M.A. and I.K.; visualization M.A.; supervision M.A; project administration M.A.; funding acquisition M.A. All authors have read and agreed to the published version of the manuscript.

Funding: This paper has been financed by the funding program "MEDICUS", of the University of Patras (Number: 81654).

Data Availability Statement: Data are contained within the article.

Acknowledgments: This paper has been financed by the funding program "MEDICUS", of the University of Patras. The authors would like to thank the Laboratory of Instrumental Analysis of the University of Patras for UHPLC-MS analysis.

Conflicts of Interest: The authors declare no conflict of interest.

References

1. Calisto, V.; Esteves, V.I. Psychiatric Pharmaceuticals in the Environment. *Chemosphere* **2009**, *77*, 1257–1274. [CrossRef] [PubMed]
2. Fatta-Kassinos, D.; Meric, S.; Nikolaou, A. Pharmaceutical Residues in Environmental Waters and Wastewater: Current State of Knowledge and Future Research. *Anal. Bioanal. Chem.* **2011**, *399*, 251–275. [CrossRef] [PubMed]
3. Kanakaraju, D.; Glass, B.D.; Oelgemöller, M. Titanium Dioxide Photocatalysis for Pharmaceutical Wastewater Treatment. *Environ. Chem. Lett.* **2014**, *12*, 27–47. [CrossRef]
4. Rivera-Utrilla, J.; Sánchez-Polo, M.; Ferro-García, M.Á.; Prados-Joya, G.; Ocampo-Pérez, R. Pharmaceuticals as Emerging Contaminants and Their Removal from Water. A Review. *Chemosphere* **2013**, *93*, 1268–1287. [CrossRef] [PubMed]
5. aus der Beek, T.; Weber, F.A.; Bergmann, A.; Hickmann, S.; Ebert, I.; Hein, A.; Küster, A. Pharmaceuticals in the Environment-Global Occurrences and Perspectives. *Environ. Toxicol. Chem.* **2016**, *35*, 823–835. [CrossRef] [PubMed]
6. Rapti, I.; Kosma, C.; Albanis, T.; Konstantinou, I. Solar Photocatalytic Degradation of Inherent Pharmaceutical Residues in Real Hospital WWTP Effluents Using Titanium Dioxide on a CPC Pilot Scale Reactor. *Catal. Today* **2022**, *in press*. [CrossRef]
7. Gros, M.; Williams, M.; Llorca, M.; Rodriguez-Mozaz, S.; Barceló, D.; Kookana, R.S. Photolysis of the Antidepressants Amisulpride and Desipramine in Wastewaters: Identification of Transformation Products Formed and Their Fate. *Sci. Total Environ.* **2015**, *530–531*, 434–444. [CrossRef]
8. Konstas, P.S.; Kosma, C.; Konstantinou, I.; Albanis, T. Photocatalytic Treatment of Pharmaceuticals in Real Hospital Wastewaters for Effluent Quality Amelioration. *Water* **2019**, *11*, 2165. [CrossRef]
9. Alygizakis, N.A.; Gago-Ferrero, P.; Borova, V.L.; Pavlidou, A.; Hatzianestis, I.; Thomaidis, N.S. Occurrence and Spatial Distribution of 158 Pharmaceuticals, Drugs of Abuse and Related Metabolites in Offshore Seawater. *Sci. Total Environ.* **2016**, *541*, 1097–1105. [CrossRef]

10. Gago-Ferrero, P.; Bletsou, A.A.; Damalas, D.E.; Aalizadeh, R.; Alygizakis, N.A.; Singer, H.P.; Hollender, J.; Thomaidis, N.S. Wide-Scope Target Screening of >2000 Emerging Contaminants in Wastewater Samples with UPLC-Q-ToF-HRMS/MS and Smart Evaluation of Its Performance through the Validation of 195 Selected Representative Analytes. *J. Hazard. Mater.* **2020**, *387*, 121712. [CrossRef]
11. Cappelli, F.; Longoni, O.; Rigato, J.; Rusconi, M.; Sala, A.; Fochi, I.; Palumbo, M.T.; Polesello, S.; Roscioli, C.; Salerno, F.; et al. Suspect Screening of Wastewaters to Trace Anti-COVID-19 Drugs: Potential Adverse Effects on Aquatic Environment. *Sci. Total Environ.* **2022**, *824*, 153756. [CrossRef]
12. Bollmann, A.F.; Seitz, W.; Prasse, C.; Lucke, T.; Schulz, W.; Ternes, T. Occurrence and fate of amisulpride, sulpiride, and lamotrigine in municipal wastewater treatment plants with biological treatment and ozonation. *J. Hazard. Mater.* **2016**, *320*, 204–215. [CrossRef] [PubMed]
13. Spyrou, A.; Tzamaria, A.; Dormousoglou, M.; Skourti, A.; Vlastos, D.; Papadaki, M.; Antonopoulou, M. The overall assessment of simultaneous photocatalytic degradation of Cimetidine and Amisulpride by using chemical and genotoxicological approaches. *Sci. Total Environ.* **2022**, *838*, 156140. [CrossRef] [PubMed]
14. Kosma, C.I.; Kapsi, M.G.; Konstas, P.S.G.; Trantopoulos, E.P.; Boti, V.I.; Konstantinou, I.K.; Albanis, T.A. Assessment of Multiclass Pharmaceutical Active Compounds (PhACs) in Hospital WWTP Influent and Effluent Samples by UHPLC-Orbitrap MS: Temporal Variation, Removals and Environmental Risk Assessment. *Environ. Res.* **2020**, *191*, 110152. [CrossRef] [PubMed]
15. Parry, E.; Young, T.M. Comparing Targeted and Non-Targeted High-Resolution Mass Spectrometric Approaches for Assessing Advanced Oxidation Reactor Performance. *Water Res.* **2016**, *104*, 72–81. [CrossRef]
16. Rapti, I.; Bairamis, F.; Konstantinou, I. g-C_3N_4/MoS_2 Heterojunction for Photocatalytic Removal of Phenol and Cr(VI). *Photochem* **2021**, *1*, 358–370. [CrossRef]
17. Bairamis, F.; Konstantinou, I. WO_3 Fibers/g-C_3N_4 z-Scheme Heterostructure Photocatalysts for Simultaneous Oxidation/Reduction of Phenol/Cr (Vi) in Aquatic Media. *Catalysts* **2021**, *11*, 792. [CrossRef]
18. Moreira, N.F.F.; Sampaio, M.J.; Ribeiro, A.R.; Silva, C.G.; Faria, J.L.; Silva, A.M.T. Metal-Free g-C_3N_4 Photocatalysis of Organic Micropollutants in Urban Wastewater under Visible Light. *Appl. Catal. B Environ.* **2019**, *248*, 184–192. [CrossRef]
19. Mamba, G.; Mishra, A.K. Graphitic Carbon Nitride (g-C_3N_4) Nanocomposites: A New and Exciting Generation of Visible Light Driven Photocatalysts for Environmental Pollution Remediation. *Appl. Catal. B Environ.* **2016**, *198*, 347–377. [CrossRef]
20. Antonopoulou, M.; Hela, D.; Konstantinou, I. Photocatalytic degradation kinetics, mechanism and ecotoxicity assessment of tramadol metabolites in aqueous TiO_2 suspensions. *Sci. Total Environ.* **2016**, *545-546*, 476–485. [CrossRef]
21. Calza, P.; Hadjicostas, C.; Sakkas, V.A.; Sarro, M.; Minero, C.; Medana, C.; Albanis, T.A. Photocatalytic transformation of the antipsychotic drug risperidone in aqueous media on reduced graphene oxide—TiO_2 composites. *Appl. Catal. B Environ.* **2016**, *183*, 96–106. [CrossRef]
22. Chen, C.; Li, X.; Ma, W.; Zhao, J.; Hidaka, H.; Serpone, N. Effect of Transition Metal Ions on the TiO_2-Assisted Photodegradation of Dyes under Visible Irradiation: A Probe for the Interfacial Electron Transfer Process and Reaction Mechanism. *J. Phys. Chem. B* **2002**, *106*, 318–324. [CrossRef]
23. Palominos, C.R.; Freer, J.; Mondaca, M.A.; Mansilla, H. Evidence for hole participation during the photocatalytic oxidation of the antibiotic flumequine. *J. Photochem. Photobiol. Chem.* **2008**, *193*, 139–145. [CrossRef]
24. Antonopoulou, M.; Konstantinou, I. Photocatalytic degradation and mineralization of tramadol pharmaceutical in aqueous TiO_2 suspensions: Evaluation of kinetics, mechanisms and ecotoxicity. *Appl. Catal. Gen. C* **2016**, *515*, 136–143. [CrossRef]
25. Hazime, R.; Ferronato, C.; Fine, L.; Salvador, A.; Jaber, F.; Chovelon, J.-M. Photocatalytic degradation of imazalil in an aqueous suspension of TiO_2 and influence of alcohols on the degradation. *Appl. Catal. B Environ.* **2012**, *126*, 90–99. [CrossRef]
26. Papailias, I.; Todorova, N.; Giannakopoulou, T.; Ioannidis, N.; Boukos, N.; Athanasekou, C.P.; Dimotikali, D.; Trapalis, C. Chemical vs thermal exfoliation of g-C_3N_4 for NOx removal under visible light irradiation. *Appl. Cat. B Environ.* **2018**, *239*, 16–26. [CrossRef]
27. Bairamis, F.; Konstantinou, I.; Petrakis, D.; Vaimakis, T. Enhanced Performance of Electrospun Nanofibrous TiO_2/g-C_3N_4 Photocatalyst in Photocatalytic Degradation of Methylene Blue. *Catalysts* **2019**, *9*, 880. [CrossRef]
28. Xu, T.; Wang, D.; Dong, L.; Shen, H.; Lu, W.; Chen, W. Graphitic carbon nitride co-modified by zinc phthalocyanine and graphene quantum dots for the efficient photocatalytic degradation of refractory contaminants. *Appl. Catal. B Environ.* **2019**, *244*, 96–106. [CrossRef]
29. Skibiński, R. Identification of photodegradation product of amisulpride by ultra-high-pressure liquid chromatography–DAD/ESI-quadrupole time-of-flight-mass spectrometry. *J. Pharm. Biomed. Anal.* **2011**, *56*, 904–910. [CrossRef]
30. Gomes, J.F.; Lopes, A.; Gmurek, M.; Quinta-Ferreira, R.M.; Martins, R.C. Study of the influence of the matrix characteristics over the photocatalytic ozonation of parabens using Ag-TiO2. *Sci. Total Environ.* **2019**, *646*, 1468–1477. [CrossRef] [PubMed]
31. Martins, R.C.; Rossi, A.F.; Quinta-Ferreira, R.M. Fenton's oxidation process for phenolic wastewater remediation and biodegradability enhancement. *J. Hazard. Mater.* **2010**, *180*, 716–721. [CrossRef] [PubMed]
32. Konstas, P.-S.; Konstantinou, I.; Petrakis, D.; Albanis, T. Synthesis, Characterization of g-C3N4/SrTiO3 Heterojunctions and Photocatalytic Activity for Organic Pollutants Degradation. *Catalysts* **2018**, *8*, 554. [CrossRef]
33. Baxendale, J.H.; Bridge, N.K. The photoreduction of some ferric compounds in aqueous solution. *J. Phys. Chem.* **1955**, *59*, 783–788. [CrossRef]

34. Calvert, J.; Pitts, J.N. *Liquid–Phase Chemical Actinometry Using Potassium Ferrioxalate. Photochemistry*; John Wiley: New York, NY, USA, 1966; pp. 783–786.
35. Organization for the Economic Cooperation and Development. *Test No. 201: Freshwater Alga and Cyanobacteria, Growth Inhibition Test*; OECD Guidelines for the Testing of Chemicals, Section 2; OECD Publishing: Paris, France, 2011. [CrossRef]

Disclaimer/Publisher's Note: The statements, opinions and data contained in all publications are solely those of the individual author(s) and contributor(s) and not of MDPI and/or the editor(s). MDPI and/or the editor(s) disclaim responsibility for any injury to people or property resulting from any ideas, methods, instructions or products referred to in the content.

Article

Photocatalytic Degradation of Psychiatric Pharmaceuticals in Hospital WWTP Secondary Effluents Using g-C$_3$N$_4$ and g-C$_3$N$_4$/MoS$_2$ Catalysts in Laboratory-Scale Pilot

Ilaeira Rapti [1], Vasiliki Boti [1], Triantafyllos Albanis [1,2] and Ioannis Konstantinou [1,2,*]

[1] Laboratory of Industrial Chemistry, Department of Chemistry, University of Ioannina, 45110 Ioannina, Greece
[2] Institute of Environment and Sustainable Development, University Research Center of Ioannina (URCI), 45110 Ioannina, Greece
* Correspondence: iokonst@uoi.gr; Tel.: +30-26510-08349

Abstract: Today, the pollution caused by a multitude of pharmaceuticals used by humans has been recognized as a major environmental problem. The objective of this study was to evaluate and compare the photocatalytic degradation of ten target psychiatric drugs in hospital wastewater effluents using g-C$_3$N$_4$ and 1%MoS$_2$/g-C$_3$N$_4$ (1MSCN) as photocatalytic materials. The experiments were performed using real wastewater samples collected from hospital wastewater treatment plant (WWTP) secondary effluent in spiked and inherent pharmaceutical concentration levels. The photocatalytic experiments were performed in a laboratory-scale pilot plant composed of a stainless-steel lamp reactor (46 L) equipped with ten UVA lamps and quartz filters connected in series with a polypropylene recirculation tank (55–100 L). In addition, experiments were carried out in a solar simulator apparatus Atlas Suntest XLS+ at a 500 Wm^{-2} irradiation intensity. The analysis of the samples was accomplished by solid-phase extraction, followed by liquid chromatography-Orbitrap high-resolution mass spectrometry. Results showed that the photocatalytic degradation of pharmaceutical compounds followed first-order kinetics. In all cases, 1MSCN presented higher photocatalytic performance than g-C$_3$N$_4$. The removal rates of the pharmaceutical compounds were determined above 30% and 54% using g-C$_3$N$_4$ and 1MSCN, respectively. Parallel to kinetic studies, the transformation products (TPs) generated during the treatment were investigated.

Keywords: pharmaceuticals; psychiatric drugs; hospital wastewaters; solar photocatalysis; g-C$_3$N$_4$; 1MoS$_2$/g-C$_3$N$_4$; transformation products

1. Introduction

One of the emerging concerns in environmental science is the occurrence of pharmaceuticals and their metabolites in the environment. Pharmaceutical compounds are regarded as frustrating compounds, as they are released into the environment continuously from various scattered points [1], due to the large amounts of medical compounds, human and veterinary, that are consumed each year [2,3]. Pharmaceuticals are designed with high stability and can maintain their chemical form long enough to be retained in the human body and to remain in the environment in their original structure [4]. Pharmaceutical compounds reach waters with concentrations from ng L^{-1} to µg L^{-1}, but even at these very low concentrations, they can cause toxicological risk to living organisms due to their chemical and physical properties [2,3,5,6].

The appearance of mental health problems is associated with many social and economic determinants, such as poverty, deprivation, and inequalities. Given that there is an increase in these factors in times of economic crisis, it is to be expected that the mental health of the population is also at higher risk. According to the World Health Organization, the COVID-19 pandemic has also created a global mental health crisis, undermining the mental health of millions of people, although even before it, one in eight people worldwide were already living with a mental disorder [7]. There has been special interest in psychiatric drugs, as contaminants, because of their continuously increasing use, on the one hand, and because many of them affect phylogenetically conserved neuroendocrine systems, on the other hand, which may create problems to other non-target organisms. The rapid increase in the use of psychiatric drugs has resulted in their strong presence in the environment. As to their therapeutic use, psychiatric drugs are categorized into six classes: (i) anxiolytics, hypnotics, and sedatives; (ii) antidepressants; (iii) stimulants; (iv) antipsychotics; (v) mood stabilizers; (vi) drugs to treat drug dependence [8].

The psychiatric drugs enter the wastewater treatment plants from the disposal of unused or expired medicines and from human excretions. Although psychiatric drugs are excreted by humans mainly as metabolites, biologically active or not, part of the drug is excreted without being metabolized. Hospital and municipal WWTPs represent the major source for the presence of this class of micropollutants in the environment due to their limited capacity to remove such micropollutants [2,9].

Pharmaceuticals may have diverse consequences on the environment, as these compounds can be persistent, bioaccumulative, and can cause both serious and chronic human and ecotoxicological harm. The presence of psychiatric drugs in water reservoirs worldwide, as well as with their potential bioaccumulation in aquatic organisms, has been confirmed in several studies. For instance, in hospital WWTP effluent of Ioannina (northwestern Greece), venlafaxine has been detected with a maximum mean concentration of 550.3 ngL^{-1} and bupropion with a minimum mean concentration of 39.9 ngL^{-1} [9]. In addition, in a monitoring study of psychiatric drugs in rivers of Portugal, the most frequent psychiatric drugs found were antidepressants such as carbamazepine, citalopram, fluoxetine, sertraline, and trazodone in concentrations of up to 2.0 ng L^{-1} [10]. Moreover, Vasskog et al. reported sertraline detection in wastewater treatment plants in Norway at concentrations ranging from 0.9 to 6.3 ngL^{-1} [11]. The drugs used in neurology and psychiatry affect humans as well as every other living organism. Telles-Correia et al. found that venlafaxine, bupropion, quetiapine, and other pharmaceuticals cause severe liver diseases [12]. Antidepressants are considered toxic to very toxic to algae [13]. In addition, Grzesiuk et al. (2018) demonstrated that, upon chronic exposure to low concentrations, fluoxetine affected the ecophysiology of two species of microalgae, *Acutodesmus obliquus* and *Nannochloropsis limnetica* [14]. Finally, Best et al. (2014) demonstrated that Venlafaxine at environmentally relevant concentrations affects the metabolic capacities and may compromise the adaptive responses of rainbow trout to an acute stressor [15].

Another important issue to be considered is the production of transformation products (TPs) under environmental conditions or during wastewater treatment. TPs are products generated in the environment and in the wastewater treatment plants by chemical, physical, and biological processes. Some studies have noticed the presence of TPs in different environmental matrices, including wastewaters, surface waters, and soil. The mostly unknown ecotoxicological effects of TPs can affect human health and aquatic biota [16].

Different methods, such as adsorption, and biological and chemical processes such as chlorination have been used for water treatment and environmental protection. These methods have shown limited success and usually do not reach complete removal of pharmaceutical compounds, so more powerful and efficient technologies are required for application in treatment of pharmaceuticals. On the other hand, Advanced Oxidation Processes (AOPs) are promising, modern, and environmentally friendly methods used to remove organic pollutants, such as pharmaceutical compounds, prior to discharge into aquatic systems [3,17]. AOPs can attain better quality of treated water as they improve overall removal efficiencies of contaminants [18]. Among AOPs, heterogeneous photocatalysis using semiconductors has proven to be a promising technique due to its high efficiency, photostability, and non-toxic properties of the catalysts.

Graphitic carbon nitride (g-C_3N_4) has semiconductor characteristics with a narrow band gap of ~2.7 eV. g-C_3N_4 expedited interest as a non-toxic, environmentally friendly, thermally and chemically stable, and facilely available inexpensive material, due to low-cost precursors (e.g., urea, thiourea, and melamine), for a broad variety of photocatalytic applications [19,20]. However, its photocatalytic efficiency is limited because of the recombination process, which consists of the transport of electrons back to the valence band. This phenomenon competes with the electron–hole pair formation, which diminishes the efficiency of the pollutant degradation. To circumvent this problem, several strategies have been established by different research groups. One approach is to combine two or more semiconductors. Heterostructuring can separate the electrons and holes and change their band structures, decreasing the recombination effects. Molybdenum sulfide (MoS_2) has also attracted great attention as a co-catalyst as it possesses a narrow band gap (1.2–1.9 eV), high stability, and proper band edge potential for the interfacial charge transfer in heterostructures [21]. The coupling of MoS_2 and $g-C_3N_4$ could significantly reduce the charge carrier recombination and highly improve photocatalytic activity compared to bare materials.

This study focuses on the application of photocatalysis for the removal of psychiatric drugs present in HWW secondary effluents at spiked and inherent concentrations by different photocatalytic semiconductors and different irradiation sources. $g-C_3N_4$ and 1MSCN were compared for the photocatalytic degradation of target compounds. According to our knowledge, photocatalytic degradation of psychiatric drugs by $g-C_3N_4$ and 1MSCN in real hospital WWTP secondary effluents and environmentally relevant concentrations has not yet been published in the scientific literature. Experiments have also been performed in order to investigate the influence of different irradiation types (UV and simulated solar radiation) on the process performance as a limited number of articles deal with the comparison among the commonly used light sources.

2. Results and Discussion

2.1. Degradation of Psychiatric Drugs

The efficiency of photocatalysts $g-C_3N_4$ and 1MSCN in the photocatalytic degradation of the target compounds was investigated in two photocatalytic systems, a small lab-scale reactor and solar simulator (suntest), as well as lab-scale pilot plant (L-PP) and UV-lamps. The degradation of the psychiatric drugs followed first-order kinetics in both cases. Tables 1 and 2 show the photocatalytic degradation rate constants, correlation coefficients, and % removal of the target compounds, while Figure 1 shows the degradation kinetics.

Table 1. Kinetic parameters (first-order kinetic constants (k, min^{-1}), correlation coefficients (R^2), and % degradation of pharmaceuticals) after photocatalytic treatment at the solar simulator.

Pharmaceutical	$g-C_3N_4$			1MSCN		
	k (min^{-1})	R^2	% Degradation (Time, min)	k (min^{-1})	R^2	% Degradation (Time, min)
Amisulpride	0.029	0.9944	100 (240)	0.03	0.9901	100 (180)
O-desmethyl-venlafaxine	0.012	0.9394	98 (300)	0.013	0.9728	99 (300)
Venlafaxine	0.015	0.9989	99 (300)	0.015	0.9791	99 (300)
Clozapine	0.003	0.9814	60 (300)	0.003	0.9699	66 (360)
Citalopram	0.002	0.9882	53 (300)	0.003	0.9594	62 (300)
Quetiapine	0.015	0.99	99 (300)	0.04	0.9926	99 (120)
Carbamazepine	0.009	0.9677	96 (300)	0.012	0.9836	97 (300)
Bupropion	0.004	0.9772	76 (300)	0.004	0.9476	79 (360)
Fluoxetine	0.017	0.9771	99 (300)	0.02	0.9934	100 (300)
Amitriptyline	0.001	0.9265	30 (300)	0.002	0.9915	54 (300)

As can be seen in Figure 1, photocatalytic degradation of target compounds using $1MoS_2/g-C_3N_4$ proceeded faster than the $g-C_3N_4$ photocatalyst in all experiments. Photocatalytic degradation experiments at the lab-scale pilot plant using 1MSCN showed removal percentages at the end of the process that ranged between 91% and 100%. Namely, in the presence of 1MSCN, after 360 min, photocatalytic removal by 100% for amisulpride, venlafaxine, and metabolite O-desmethyl venlafaxine and 99% for quetiapine, fluoxetine, carbamazepine, and amitriptyline were accomplished; while for clozapine, citalopram, and bupropion, 98%, 97%, and 91% were recorded, respectively. On the other hand, the degradation efficiency of target compounds differed to a lesser extent in the presence of g-C_3N_4 (78–100%). Using g-C_3N_4 as a photocatalyst, the lowest degradation performance of 78% was recorded in the case of bupropion. Amitriptyline showed the highest degradation rate constant in both experiments (0.039 min^{-1} and 0.025 min^{-1} for 1MoS_2/g-C_3N_4 and g-C_3N_4, respectively). The 1MSCN photocatalyst showed a higher photocatalytic degradation of the psychiatric drugs also in experiments carried out at the solar simulator apparatus. Amisulpride, fluoxetine, and metabolite O-desmethyl venlafaxine were completely degraded. Using g-C_3N_4, the photocatalytic efficiency decreased by 6% and 9% for clozapine and citalopram, respectively. The lowest degradation performances of 54% and 30% were recorded for amitriptyline in the case of 1MSCN and g-C_3N_4, respectively.

Regarding the influence of different types of irradiation sources on the efficiency of psychiatric drugs removal, lab-scale pilot plant experiments using 1MSCN photocatalyst led to higher photocatalytic rate constants by 3.45% (for citalopram) and 56% (for amitriptyline) compared to the efficiency of g-C_3N_4 photocatalyst. In the case of simulated solar simulator experiments, the photocatalytic process led to the complete degradation of amisulpride, metabolite O-desmethyl venlafaxine, fluoxetine, and quetiapine. Using 1MSCN photocatalyst instead of g-C_3N_4 photocatalyst, the rate constants increased by 3.45% for amisulpride and metabolite O-desmethyl venlafaxine, 17.65% for fluoxetine, and 166.7% for quetiapine.

For comparison to previously reported results dealing with the degradation of psychoactive pharmaceuticals using g-C_3N_4 photocatalysts, Kane et al. [22] studied the degradation of carbamazepine under UV light irradiation using g-C_3N_4 and g-C_3N_4/TiO_2 photocatalysts, reporting that composite materials possess higher degradation activity. More specifically, removal percentages were 71.41% and 5.97% for 10% g-C_3N_4/TiO_2 and g-C_3N_4, respectively. In addition, Moreira et al. [23] studied the photocatalytic degradation of nine organic pollutants (including three psychiatric drugs, namely carbamazepine, venlafaxine, and fluoxetine) found in biologically treated effluents of an urban wastewater treatment plant (North Portugal) using exfoliated g-C_3N_4. At the end of the photocatalytic process, an almost complete removal of all compounds was observed.

Table 2. Kinetic parameters (rate constants (k, min^{-1}), correlation coefficients (R^2), and % degradation of pharmaceuticals) after photocatalytic treatment at laboratory pilot plant.

Pharmaceutical	g-C_3N_4			1MoS_2/g-C_3N_4		
	k (min^{-1})	R^2	% Degradation (Time, min)	k (min^{-1})	R^2	% Degradation (Time, min)
Amisulpride	0.017	0.9958	100 (360)	0.018	0.9028	100 (360)
O-desmethyl-venlafaxine	0.009	0.9831	97 (360)	0.009	0.9259	100 (360)
Venlafaxine	0.014	0.9837	100 (360)	0.016	0.9956	100 (360)
Clozapine	0.01	0.9786	98 (360)	0.011	0.9892	98 (360)
Citalopram	0.007	0.9371	93 (360)	0.008	0.9669	97 (360)
Quetiapine	0.015	0.9915	99 (360)	0.02	0.9953	99 (240)
Carbamazepine	0.013	0.9877	99 (360)	0.015	0.9922	99 (360)
Bupropion	0.004	0.9751	78 (360)	0.005	0.9772	91 (360)
Fluoxetine	0.011	0.9869	99 (360)	0.011	0.9894	99 (360)
Amitriptyline	0.025	0.9945	99 (180)	0.039	0.9961	99 (120)

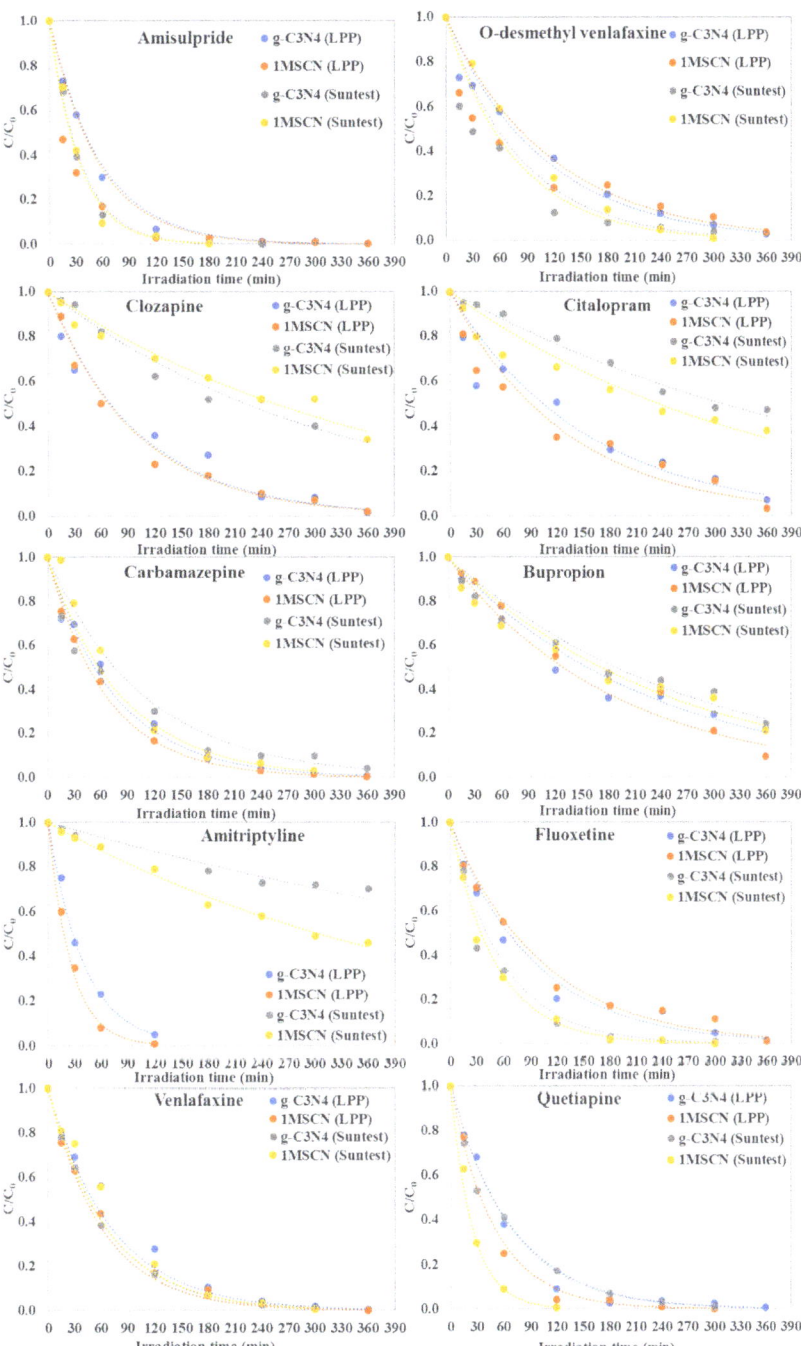

Figure 1. Photocatalytic degradation of psychiatric drugs using g-C$_3$N$_4$ and 1MoS$_2$/g-C$_3$N$_4$ catalysts in lab-scale pilot plant (LPP) and in solar simulator (Suntest) as a function of irradiation time.

The occurrence and photocatalytic removal of the inherent concentration of target psychiatrics in the effluent using g-C$_3$N$_4$ and 1MSCN photocatalysts (100 mgL^{-1}) under simulated solar radiation were also determined. As we are dealing with the inherent concentration of pharmaceutical compounds, the presence of all ten psychiatric drugs was not detected in every case. More specifically, using g-C$_3$N$_4$, a total of five compounds (amisulpride, venlafaxine, O-desmethyl venlafaxine, and quetiapine) were detected in concentrations higher than LOQ, while fluoxetine was found in trace levels. Amisulpride's concentration in the effluent was 42.2 ngL^{-1}. Venlafaxine and metabolite O-desmethyl venlafaxine was found at concentrations levels of 24.6 ngL^{-1} and 38.1 ngL^{-1}, respectively. Quetiapine presented the lowest observed concentration at 8.2 ngL^{-1}. After 120 min, amisulpride presented 44% degradation, while venlafaxine and metabolite O-desmethyl venlafaxine presented 43% and 37% elimination, respectively. The highest degradation rate, 85%, was noticed for quetiapine after 60 min of irradiation. Amisulpride, venlafaxine, O-desmethyl venlafaxine, carbamazepine, and fluoxetine were also found when 1MSCN photocatalyst was used. Concentrations ranging between 5.04 ngL^{-1} for fluoxetine and 578.5 ngL^{-1} for amisulpride. Amisulpride presented the highest degradation rate, 99%, after 180 min of irradiation, while the lowest degradation rate was noticed for fluoxetine, 47%, after 300 min of irradiation. At the end of the photocatalytic reaction (after 300 min), venlafaxine presented 97% degradation, while O-desmethyl venlafaxine and carbamazepine presented 90% and 80% degradation, respectively.

Additionally, BOD$_5$ and COD measurements were performed before and after the photocatalytic treatment at the lab-scale pilot plant using g-C$_3$N$_4$ and 1MSCN. The BOD$_5$/COD ratio is considered to be a suitable criterion for biodegradability, as it is not affected by the amount or the oxidation state of organic matter [24–26]. Table 3 shows the BOD$_5$, COD, and BOD$_5$/COD ratio measured in secondary effluent of hospital WWTP and after the photocatalytic processes. The ratio of BOD$_5$/COD after 360 min of irradiation time decreased from the initial 0.55 to 0.38 for g-C$_3$N$_4$ and 0.13 to 0.07 for 1MSCN. The variation in COD before and after the photocatalytic treatment may also be affected by the initial suspended solid concentrations in the treated effluents or the interference of produced H$_2$O$_2$ from the catalysts during the treatment.

Table 3. BOD$_5$ (mgL^{-1}), COD (mgL^{-1}), and BOD$_5$/COD measured in secondary effluent of hospital WWTP before and after photocatalytic treatment.

	g-C$_3$N$_4$		1MoS$_2$/g-C$_3$N$_4$	
	Before Treatment (t = 0 min)	After Treatment (t = 360 min)	Before Treatment (t = 0 min)	After Treatment (t = 360 min)
BOD$_5$ (mgL^{-1})	15.5	14	4.3	1.4
COD (mgL^{-1})	28	37	33	19
BOD$_5$/COD	0.55	0.38	0.13	0.07

2.2. Transformation Products

Most studies dealing with the treatment of real wastewaters examined usually only the removal of parent drug compounds, whereas our study coped with the challenge to identify transformation products at very low concentration in spiked and unspiked real effluent samples. LC–HRMS data on the identification of target psychiatric compounds TPs as well as their structures are summarized in Table 4. Figure 2 presents the evolution profiles of the identified transformation products after the photocatalytic process at the laboratory-scale pilot plant using 1MSCN.

Regarding venlafaxine (VNX), two TPs were found. The first with m/z 262.1802 and elemental composition C$_{16}$H$_{24}$NO$_2$$^+$ could be generated from further oxidation of metabolite ODV (m/z 264.1958). Second, TP yielded an accurate mass of 250.1802 (C$_{15}$H$_{24}$NO$_2$$^+$). VNF-TP1 and VNX-TP2 maximized at 15 min. Both TPs were identified as intermediate products of venlafaxine after the photocatalytic process by Konstas et al. (2019) [27].

Table 4. Retention time (R_t), elemental formula, mass error, and double-bond–ring-equivalent number (RDB) of parent compounds and transformation products (TPs) identified during the photocatalytic treatment by UHPLC-Orbitrap MS.

Parent Compounds/TPs	R_t (min)	Elemental Formula $[M + H]^+$	Δ (ppm)	RDB	Chemical Structure
Venlafaxine	3.60	$C_{17}H_{28}NO_2$	2.644	4.5	
VNX–TP1	3.47	$C_{16}H_{24}NO_2$	0.402	5.5	
VNX–TP2	3.30	$C_{15}H_{24}NO_2$	0.098	4.5	
O-desmethyl-venlafaxine	3.35	$C_{16}H_{26}NO_2$	3.011	4.5	
ODV-TP1	3.05	$C_{16}H_{24}NO$	1.629	5.5	
Quetiapine	3.75	$C_{21}H_{25}N_3O_2S$	1.391	10.5	
QTP-TP1	3.36	$C_7H_{15}O_2N_2$	1.598	1.5	
Fluoxetine	3.93	$C_{17}H_{19}NOF_3$	4.628	7.5	
FLX-TP1	3.50	$C_{10}H_{16}NO$	1.991	3.5	
FLX-TP2	3.75	$C_{10}H_{13}NO$	0.187	4.5	

According to previous published data, O-desmethyl venlafaxine is the main product of human metabolism at a rate of 90%, while N-desmethyl venlafaxine is also produced but only at a rate of 10%. Both TPs have been identified elsewhere during the photocatalytic process with O-desmethyl venlafaxine to represent the main TP, while degradation pathways were also discussed [28]. In our study, O-desmethyl venlafaxine with m/z 264.1954

($C_{16}H_{25}NO_2$) was identified. Only one TP of ODV was observed after 15 min with m/z 246.1852, corresponded to H_2O loss, yielded an elemental structure of $C_{16}H_{24}NO$, and was completely removed after 120 min of the photocatalytic process.

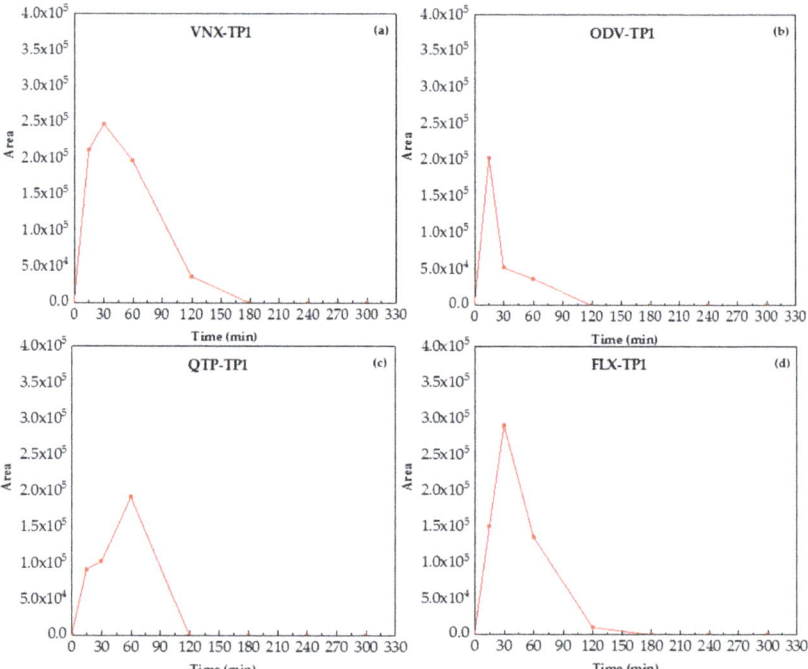

Figure 2. Evolution profiles of (**a**) venlafaxine TP1, (**b**) O-desmethyl venlafaxine TP1, (**c**) quetiapine TP1, and (**d**) fluoxetine TP1, formed by photocatalytic oxidation process at the laboratory-scale pilot plant using $1MoS_2/g$-C_3N_4.

Concerning quetiapine, one TP was observed. This TP exhibited an m/z of 159.1120 and peak up at 60 min. This TP has been reported by Skibinski as a TP formed after a photodegradation study of quetiapine [29].

As for fluoxetine, two TPs were identified. TP1 was m/z 166.1225 with the chemical formula $C_{10}H_{15}NO$. Moreira et al. (2020) also reported the finding of this TP in the photocatalytic degradation of Prozac by TiO_2 nanoparticles [30]. TP2 with molecular formula $C_{10}H_{13}NO$ provided a peak at m/z 164.1069, which indicates the loss of the trifluorotoluene and the oxidation of the alcohol produced in the corresponding ketone. Both TPs were observed after 15 min, maximized at 30 min, and completely degraded after 120 min.

Photocatalytic experiments in the presence of scavengers have been previously studied [19] for the degradation of phenolics using the same catalysts. Isopropanol (IPA), superoxide dismutase (SOD), and triethanolamine (TEOA) were used to scavenge $^{\bullet}OH$, $O_2^{\bullet-}$, and h^+, respectively. Based on the previous results, kinetics retardation follows the trend TEOA > SOD > IPA, suggesting mainly the participation of positive holes (h^+) and $O_2^{\bullet-}$ with $^{\bullet}OH$ being formed secondary in a minor extent. The critical role of holes in contaminant degradation using g-C_3N_4 catalysts was also elucidated previously [31]. Based on the above, the formation of the detected TPs or the absence of TPs related to $^{\bullet}OH$ attack such as hydroxyderivatives is consistent with the formation of major oxidant species.

3. Materials and Methods

3.1. Reagents and Chemicals

All reagents used in the experiments were of high purity grade (>97%). Bupropion hydrochloride (BUP) was obtained from LGC (Wesel, Germany). Amisulpride (AMS), amitriptyline (AMT), fluoxetine hydrochloride (FLX), and venlafaxine hydrochloride (VNX) were purchased from TCI (Zwijndrecht, Belgium). Carbamazepine (CBZ), citalopram (CIT), clozapine (CZP), O-desmethyl venlafaxine (ODV), and quetiapine (QTP) were purchased from Sigma-Aldrich (Darmstadt, Germany). Carbamazepine-d10 and fluoxetine-d5 hydrochloride were supplied from A2S (Saint Jean d'Illac, France). Individual stock solutions of each compound, as well as isotopically labeled internal standard solution, were prepared in methanol and stored at −20 °C. LC-MS-grade methanol, LC-MS-grade water, and Na_2EDTA grade were purchased from Fisher Scientific (Leicestershire, UK). Folin–Ciocalteu's phenol reagent and formic acid (purity, 98–100%) were obtained from Merck KGaA (Darmstadt, Germany). Sodium carbonate and p-hydroxy benzoic acid were supplied from Sigma Aldrich (St. Louis, MO, USA). Oasis HLB (200 mg, 6 cm^3) cartridges used for solid-phase extraction were bought from Waters Corporation (Milford, MA, USA).

3.2. Collection of Hospital Wastewater Treatment Plant Samples

Samples of real hospital wastewater treatment plant effluent were collected from the University hospital WWTP of Ioannina (Northwestern Greece) and used for all experiments. The hospital is an academic medical center that interrelates with Ioannina's University's School of Medicine and School of Nursing. It has a capacity of 800 beds and almost 45,000 people are treated in the hospitals' care clinics, while almost 130,000 people use the hospitals' casualty department every year. The HWWTP consists of a pretreatment step (grit-removal), flow equilibration tank, and a biological secondary treatment, with the final step being the disinfection with the addition of NaClO (15% solution). The hydraulic retention time (HRT) of the WWTP is 6 h, while the solid retention time (SRT) is 1.5 days. This plant discharges its effluent wastewater into the urban network, which results in the municipal WWTP; therefore, the assessment of its efficiency has substantial interest.

Secondary effluent, for all lab-scale experiments, was collected in jerrycans and transported immediately to the laboratory. Physicochemical parameters of the samples were determined by applying standard methods. The chemical oxygen demand (COD) was measured by a WTW Thermoreactor 3200 and a WTW pHotoFlex portable photometer by following the corresponding set test for each application (WTW, Weilheim, Germany). Five-day biochemical oxygen demand (BOD_5) was determined by means of a WTW OxiTop OC 110 system and a WTW TS 606-G/2-i thermostat cabinet (WTW, Weilheim, Germany).

3.3. Photocatalytic Materials

Two different semiconductor materials were used, i.e., graphitic carbon nitride (g-C_3N_4) (specific surface area: 35 m^2g^{-1}, particle size of 25 nm, E_g = 2.82 eV) prepared using urea as a precursor compound, and a composite catalyst 1%MoS_2/g-C_3N_4 (1MSCN) (specific surface area: 62.2 m^2g^{-1}, particle size of 9.5 nm, E_g = 2.66 eV). The synthesis and characterization of the semiconductor materials are described elsewhere [19]. The selection of 1MSCN material is based on our previous publication [19] that studied a series of composite g-C_3N_4/MoS_2 materials with different percentages (1, 2, 5, and 10%) of MoS_2 loadings, concluding that the catalyst with 1% MoS_2 loading was the most active toward the photocatalytic degradation of phenolics.

3.4. Photocatalytic Experiments

3.4.1. Solar Simulator

Photocatalytic experiments were carried out in a solar simulator apparatus Atlas Suntest XLS+ (Atlas, Germany) equipped with a xenon lamp (2.2 kW) and special filters in place to prevent the transmission of wavelengths below 290 nm. During the experiments, the irradiation intensity was maintained at 500 Wm^{-2}. In the wastewater photocatalytic treatment process, aqueous solutions (250 mL) and the catalyst (100 mgL^{-1}) were transferred into a double-walled Pyrex glass reactor, thermostated by water flowing in the double-skin of the reactor. The suspension was spiked with a mixed standard solution of target psychiatric drugs at a concentration of 250 ngL^{-1}. To ensure the establishment of adsorption–desorption equilibrium onto the catalyst surface, the suspension was stirred by a magnetic stirrer in the dark for 30 min, before exposure to light. The wastewater samples were irradiated for 300 min. Samples were withdrawn at different time intervals and were centrifuged (Thermo Scientific, HERAUS Megafuge 8, Shanghai, China; 4400 rpm) for 20 min for the separation of the catalyst particles. The samples were filtered by 0.45 μm filters before extraction.

3.4.2. Laboratory-Scale Pilot Plant

Photocatalytic experiments were carried out also in a laboratory-scale pilot reactor (Ecosystem S.A., Barcelona, Spain) composed of a stainless-steel reactor of total volume 46 L equipped with ten UVA lamps (Philips PL-L 36W. UVA radiation 8.5W, UVA range 340–400 nm, λ_{peak} = 375 nm) and quartz filters. A polypropylene recirculation tank of a working volume between 55 L and 100 L was connected in series. Hence, the pilot plant behaved as a plug-flow reactor where water is circulating using a circulation pump. The experiment started with the addition of secondary effluent to the circulation tank spiked with a mixed standard solution of target psychiatric drugs at a concentration of 250 ngL^{-1} (amisulpride, venlafaxine, O-desmethyl venlafaxine, clozapine, citalopram, quetiapine, carbamazepine, bupropion, fluoxetine, and amitriptyline) following 15 min of homogenization by mechanical stirrer. Then, 100 mgL^{-1} of catalyst was added and for ensuring homogenization, the suspension was mechanically stirred in the dark for 15 min followed by recirculation for 30 min before exposure to light. The wastewater samples were irradiated for 360 min. Samples were withdrawn at different time intervals and were centrifuged for 20 min for the separation of the catalyst particles. The samples were filtered by 0.45 μm filters.

3.5. Extraction of Wastewater Samples

Concentration levels of psychiatric drugs in the raw and treated samples were determined using Ultra-High-Performance Liquid Chromatography-Orbitrap-Mass Spectrometry (UHPLC-Orbitrap-MS), after the Solid-Phase Extraction (SPE) procedure. The Oasis HLB (200 mg, 6 cm^3) cartridges were selected for the determination of the target analytes. After the filtration of the samples, the pH value was regulated to ~7. An appropriate volume of 5% Na$_2$EDTA solution was added (final concentration of 0.1% in the sample). The samples were spiked with the appropriate volume of internal standard mixture. The preconditioning of the cartridges was performed with 5 mL of LC-MC-grade methanol and 5 mL of LC-MS-grade water and then the samples were loaded into the cartridges and percolated with a flow rate of 6 mL/min. After the extraction, the cartridges were washed with 5 mL of LC-MS-grade water and dried for 20 min. Elution of the analytes was performed twice with 5 mL of LC-MS-grade methanol at 1 mL/min and the extracts were evaporated to dryness under a gentle stream of nitrogen by means of a Techne Dri-Block heater Model DB-3D. The final step was the reconstitution, which was performed with 500 μL of methanol: water 20:80 (v/v) with 0.1% formic acid, and the samples were stored at −20 °C until analysis.

3.6. LTQ-FT Orbitrap Instrument Operational Parameters

The analysis of the samples was performed by a UHPLC Accela LC system, connected with a hybrid LTQ-FT Orbitrap XL 2.5.5 SP1 mass spectrometer, equipped with an electrospray ionization source (ESI) (Thermo Fisher Scientific, Inc., GmbH, Bremen, Germany) as reported in our previous work. Identification and quantification of pharmaceuticals were acquired using full scan mode in positive and negative ionization. Collision-induced dissociation (CID) was performed for the data-dependent acquisition (full MS/dd-MS2) and the mass tolerance window was set to 5 ppm. To process the data, Thermo Xcalibur 2.1 software (Thermo Electron, San Jose, CA, USA) was used. Chromatographic separation was performed on a reversed phase Hypersil Gold C18 (Thermo Fisher Scientific) analytical column (100 × 2.1 mm, 1.9 μm). In positive ionization (PI), two mobile phases, A: 0.1% formic acid in LC-MS-grade water, and B: 0.1% formic acid in LC-MS-grade methanol, were used. In negative ionization (NI), LC-MS-grade water and LC-MS-grade methanol were used as mobile phases A and B, respectively. The elution gradient in PI started at 95% A and remained for 1 min, progressed to 30% in 3 min, and then to 0% in 6 min and returned to 95% A after 3 min with re-equilibration of the column set at 1 min. The elution gradient in NI started with 90% A, remained for 0.5 min, progressed to 30% in 2 min, reached 10% in 3 min, decreased to 5% at 3.9 min, decreased to 0% at 4.5 min, and remained for 0.5 min. After 1 min, it returned to 90% A with re-equilibration of the column set at 2 min. The total run time for PI and NI was 10 and 8 min, respectively. In both cases, the injection volume was 20 μL and the flow rate was 0.4 mL/min. The oven temperature was maintained at 27 °C.

4. Conclusions

The results of the present study point out that heterogeneous photocatalysis using g-C_3N_4 catalysts is an effective treatment process for the elimination of psychiatric drugs from real hospital effluent wastewater samples. The photocatalytic pattern of the targeted compounds followed the pseudo-first-order kinetics. 1MoS_2/g-C_3N_4 presented higher photocatalytic performance than g-C_3N_4 in all experiments. All psychiatric drugs after the photocatalytic process were removed in percentages higher than 30% and 54% using g-C_3N_4 and 1MSCN, respectively. Five psychiatric compounds (amisulpride, venlafaxine, O-desmethyl venlafaxine, and quetiapine) were detected in inherent concentrations higher than LOQ, whereas fluoxetine was found in trace levels. Six transformation products were identified during the photocatalytic treatment; however, all of them were totally degraded at the end of the treatment.

Future studies should focus on the reuse of the catalyst and the monitoring of the toxicity along the treatment, as well as on the effective separation of the catalyst and its application on a larger pilot scale using natural solar irradiation.

Author Contributions: Conceptualization, I.K. and T.A.; methodology, I.K.; validation, formal analysis, I.R. and V.B.; investigation, I.R. and I.K.; resources, I.K.; data curation, I.R. and V.B.; writing—original draft preparation, I.R. and I.K.; writing—review and editing, I.R. and I.K.; visualization, I.R.; supervision, I.K.; project administration, T.A. and I.K.; funding acquisition, T.A. and I.K. All authors have read and agreed to the published version of the manuscript.

Funding: This research received no external funding.

Data Availability Statement: Data are contained within the article.

Acknowledgments: The authors would like to thank the Unit of Environmental, Organic and Biochemical high-resolution analysis—Orbitrap-LC–MS—of the University of Ioannina for providing access to the facilities.

Conflicts of Interest: The authors declare no conflict of interest.

References

1. Deegan, A.M.; Shaik, B.; Nolan, K.; Urell, K.; Oelgemöller, M.; Tobin, J.; Morrissey, A. Treatment Options for Wastewater Effluents from Pharmaceutical Companies. *Int. J. Environ. Sci. Technol.* **2011**, *8*, 649–666. [CrossRef]
2. Iqbal, J.; Shah, N.S.; Khan, Z.U.H.; Rizwan, M.; Murtaza, B.; Jamil, F.; Shah, A.; Ullah, A.; Nazzal, Y.; Howari, F. Visible Light Driven Doped CeO_2 for the Treatment of Pharmaceuticals in Wastewater: A Review. *J. Water Process. Eng.* **2022**, *49*, 103130. [CrossRef]
3. Tang, L.; Wang, J.; Jia, C.; Lv, G.; Xu, G.; Li, W.; Wang, L.; Zhang, J.; Wu, M. Simulated Solar Driven Catalytic Degradation of Psychiatric Drug Carbamazepine with Binary $BiVO_4$ Heterostructures Sensitized by Graphene Quantum Dots. *Appl. Catal. B Environ.* **2017**, *205*, 587–596. [CrossRef]

4. Kanakaraju, D.; Glass, B.D.; Oelgemöller, M. Titanium Dioxide Photocatalysis for Pharmaceutical Wastewater Treatment. *Environ. Chem. Lett.* **2014**, *12*, 27–47. [CrossRef]
5. aus der Beek, T.; Weber, F.A.; Bergmann, A.; Hickmann, S.; Ebert, I.; Hein, A.; Küster, A. Pharmaceuticals in the Environment-Global Occurrences and Perspectives. *Environ. Toxicol. Chem.* **2016**, *35*, 823–835. [CrossRef] [PubMed]
6. Rapti, I.; Kosma, C.; Albanis, T.; Konstantinou, I. Solar Photocatalytic Degradation of Inherent Pharmaceutical Residues in Real Hospital WWTP Effluents Using Titanium Dioxide on a CPC Pilot Scale Reactor. *Catal. Today* **2022**, in press. [CrossRef]
7. Jacob, L.; Bohlken, J.; Kostev, K. What Have We Learned in the Past Year? A Study on Pharmacy Purchases of Psychiatric Drugs from Wholesalers in the Days Prior to the First and Second COVID-19 Lockdowns in Germany. *J. Psychiatr. Res.* **2021**, *140*, 346–349. [CrossRef]
8. Lopez-Leon, S.; Lopez-Gomez, M.I.; Warner, B.; Ruiter-Lopez, L. Psychotropic Medication in Children and Adolescents in the United States in the Year 2004 vs 2014. *DARU J. Pharm. Sci.* **2018**, *26*, 5–10. [CrossRef]
9. Kosma, C.I.; Kapsi, M.G.; Konstas, P.S.G.; Trantopoulos, E.P.; Boti, V.I.; Konstantinou, I.K.; Albanis, T.A. Assessment of Multiclass Pharmaceutical Active Compounds (PhACs) in Hospital WWTP Influent and Effluent Samples by UHPLC-Orbitrap MS: Temporal Variation, Removals and Environmental Risk Assessment. *Environ. Res.* **2020**, *191*, 110152. [CrossRef]
10. Fernandes, M.J.; Paíga, P.; Silva, A.; Llaguno, C.P.; Carvalho, M.; Vázquez, F.M.; Delerue-Matos, C. Antibiotics and Antidepressants Occurrence in Surface Waters and Sediments Collected in the North of Portugal. *Chemosphere* **2020**, *239*, 124729. [CrossRef]
11. Vasskog, T.; Anderssen, T.; Pedersen-Bjergaard, S.; Kallenborn, R.; Jensen, E. Occurrence of Selective Serotonin Reuptake Inhibitors in Sewage and Receiving Waters at Spitsbergen and in Norway. *J. Chromatogr. A* **2008**, *1185*, 194–205. [CrossRef] [PubMed]
12. Telles-Correia, D.; Barbosa, A.; Cortez-Pinto, H.; Campos, C.; Rocha, N.B.F.; Machado, S. Psychotropic Drugs and Liver Disease: A Critical Review of Pharmacokinetics and Liver Toxicity. *World J. Gastrointest. Pharmacol. Ther.* **2017**, *8*, 26. [CrossRef] [PubMed]
13. Minguez, L.; Bureau, R.; Halm-Lemeille, M.P. Joint Effects of Nine Antidepressants on Raphidocelis Subcapitata and Skeletonema Marinoi: A Matter of Amine Functional Groups. *Aquat. Toxicol.* **2018**, *196*, 117–123. [CrossRef] [PubMed]
14. Grzesiuk, M.; Spijkerman, E.; Lachmann, S.C.; Wacker, A. Environmental Concentrations of Pharmaceuticals Directly Affect Phytoplankton and Effects Propagate through Trophic Interactions. *Ecotoxicol. Environ. Saf.* **2018**, *156*, 271–278. [CrossRef] [PubMed]
15. Best, C.; Melnyk-Lamont, N.; Gesto, M.; Vijayan, M.M. Environmental Levels of the Antidepressant Venlafaxine Impact the Metabolic Capacity of Rainbow Trout. *Aquat. Toxicol.* **2014**, *155*, 190–198. [CrossRef] [PubMed]
16. Ibáñez, M.; Bijlsma, L.; Pitarch, E.; López, F.J.; Hernández, F. Occurrence of Pharmaceutical Metabolites and Transformation Products in the Aquatic Environment of the Mediterranean Area. *Trends Environ. Anal. Chem.* **2021**, *29*, e00118. [CrossRef]
17. Sousa, M.A.; Gonçalves, C.; Vilar, V.J.P.; Boaventura, R.A.R.; Alpendurada, M.F. Suspended TiO$_2$-Assisted Photocatalytic Degradation of Emerging Contaminants in a Municipal WWTP Effluent Using a Solar Pilot Plant with CPCs. *Chem. Eng. J.* **2012**, *198–199*, 301–309. [CrossRef]
18. Johnson, G.J.; Bullen, L.A.; Akil, M. Review of emerging pollutants and advanced oxidation processes. *Int. J. Adv. Res.* **2017**, *5*, 2315–2324. [CrossRef] [PubMed]
19. Rapti, I.; Bairamis, F.; Konstantinou, I. g-C$_3$N$_4$/MoS$_2$ Heterojunction for Photocatalytic Removal of Phenol and Cr(VI). *Photochem* **2021**, *1*, 358–370. [CrossRef]
20. Bairamis, F.; Konstantinou, I. WO$_3$ Fibers/ g-C$_3$N$_4$ z-Scheme Heterostructure Photocatalysts for Simultaneous Oxidation/Reduction of Phenol/Cr(VI) in Aquatic Media. *Catalysts* **2021**, *11*, 792. [CrossRef]
21. Zhang, X.; Zhang, R.; Niu, S.; Zheng, J.; Guo, C. Enhanced Photo-Catalytic Performance by Effective Electron-Hole Separation for MoS$_2$ Inlaying in g-C$_3$N$_4$ Hetero-Junction. *Appl. Surf. Sci.* **2019**, *475*, 355–362. [CrossRef]
22. Kane, A.; Chafiq, L.; Dalhatou, S.; Bonnet, P.; Nasr, M.; Gaillard, N.; Dikdim, J.M.D.; Monier, G.; Assadie, A.A.; Zeghioud, H. g-C$_3$N$_4$/TiO$_2$ S-Scheme Heterojunction Photocatalyst with Enhanced Photocatalytic Carbamazepine Degradation and Mineralization. *J. Photochem. Photobiol. A Chem.* **2022**, *430*, 113971. [CrossRef]
23. Moreira, N.F.F.; Sampaio, M.J.; Ribeiro, A.R.; Silva, C.G.; Faria, J.L.; Silva, A.M.T. Metal-Free g-C$_3$N$_4$ Photocatalysis of Organic Micropollutants in Urban Wastewater under Visible Light. *Appl. Catal. B Environ.* **2019**, *248*, 184–192. [CrossRef]
24. Oller, I.; Malato, S.; Sánchez-Pérez, J.A. Combination of Advanced Oxidation Processes and Biological Treatments for Wastewater Decontamination-A Review. *Sci. Total Environ.* **2011**, *409*, 4141–4166. [CrossRef] [PubMed]
25. Colina-Márquez, J.; Machuca-Martínez, F.; Salas, W. Enhancement of the Potential Biodegradability and the Mineralization of a Pesticides Mixture after Being Treated by a Coupled Process of TiO$_2$-Based Solar Photocatalysis with Constructed Wetlands. *Ing. Compet.* **2013**, *15*, 181–190.
26. Amat, A.M.; Arques, A.; García-Ripoll, A.; Santos-Juanes, L.; Vicente, R.; Oller, I.; Maldonado, M.I.; Malato, S. A Reliable Monitoring of the Biocompatibility of an Effluent along an Oxidative Pre-Treatment by Sequential Bioassays and Chemical Analyses. *Water Res.* **2009**, *43*, 784–792. [CrossRef]
27. Konstas, P.S.; Kosma, C.; Konstantinou, I.; Albanis, T. Photocatalytic Treatment of Pharmaceuticals in Real Hospital Wastewaters for Effluent Quality Amelioration. *Water* **2019**, *11*, 2165. [CrossRef]
28. Lambropoulou, D.; Evgenidou, E.; Saliverou, V.; Kosma, C.; Konstantinou, I. Degradation of Venlafaxine Using TiO$_2$/UV Process: Kinetic Studies, RSM Optimization, Identification of Transformation Products and Toxicity Evaluation. *J. Hazard. Mater.* **2017**, *323*, 513–526. [CrossRef]

29. Skibiński, R. A Study of Photodegradation of Quetiapine by the Use of LC-MS/MS Method. *Cent. Eur. J. Chem.* **2012**, *10*, 232–240. [CrossRef]
30. Moreira, A.J.; Campos, L.O.; Maldi, C.P.; Dias, J.A.; Paris, E.C.; Giraldi, T.R.; Freschi, G.P.G. Photocatalytic Degradation of Prozac® Mediated by TiO_2 Nanoparticles Obtained via Three Synthesis Methods: Sonochemical, Microwave Hydrothermal, and Polymeric Precursor. *Environ. Sci. Pollut. Res.* **2020**, *27*, 27032–27047. [CrossRef]
31. Zheng, Q.; Xu, E.; Park, E.; Chen, H.; Shuai, D. Looking at the overlooked hole oxidation: Photocatalytic transformation of organic contaminants on graphitic carbon nitride under visible light irradiation. *Appl. Catal. B Environ.* **2019**, *240*, 262–269. [CrossRef]

Disclaimer/Publisher's Note: The statements, opinions and data contained in all publications are solely those of the individual author(s) and contributor(s) and not of MDPI and/or the editor(s). MDPI and/or the editor(s) disclaim responsibility for any injury to people or property resulting from any ideas, methods, instructions or products referred to in the content.

Review

Surface Modification of Biochar for Dye Removal from Wastewater

Lalit Goswami, Anamika Kushwaha, Saroj Raj Kafle and Beom-Soo Kim *

Department of Chemical Engineering, Chungbuk National University, Cheongju 28644, Korea; lalitgoswami660323@gmail.com (L.G.); kushwaha.anamika@gmail.com (A.K.); 100sarojraj@gmail.com (S.R.K.)
* Correspondence: bskim@chungbuk.ac.kr

Abstract: Nowadays, biochar is being studied to a great degree because of its potential for carbon sequestration, soil improvement, climate change mitigation, catalysis, wastewater treatment, energy storage, and waste management. The present review emphasizes on the utilization of biochar and biochar-based nanocomposites to play a key role in decontaminating dyes from wastewater. Numerous trials are underway to synthesize functionalized, surface engineered biochar-based nanocomposites that can sufficiently remove dye-contaminated wastewater. The removal of dyes from wastewater via natural and modified biochar follows numerous mechanisms such as precipitation, surface complexation, ion exchange, cation–π interactions, and electrostatic attraction. Further, biochar production and modification promote good adsorption capacity for dye removal owing to the properties tailored from the production stage and linked with specific adsorption mechanisms such as hydrophobic and electrostatic interactions. Meanwhile, a framework for artificial neural networking and machine learning to model the dye removal efficiency of biochar from wastewater is proposed even though such studies are still in their infancy stage. The present review article recommends that smart technologies for modelling and forecasting the potential of such modification of biochar should be included for their proper applications.

Keywords: post-processing modification; surface-engineered biochar; dye removal; machine learning; artificial neural network

1. Introduction

Unprecedented rising globalization, industrialization, urbanization, and anthropogenic human activities have led to a worldwide shortage of clean water [1–3]. In this regard, wastewater contaminated by water-soluble dyes is one of the prime environmental issues [4–6]. Nowadays, numerous dyes and additives are being utilized enormously in industrial applications such as textile, paper, printing, paint, laundry, cosmetics, carpet, leather, food, and rubber [4]. Globally, more than 10,000 different types of natural and synthetic dyes are produced annually, weighing in the range of 7×10^5–1×10^6 tons [7]. The chemical complexity, stability, and poor biodegradability of these dye-contaminated wastewaters are of prime concern and continue to limit the clean water resources available. The rising demand for dyes is simultaneously leading to the malefactors of inadvertent discharge of dye-contaminated wastewater into water streams; it directly affects the life of aquatic flora and fauna along with the food chain and is indirectly deleterious to human health [8–10]. Henceforth, it is highly significant to develop sustainable remediation solutions to remove soluble dyes and other contaminants from water.

It typically costs about $1 billion annually to treat 640 million m³ of textile and dyeing wastewater [11]. Several conventional dye-contaminated wastewater treatment technologies are used for dye removal from effluents, such as chemisorption, electrochemical oxidation, ozonation, ion exchange, membrane filtration, anaerobic lagoons, sedimentation, oxidation ponds, coagulation flocculation, photocatalytic degradation, and gamma irradiation [4,11–13]. Among these treatments, adsorption is one of the most sustainable and cost-effective techniques because other technologies require a lot of chemicals, high energy, and are not cost effective. Carbonaceous materials possessing a high specific surface area (SSA) are widely used as adsorbing agents in wastewater treatment for dye removal [14,15]. Inexpensive adsorbents, preferably derived from natural materials and industrial wastes/by-products, such as biochar, cellulose, aerogels, activated charcoal, bentonite, fly ash, and silica, can be utilized for wastewater treatment [16–20].

Biochar (BC) is a highly stable carbonaceous material that is aromatized and amorphous in nature. It is usually formed after thermochemical conversion of organic matter and wastes at temperatures of 350–750 °C under limited oxygen conditions [21–23]. Its high SSA, pore volume, hydrophobicity, etc., enable its use as an efficient biomaterial for carbon sequestration, soil improvement, climate change mitigation, catalysis, wastewater treatment, energy storage, and waste management [4]. In addition, it has been well recognized as a potentially highly efficient, low-cost, and eco-friendly adsorbent for the removal of organic and inorganic pollutants, particularly heavy metals and dyes from wastewater.

The physicochemical properties and primary composition of BC are considerably altered according to the biomass feedstock, carbonization procedure, degree of pyrolysis, activation, and functionalization techniques [23]. Indulging in modification techniques, BC depicts multiscale porous structures, extensive surface functional groups, and high surface areas, utilizing various organic feedstocks. In addition, inherent functional sites (hydroquinone, defects, etc.) and minerals (silica, transition metals) add BC as a promising ingredient to tailor heterojunctions/composites. The facile metal impregnation, gas activation, sulfonation, and ionic liquid grafting of BC enable many advantages in catalytic processes owing to enhanced accessibility to active sites, excellent $\pi-\pi$ interactions, high active surface area, and enhanced charge transfer [24].

In this review, we focused on surface modification and alteration of BC to enhance the efficiency of dye removal from wastewater. A network visualization of terms associated with biochar and dye with a minimum number of 10 occurrences of 3589 associated keywords is represented in Figure 1 (assessed 30 June 2022). It represents the current trends in research related to the application of biochar in association with dye in the Web of Science. Here, the various colors of the nodes represent different clusters, while the size of each bubble depicts its frequency of occurrence. A literature review of BC reports that thermochemical conversion often exhibits low reactivity and selectivity for dye removal [25]. To eliminate these limitations, BC is tailored via numerous techniques to achieve the desired selectivity and reactivity to enhance surface sites and hydrophobicity for regulating the dye removal kinetics [26–28]. Factors affecting BC properties and various post-processing modifications, functionalization, and BC activation for dye removal are included. Literature review reports on surface modification of BC for dye removal are very scarce. The application of smart digital technologies such as machine learning and artificial neural networks (ANN) is also covered to further enhance dye removal via BC.

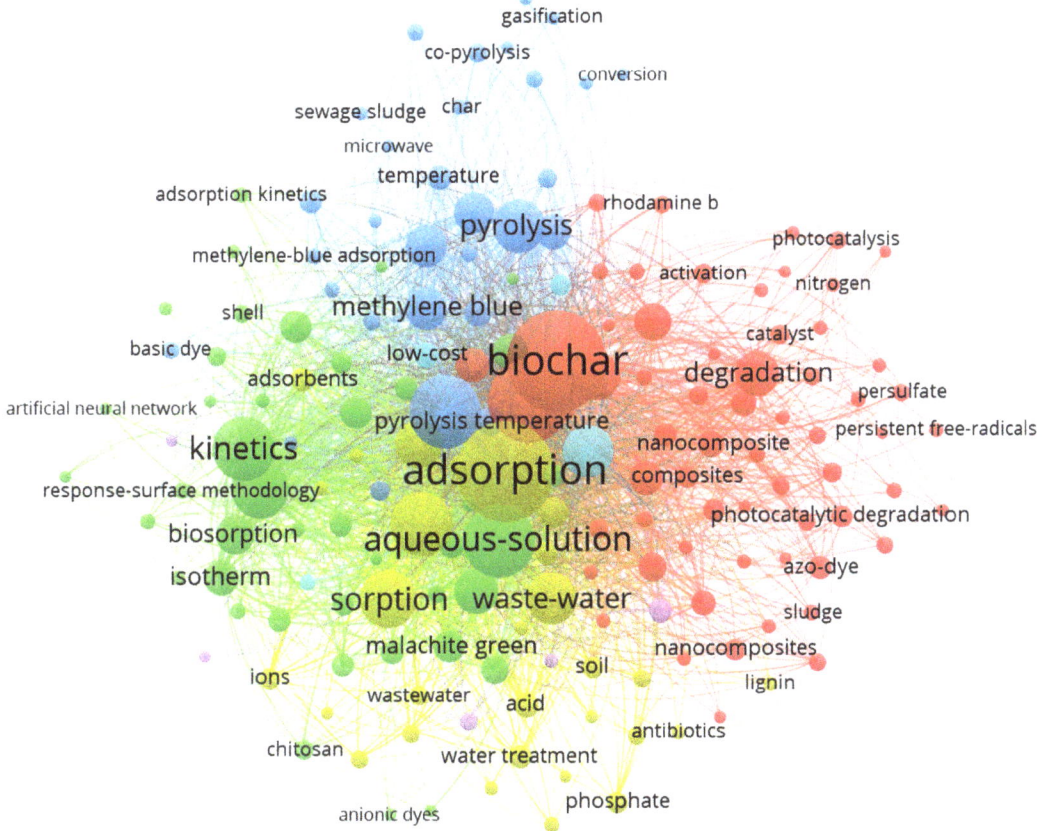

Figure 1. Network visualization of terms associated with biochar and dye.

2. Biochar

2.1. Production of Biochar

Various methods have been employed for BC production, such as combustion, torrefaction, gasification, and pyrolysis (slow and fast types) [29]. Each method results in a different char yield and carbon level. These processes differ based on temperature, char yield, heating rate, feedstock residence time, and carbon content and yield. Earlier, combustion was used to produce charcoal from woody biomass; however, it resulted in low yields and extreme air pollution. With time, the advancement in technology (endothermic and exothermic processes) allows the maximum energy extraction from organic matter. Combustion and gasification generate heat and gas, respectively, by thermally decomposing organic matter in an oxic environment [30]. However, these processes are less effective and satisfactory in decreasing emissions while meeting energy requirements. On the other hand, pyrolysis is the most common, oldest, and the most effective process for meeting energy requirements and reducing emissions. Figure 2 depicts the mechanism of BC formation from cellulose, hemicellulose, and lignin [31]. Pyrolysis leads to the production of BC (solid), bio-oil (liquid), and syngas (gas). The generated by-products can also be utilized as energy substances to meet the energy crisis.

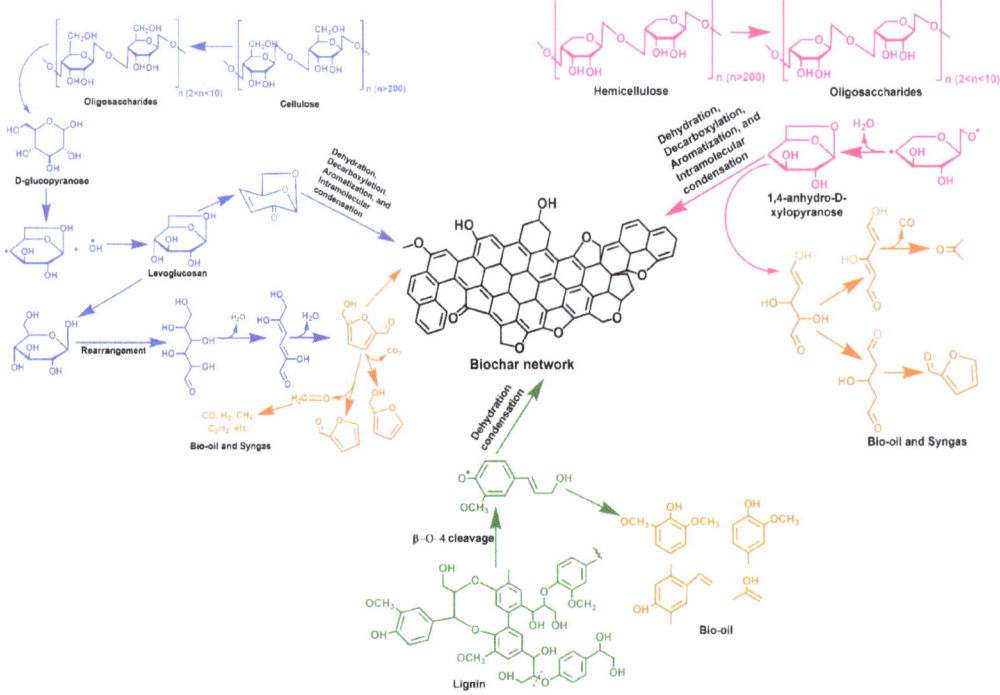

Figure 2. Mechanism of biochar formation from cellulose, hemicellulose, and lignin.

Here, pyrolysis can be classified into slow, intermediate, and fast pyrolysis in accordance with the heating rate, temperature, pressure, and residence time [23]. For high yield of BC, slow pyrolysis should meet heating rate of 0.1–1 °C min^{-1}, whereas fast pyrolysis and high temperature can result in generation of more liquid and syngas. In addition, the mass fraction range of pyrolysis solid increases with increasing temperature. In case of intermediate pyrolysis where the conditions are between slow and fast pyrolysis, there is a balance yield of BC and bio-oil yield, but it is rarely approved from a BC production perspective [32,33]. Further, the properties of BC depend on numerous factors and parameters such as pyrolysis temperature and type of feedstock.

Some conceivable approaches for more efficient BC production are proposed as follows: (i) improving energy efficiency via utilization of batch reactors through continuous feeding pyrolyzers; (ii) enhancing energy efficiency and BC yield using exothermic operation; (iii) tuning the BC properties (texture feature and surface chemistry) by adjusting the operating conditions; (iv) improving process economics and reducing pollutant emissions; (v) exploring new biomass-based feedstocks [34].

2.2. Characteristics of Biochar

2.2.1. Physical Properties

The physical properties of BC are directly dependent on the biomass feedstock and pyrolysis conditions, including biomass pre-treatment and handling. In particular, the attrition, crack formation, and microstructural rearrangement generated under different pyrolysis conditions change the native structure of the biomass feedstock to varying degrees [34]. As the pyrolysis process progresses, the biomass feedstock is reduced, converting macropores into mesopores and micropores. The ideal BC physical structure can be enhanced by increasing the pyrolysis temperatures to the point where deformation occurs [35]. Thermal decomposition of organic materials begins at >120 °C. More specifically, hemicellulose, cellulose, and lignin decompose at 200–260 °C, 240–350 °C, and 280–500 °C, respectively.

Nano and Macroporosity

The pore size distribution is one of the most important physical properties. Macropores (>50 nm), mesopores (2–50 nm), and micropores (<2 nm) constitute the total pore volume of BC. Here, micropores are the prime contributors to the BC surface area and are also responsible for efficient adsorbents, whereas mesopores play a noteworthy role in the liquid–solid adsorption process [24,34]. Further, micropores are amended via prolonging pyrolysis at higher temperatures which simultaneously release the volatile content of the feedstock along with the creation of numerous pores. Slow pyrolysis enhances the porosity of BC via slow decomposition of lignin content. The surface area of BC increases quickly when the pyrolysis temperature exceeds >400 °C, owing to thermal condensation of the biomass feedstock and micropore formation [35]. Biomass possessing aromatic lignin cores, aliphatic alkyl, and ester groups produces a higher surface area when pyrolyzed at higher temperatures.

Particle Size Distribution

BC particle size is determined by biomass rigidity against shrinkage and slow destruction during the pyrolysis process. Usually, the BC particle sizes are smaller compared to the un-pyrolyzed biomass. However, agglomeration may occur during pyrolysis, resulting in BC with a larger particle size [34]. Post-mechanical stresses may also occur, crumbling BC more susceptible than the original biomass. Slow pyrolysis (5–30 °C/min) results in a smaller size of BC. BC particle size has an inverse correlation with reaction temperature. For instance, an increase in reaction temperature (450 °C to 700 °C) results in the formation of smaller size particles [23]. This could be due to the augmented susceptibility to attrition because of the lower biomass/BC tensile strength [36].

Density

Apparent/bulk/solid density can be evaluated relating to determining the physical properties of BC. Usually, there is an antagonistic relationship between bulk density and solid density. In general, the solid density of BC is higher than that of the original biomass as a result of the release of volatile and condensable compounds and the formation of graphitic crystallites [34,36]. In contrast, BC has a lower bulk density compared to wood precursor owing to biomass drying and carbonization.

Mechanical Strength

This property correlates directly with density and inversely with porosity. Few studies have reported the evaluation of mechanical strength of BC [37]. In comparison to the virgin wood, monolithic carbonized wood BC has lower stiffness (37%) and higher strength (28%). Here, both the reduced modulus and hardness of pyrolyzed wood BC incessantly increases with increasing temperature (between 700 and 2000 °C) [38]. A negative effect on these properties follows with a further increase in the pyrolysis temperature. In addition, compared to harder native woods (low ash and high lignin content), fruit stones/pits and nutshells are more appreciated for BC production depicting admirable mechanical properties [34]. Further, it was found that the pyrolytic temperature is more dominant than the residence time in determining the mechanical properties of BC.

2.2.2. Chemical Properties

BC properties depend on the biomass type and pyrolysis temperature. In fact, data on synthesis techniques can be used to predict the properties and functions of BC [39]. In general, with a surge in pyrolysis temperature, BC yield declines exponentially while BC alkalinity (pH) displays a linear rise [34,39]. The increase in pH can be ascribed to thermal decomposition of hydroxyl groups and other weak bonds within the BC structure at high temperatures. In comparison, due to the removal of acidic functional groups, the cation exchange capacity of BC is inversely proportional to the pyrolysis temperature [34,39]. BC derived from biosolids display the highest cation exchange capacity due to the presence of minerals (P, Mg, Ca, Na, and K) in the biosolids, which encourage the generation of oxygen-containing functional groups on the surface of BC [40]. Unlike ash, the volatile matter content of BC decreases linearly with increasing pyrolysis temperature. Ash formation is due to inorganic minerals remaining following the decomposition of H, O, and C of biomass [39]. The destruction of hydroxyl, azanide, and other weakly bonded groups results in the reduction of these elements with increasing temperature. However, the C content gradually declines, resulting in a higher C content in the BC at the enhanced pyrolysis temperatures [39].

2.2.3. Microchemical Characteristics

BC microchemical properties can influence its superficial sorption characteristics. The microchemical properties vary significantly depending on the nature and composition of solid phase, entrapped oils and their arrangement, and indicate the electrochemical properties and functional groups on the BC surface [34,41]. Dried biomass pyrolysis is activated by hemolytic cleavage of covalent bonds to release free radicals and structural O [41]. In the initial phases of pyrolysis, free radicals are generated from low atmospheric O_2 levels [34,40,41]. This process proceeds with the formation of carboxyl and carbonyl groups, followed by their cleavage into CO_2 and CO. Finally, BC residues are produced from free radical fragments that recombine with the substrate in various ways [41].

2.2.4. Organo-Chemical Characteristics

In general, the H-to-C ratio declines from ~1.5 to ≤0.5 in lignocellulosic biomass with pyrolysis temperatures of >400 °C. This reduction in the H-to-C ratio could be due to changes in the elemental (N, O, H, and C) content during thermal decomposition of the biomass. O-to-C and H-to-C ratios (displaying the degree of aromaticity and maturation) in BC decrease at higher reaction temperatures [23,34]. Burnt peat displayed an H-to-C ratio of 1.3 when the maximum C content was either associated through a hydroxyl group or directly bonded to a proton. If the H-to-C ratio drops to 0.4–0.6, it indicates that every second to third C is associated with a proton [34].

2.3. Factors Influencing Biochar Sorption Efficiency

BC is a carbonaceous material derived via biomass thermal conversion (pyrolysis, gasification, hydrothermal carbonization, and torrefaction) under oxygen-limited conditions [4]. Typically, cellulose (40–60%), hemicellulose (20–40%), and lignin (10–25%) are included within biomass [5]. Biomass structural building blocks undergo a sequence of reactions during thermal decomposition, such as dehydration, de-polymerization, rearrangement, re-polymerization, condensation, and carbonization at various temperatures, producing BC, bio-oil, and syngas [22]. Based on the feedstock, the desired product and its application vary. Biomass characteristics include elemental composition, size, and ash mineral content, whereas thermal conditions include temperature, heating rate, pressure, and residence time [11,23]. Several parameters affect BC performance during adsorption. BC physisorption isotherm characteristics govern properties such as morphological structure and reactivity depending on pyrolytic conditions.

2.3.1. Temperature

Reaction temperature plays a vital role in the process and reaction rate. Huang et al. [42] suggested that the adsorption is chemical rather than physical when the equilibrium temperature does not significantly affect the adsorption process. Based on the pore size, adsorbent materials of various sizes are grouped into micropores, mesopores, and macropores, which are highly dependent upon the production conditions of BC. BC produced by pyrolysis at high temperatures contains high micropore volumes ranging from 50% to 70% of the total pores. Adsorption of contaminants by BC is an endothermic process and the adsorption capacity increases with increasing temperature.

An experimental study was conducted by Tan et al. [43] on Cu (II) adsorption and textile dye adsorption into food waste BC, and thermodynamics parameters were calculated. In both cases, the heat of reaction was positive and Gibbs's free energy was negative. According to Ambaye et al. [44], as the reaction temperature increases, the BC surface area increases as the amount of oxygen-containing functional groups on the surface decreases. Wu et al. [45] used litchi peel biomass and increased the activated temperature from 650 °C to 850 °C, which ultimately increased the surface area from 531 $m^2\ g^{-1}$ to 1006 $m^2\ g^{-1}$ and the pore volume from 0.328 $cm^{-1}\ g^{-1}$ to 0.588 $cm^{-1}\ g^{-1}$.

One experimental setup was carried out by Qambrani et al. [46] to perceive the effect of temperature on BC yield and quality. The yield, total nitrogen and organic carbon content, and cation exchange capacity values were found to decrease with increasing pyrolysis temperature. Therefore, low temperatures are highly recommended for BC production from poultry litter. This condition applies analogously to the condition for BC obtained from vegetative materials.

2.3.2. Solution pH

Solution pH plays a vital role in controlling the surface charge of the adsorbent and the degree of ionization of adsorbate [47]. At high pH, a negative charge exists on the surface with deprotonation of the phenolic and carboxylic groups. At low pH, basic functional groups become protonated and positively charged, promoting adsorption of anions. Adsorption by BC is known to be a function of pH, medium, and deprotonation of functional groups [48]. The pH at the zero-point change is the point at which the net charge on the surface of any adsorbent in solution becomes neutral. At this point, pH highly affects the adsorption efficiency of the BC active surface by providing active functional groups for a charged solution. Therefore, increasing the pH of a solution containing BC leads to a negative potential by increasing the negative charge on the BC surface [49].

Qiu et al. [50] reported that the positive charge on activated carbon at pH 3.0 was higher than at pH 6.5. In contrast, the negative charge on BC was more negative at pH 6.5. Therefore, BC was more efficient than activated carbon at adsorbing dyes from solution. Similarly, the effect of pH on methylene blue dye adsorption was studied at various pH (2–10) [51]. As the pH increases, the number of positively charged sites decreases, increasing methylene blue adsorption. Although the high adsorption of the methylene blue is favorable in alkaline solution, the adsorption did not change when the pH of the solution increased. They observed a maximum dye adsorption (69.07%) at pH 4 [51]. Lin et al. [52] tested the microalgae derivative BC to remove dye and found that the higher the pH of the aqueous solution, the greater the adsorption capacity of positively charged dyes with surface electrostatic attraction due to the negative potential effects. Similarly, Huang et al. [53] studied the removal of organic dye using BC derived from *Spirulina platensis* algae biomass residue. The maximum adsorption capacity was obtained at the alkaline pH with an adsorbent dosage of 2000 mg/L and an initial dye concentration of 90 mg/L.

2.3.3. Adsorbent Dosage

The adsorbent dosage greatly influences the sorbent-sorbate equilibrium in the adsorption system. Due to the availability of more sorption sites, the removal efficiency of organic and inorganic pollutants increases with increasing adsorbent dosage. When the dosage rate is in excess, the adsorption capacity of BC decreases [48]. The reduction in adsorption may be attributed to the overlap of adsorption sites which ultimately shield the pores [48]. Moreover, at a lower dosage of adsorbent at a given initial dye concentration, the ratio of dye to adsorbent molecules increases, leading to an increase in specific uptake. In contrast, at a higher dosage of adsorbent, the availability of dye ions is insufficient for adsorption to the sites [53].

The effect of adsorbent dosage of BC derived from eucalyptus bark to remove methylene blue dye from aqueous solution was studied [53]. They observed that the dye adsorption decreased from 56.8 to 25.5 (mg/g) as the adsorbent dosage was increased from 0.01 to 0.03 g. According to Zhang et al. [54], Congo red dye removal (82.2–83.6%) increased as the adsorbent dosage was increased from 0.2 to 1 g/100 mL. Green BC composite was used for methylene blue removal with adsorbent doses up to 2 g/L. It was found that the adsorption performance decreases with increasing dosage, reducing the number of methylene blue molecules per unit adsorbent [54].

2.3.4. Initial Dye Concentration

For an efficient and effective adsorption process, the initial concentration of adsorbate plays a vital role. As the initial dye concentration increases, the percentage removal of dye increases. The initial dye concentration provides the driving force to overcome the mass transfer resistance of dye between the aqueous and solid phases [51]. According to Chowdhury et al. [55], at high concentrations, all dye molecules in solution do not interact with the binding sites of the adsorbent because the adsorbent has a limited number of active binding sites that saturate at a certain concentration. Experimental work carried out by Dawood et al. [56] found that the percentage of methylene blue removal decreased from 80.5% to 36.8% as the initial concentration (10–100 mg/L) of the methylene blue dye solution increased. Another experimental work was also carried out in which the percentage removal of methylene blue dye drastically reduced from 85.93% to 41.40% with increasing initial dye concentration (20 to 50 mg/L) [57].

2.3.5. Heating Rate

The heating rate is also one of the major factors affecting BC properties. According to Xu and Chen [58], heating rates of <20 $°C\ s^{-1}$ for BC production from pine sawdust permitted the natural porosity of sawdust to be shifted to BC without significant structural variations. In contrast, devolatilization occurred at a heating rate of 500 $°C\ s^{-1}$, resulting in deformation of the sawdust cell structure. Finally, lower pyrolysis temperature and lower heating rate along with high residence time led to BC formation. Li et al. [59] produced BC from lignin (heating rate of 10 $°C\ min^{-1}$) and the abundance of -OH started to decrease from 350 °C. They concluded that a lower heating rate with long residence time increased the thermal stability of the -OH functional groups on the BC surface. BC carbon content increased from 75.39% to 88.35%, 75.15% to 88.28%, and 77.13% to 89.70% at various heating rates 5, 10, and 15 $°C\ min^{-1}$, respectively. Li et al. [59] concluded that the smaller particle size leads to a complete pyrolysis process, and the hydrogen, sulfur, and oxygen content decreases with increasing temperature. With an upsurge in temperature, the H/C and O/C atoms gradually decrease and BC becomes more carbonaceous and aromatic.

2.3.6. Particle Size

The rate of heat and mass transfer of particles and the degree of subordinate reactions are significantly influenced by the particle size of the biomass. The particle size depends on the process being carried out and the feedstock materials. Larger particle sizes (>1.8 mm) have greater temperature gradients and thus provide higher BC in comparison with smaller particle sizes [46]. An experiment was conducted with various particle sizes of wood, and it was observed that maximum BC yield (28%) was attained with particle sizes 0.224–0.425 nm at 500 °C and 223.15 $°C\ min^{-1}$ heating rate in an inert environment [60]. Mechanical properties such as mechanical strength, yield stress, and BC density are highly dependent on biomass particle size. As a result of examining the effect of particle size on wood biomass, it was found that wood BC produced from smaller particle sizes showed higher yield stress and density compared to larger ones [61]. The smaller particle size is appropriate for fast pyrolysis due to the uniform and effective heat transfer facilitating the release of volatiles.

2.3.7. Feedstock Composition

Feedstock selections also greatly impact BC properties and elemental composition. It affects various physical properties of BC, such as pH, pore structure, surface area, adsorption efficiency, and other chemical properties [34]. Compared to cellulose, hemicellulose, lignin, and some organic compounds, a feedstock consisting of animal manure results in higher BC nutrients. As the temperature increases, the volatile compounds in the biomass decrease along with a decrease in the quantity of surface functional groups, but the surface area and ash content of BC increase [23,24]. Feedstock materials are oxygen-rich and hydrogen-deprived, for example, sugars are non-graphitizing and create robust cross-linked arrangements that immobilize the structure and tie the crystallites into a stiff mass. The various carbon and nitrogen contents in char produced from plant-derived biomass were found to be increased relative to biomass. However, the use of mineral-rich feedstock (manure) may reduce it [62].

2.4. Mechanisms of Dye Adsorption

The enhanced performance of BC (without activation) can be due to oxygen containing surface functional groups and other inorganic components that play a significant role in dye adsorption instead of only surface area and porosity. BC inorganic elements upsurge the hydrophilicity towards dyes and act as catalysts during the adsorption process [63]. However, BC activation, surface area, and surface alkalinity can also influence adsorption capacity. Figure 3 represents the adsorption process and mechanism for dye removal from wastewater. Specific adsorption mechanisms include (i) physical adsorption, (ii) ion exchange; (iii) electrostatic interactions; (iv) precipitation; and (v) surface complexation [34,35]. However, the pollutant removal process always works in conjunction with numerous mechanisms.

2.4.1. Physical Adsorption

Dye adsorption onto the BC surface primarily occurs via physical interactions such as pore filling, π stacking, and H-bonding. However, BC surface area, porosity, and aromaticity influence the physical adsorption. The enhanced surface area and pore volume favor the diffusion of contaminants [64]. The aromatic structure of BC is promising for forming π-stacking and H-bond interactions with pollutants [65]. The pore filling, π–π interactions, and H bonding between BC and dye can be improved via post-modification of BC. For example, pore filling is an illustrative of physical adsorption and can be accredited to a widespread distribution of pores via BC [66]. Hydroxyl and amine groups can be an advantage for π–π interactions due to electron-deficient functional groups on the surface of cationic dyes [7]. For instance, cationic dyes H-bonding of O and/or N center generated free hydroxyl groups on the surface of sulfur-doped tapioca peel waste BC [67], whereas fly ash and agricultural waste-derived BC alkali-fusing with O-containing surface groups resulted in H-bonding [68]. The generation of H-bonds between acid orange 7 dye and the adsorbent accounts for the adsorption process [69]. Siddiqui et al. reported the H-bonds between methylene blue and MnO_2/BC occurs due to the interaction between the -OH groups present in MnO_2/BC and the acceptor present in the methylene blue molecules [70]. Likewise, π–π interactions/π-effects/π-interactions (non-covalent) involve the π system, where positively charged molecules interact with negatively charged surfaces, similar to electrostatic interactions [34].

2.4.2. Ion Exchange

Dye adsorption on the BC surface occurs via exchange of metal ions/mineral elements or replacement of BC functional groups [64]. Furthermore, the ion exchange mechanism involves ion exchange between a liquid (dye solution) and a solid phase (adsorbent). Pirbazari et al. suggested that two principal mechanisms are involved in the removal of methylene blue dye on NaOH-treated wheat straw impregnated with Fe_3O_4, namely, surface complex formation and ion exchange between the dye molecules and adsorption surfaces [71]. According to Zheng et al., the adsorption of anionic dyes such as Congo red and methyl orange on graphene oxide-NiFe layered double hydroxide is achieved by electrostatic attraction and ion exchange phenomena [72].

2.4.3. Electrostatic Attraction

Synthetic dyes are categorized into cationic and anionic dyes. Malachite green, methylene blue, and rhodamine B are examples of cationic dyes [73], and tartrazine, sunset yellow, Congo red, and orange G are examples of anionic dyes [74]. Knowledge of dye classification aid in choosing the accurate adsorbent as a positively charged adsorbent is effective in removing negatively charged dyes due to electrostatic interactions [74]. An electrostatic attraction occurs between the dye and BC surface when the surface charges are opposite [75]. Electrostatic interaction is one of the prime adsorption mechanisms for the adsorption of synthetic dyes on BC and can influence the adsorption rate. For instance, BC derived from litchi peel was assessed for adsorption of anionic and cationic dyes. The adsorption capacity was 404.4 and 2468 mg g^{-1}, respectively [45]. However, other adsorption mechanisms were also involved in the adsorption, and since the dye and BC both have opposite charges, the dye adsorption significantly occurred via electrostatic interactions [45]. Several BCs were designed for cationic and anionic dyes that fit either pseudo-second order or Elovich kinetic models, where both are indicative of electrostatic interactions and chemisorption. Shen and Gondal reported that electrostatic and intermolecular interactions govern the adsorption of rhodamine dye on adsorbent surfaces [76].

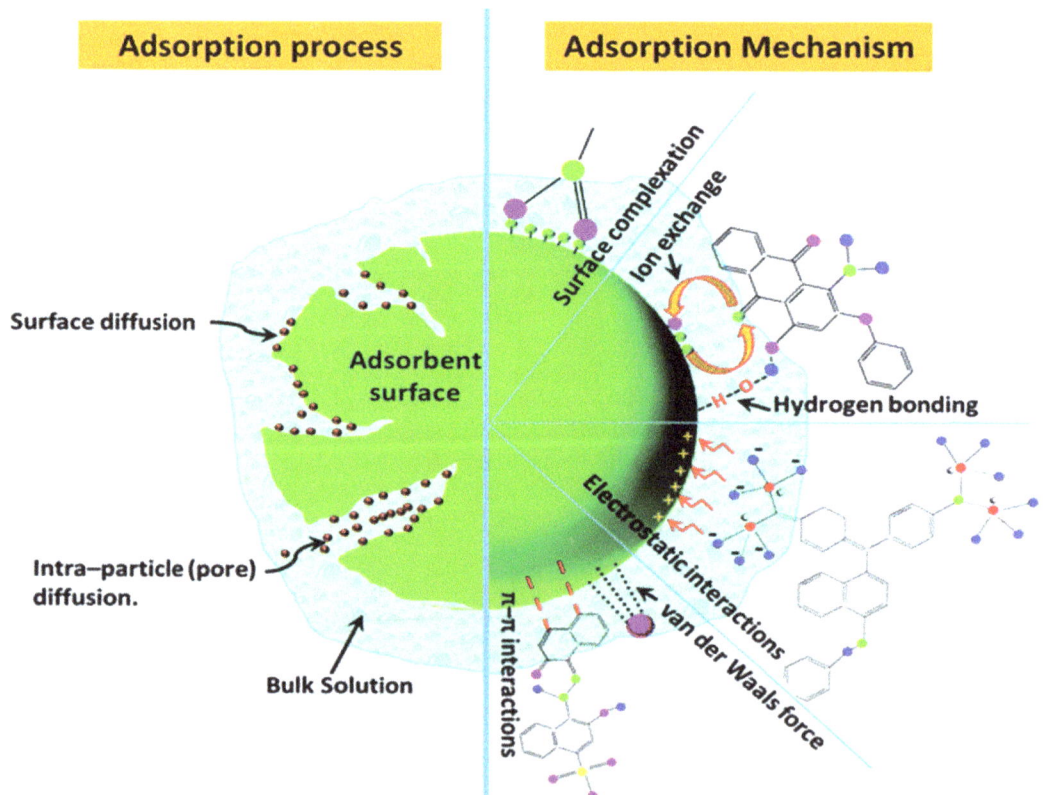

Figure 3. Adsorption process and mechanism for dye removal from wastewater [77].

3. Post-Production Modification of Biochar for Dye Removal

Biomass breaks down thermally in an oxygen-limited environment to generate BCs composed of numerous refractory oxides based on the feedstocks such as $CaCO_3$, Fe_2O_3, SiO_2, Al_2O_3, and $CaSO_4$ [78]. Moreover, pristine BC surfaces are generally negatively charged and linked with oxygen-bearing functional groups, thus displaying specific adsorption towards cations (e.g., heavy metals ions) [79]. However, adsorption of anionic species (i.e., oxyanion, anionic dyes, and organics) of BC is inadequate [80]. This restricts BC applications in various fields, instigating scientists to introduce novel metal oxides to alter BC for targeted applications [81]. Table 1 enlists few literatures available onto the treatment of dye-contaminated wastewater using BC and its nanocomposites. Table 1 enlist some literature available on the BC utilization for dye removal from wastewater.

Table 1. Literature summary on modified biochar for dye removal.

Synthetic Dye	Biochar		Adsorption Capacity	Conditions Adsorption Mechanism	Reference
	Feedstock	Method of Production			
Congo red	Litchi peel BC	Hydrothermal carbonization	404.4 mg g^{-1}	Pore filling effect π–π interaction Electrostatic interaction Hydrogen bonding	[45]
	Orange peel waste	Microwave pyrolysis using CO_2 and steam activation	136 mg g^{-1}	Electrostatic interaction	[82]
	Switchgrass	Pyrolysis (900 °C)	22.6 mg g^{-1}	Electrostatic interaction π–π interaction	[83]
Acid violet 17	Pine tree-derived BC	-	90%	Electrostatic interactions	[84]
Malachite green	Cladodes of cactus (*Opuntia ficus-indica*)	Pyrolysis at 400 °C followed by NaOH impregnation, pyrolysis at 500 °C and rinsing with HCl	1341 mg g^{-1}	Π–π EDA interaction Cation-π interaction Hydrogen bonding	[85]
	Corn straw	HNO_3 treatment, followed by washing with distilled water and drying. Then, NaOH activation and drying, followed by pyrolysis at 500 °C and $FeCl_3$ modification by precipitation technique	515.8 mg g^{-1}	Electrostatic attraction	[86]
	Frass of mealworms (*Tenebrio molitor* Linnaeus 1758)	Pyrolysis at 800 °C	1738.6 mg g^{-1}	Electrostatic interaction π–π interaction Hydrogen bonding	[87]
	Litchi peel BC	Hydrothermal carbonization at 850 °C	2468 mg g^{-1}	Pore filling effect π–π interaction Electrostatic interaction Hydrogen bonding	[45]
	Tapioca peel waste	Pyrolysis of feedstock, then mixing with thiourea and followed by pyrolysis at 800 °C to create sulfur-doped BC	30.2 mg g^{-1}	Electrostatic interaction Hydrogen bonding	[67]
	Wakame (macroalgae)	Chemical activation with KOH followed by pyrolysis at 800 °C	4066.9 mg g^{-1}	Electrostatic interaction π–π stacking Hydrogen bonding van der Waals force	[88]
Methylene blue	Macroalgae (*Undaria pinnatifida*)	Chemical activation with KOH followed by pyrolysis at 800 °C	841.6 mg g^{-1}	Electrostatic interaction π–π interaction Hydrogen bonding and van der Waals force	[88]
	Rice straw and fly ash	Alkali-fusion pre-treatment of fly ash with NaOH, followed by mixing with rice straw and pyrolysis at 700 °C	143.8 mg g^{-1}	Electrostatic interaction π–π interaction	[68]
Orange G	Switchgrass	Pyrolysis at 900 °C	38.2 mg g^{-1}	Electrostatic interaction π–π interaction	[83]

Table 1. Cont.

Synthetic Dye	Biochar		Adsorption Capacity	Conditions Adsorption Mechanism	Reference
	Feedstock	Method of Production			
Rhodamine B	Macroalgae (*Undaria pinnatifida*)	Chemical activation with KOH followed by pyrolysis at 800 °C	533.8 mg g^{-1}	Electrostatic interaction π–π interaction Hydrogen bonding van der Waals force	[88]
	Tapioca peel waste	Pyrolysis of feedstock, then mixing with thiourea and followed by pyrolysis at 800 °C to create sulfur-doped BC	33.1 mg g^{-1}	Electrostatic interaction Hydrogen bonding	[67]
Sunset yellow/ Tartrazine	Corncob	Pyrolysis at 400 °C followed by mixing with triethylenetetramine, drying and H$_2$SO$_4$ treatment to achieve a positively charged BC	77.1 mg g^{-1}	Amine groups on the surface Electrostatic interaction	[74]

Additionally, conventional pyrolysis generates BC with inefficient physicochemical properties such as surface area, surface-oxygenated groups, pore volume, and width. Different feedstock types have different BC production conditions and physicochemical properties [89]. However, the existing -COOH, -OH, and C=O functional groups are crucial in water treatment [90]. The hydrophilic or hydrophobic properties depend on the functional groups (nature and type) present on the BC surface. Various methods such as impregnation, steam activation, and chemical and heat treatments have been utilized to improve the abundant BC properties [91]. In the presence of other catalysts, BC acts as a photocatalyst due to the physicochemical properties (enhanced surface area, active site, charge separation, porous volume, functional groups, catalyzing ability, stability, and recoverability) [92]. However, photocatalyst-based BC production needs to be optimized [93]. For instance, the addition of ZnO to BC improves the nanocomposite adsorption capability and photocatalytic ability [94]. Figure 4 shows various modification and classification of engineered BC for the dye removal from wastewater.

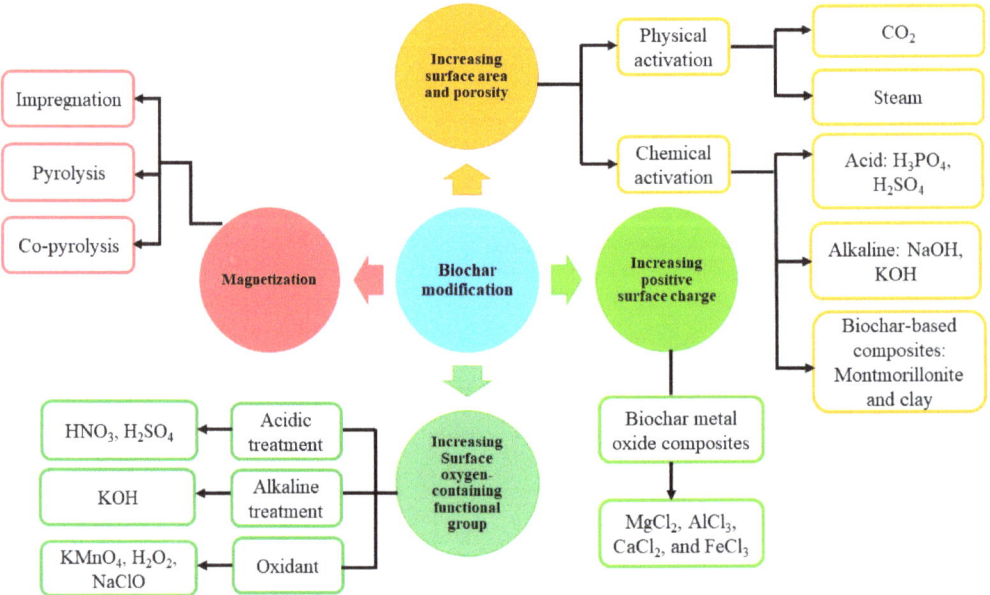

Figure 4. Modification and classification of engineered biochar.

Post-modification can be categorized into the following groups: (i) modifications to surge surface area and porosity; (ii) changes to decrease the positively charged BC surface; (iii) increase surface oxygen-containing functional groups; and (iv) magnetization of BC to enable recovery. Among all categories, any method can improve one or two BC properties simultaneously. For instance, the SSA and oxygen-containing functional groups can be increased by H_2SO_4 treatment. In the literature, these modifications are classified based on BC physicochemical characteristics or the nature of the process (chemical, physical, or composite) [95,96].

Furthermore, physical (van der Waals forces, electrostatic forces, and hydrophobic interactions) and chemical (surface complexation, ion exchange, π-interactions, co-precipitation, partition, and pore filling) interactions occur during adsorption of organic and inorganic pollutants by BC [97].

3.1. Acid-Base Activation/Decoration

Acid treatment of BC results in enhanced surface area, better porosity, increased sorption capacity, created charged and hydrophilic surface functional groups, and increased colloidal stability and mobility [98]. After BC oxidation, -OH and -COOH functional groups are enhanced to improve hydrophilicity. Acids such as HNO_3/H_2SO_4, HNO_3, HF/HNO_3, $KMnO_4$, H_2O_2, oxalic, and citric acids are used for chemical modification of BC [99].

One of the approaches for the modification of BC is HCl treatment resulting in higher adsorption of concomitant organic compounds. This is because functional groups (carboxylic, phenolic, and carbonyl groups) are added to the BC surface after treatment, the ash content is reduced, and the adsorption sites (hydrophobic) are increased [100,101]. However, this method is not well established for BC modification and more studies are needed to be performed [102].

The use of strong bases for BC modification increases functional groups, surface area, surface charge, and porosity [103]. However, the degree of improvement in BC properties varies depending on the base used. Frequently used bases include potassium oxide (KOH) and sodium hydroxide (NaOH) [104,105]. Few studies showed that the BC surface is significantly improved when using NaOH rather than KOH. For instance, Cazetta et al. [106] reported a better improvement in the surface area (49%) of coconut shell-derived BCs when using NaOH. Wakame-derived BC showed a surface area of 69.7 $m^2\ g^{-1}$ and 1156 $m^2\ g^{-1}$ for non-activated BC and KOH-activated BC, respectively [88]. Similarly, the BC surface area derived from rice husk improved from 132.9 $m^2\ g^{-1}$ for non-activated BC to 1818 $m^2\ g^{-1}$ for KOH-activated BC [107]. BC activation by KOH and $KMnO_4$ results in large pore volume, pore channels, and percentage of aromatized structures, enhancing the adsorption of methylene blue [108].

In general, modification of BC via acid is carried out in two ways: activation and decoration. BC activation via chemical activating agent increases SSA and porous structure. In contrast, decoration enhances the surface activity of BC via zero-valent iron nanoparticles [109], Fe_3O_4, and FeOOH [110]. Studies have also shown that decoration using acids and fatty acids increases the hydrophobicity of BC adsorbents [111,112].

3.2. Persulfate Activation

Lately, the activation of BC via persulfate (PS) has been extensively studied [113] owing to the enhanced contaminant removal efficacy from water by the combination of PS radicals with BC-SSA [114]. Studies have shown that the sulfate radical-based advanced oxidation processes (AOP) have higher redox potentials, extended half-life time, and extensive pH flexibility than traditional hydroxyl radical-based AOP. Liu et al. [115] studied food waste digestate-derived BC (FWDB) for peroxydisulfate (PDS) catalyst (radical-based oxidant) for the removal of azo dye (reactive brilliant red X-3B). They observed a 92.21% removal of X-3B within 30 min from solution at an initial content of 1 g/L. The reactive oxygen species $SO_4^{\bullet-}$, $^{\bullet}OH$, $O_2^{\bullet-}$, and $^1O_2$ were found to be present in the FWDB/PDS system and the generation of reactive oxygen species is due to the presence of graphitized carbon, doped-N, oxygen-containing groups, and the defective sites on the FWDB surface. One-step sol–gel pyrolysis was used to synthesize CuFe$_2$O$_4$@BC composite (CuFe$_2$O$_4$@BC) and assessed for the efficiency, stability, and mechanism of activating PS against malachite green degradation [116]. The malachite green removal efficiency in the CuFe$_2$O$_4$@BC/PS system was observed to be 98.9%, signifying that CuFe$_2$O$_4$@BC has better catalytic activity compared to other catalysts under the optimal conditions.

A composite of stalk BC (SBC) and nanoscale zero-valent iron (nZVI) was biosynthesized to remove dye. The composite displayed a synergetic role in enhancing PS catalytic activity for dye removal [117]. SBC functional groups (-OH and -COOH) stimulated PS and nZVI improved catalytic activity compared to SBC. SBC enhances catalytic activity for PS and electron transfer efficiency by increasing nZVI dispersibility and protecting it from oxidation. Studies have shown that superoxide radicals (O$_2^{\bullet-}$) and hydroxyl radicals ($^\bullet$OH) play an important part in pollutant removal, but they are not primary radicals. The primary radical involved in contaminant removal was sulfate radicals (SO$_4^{\bullet-}$). The SBC-nZVI composite also displayed better reusability, storage stability, and various dye removal applications [117]. One-step pyrolysis of loofah sponge-based BC (LSB) was prepared at various pyrolytic temperatures. LSB was further used to activate PS to remove acid orange 7 (AO7) [118]. The LSB produced at 800 °C showed the best catalytic ability for AO7 degradation, i.e., 96% degradation within 30 min [118]. The enhanced adsorption capability of LSB is due to the high surface area, carbonyl groups, and graphitized structure. The AO7 degradation mechanism in the LSB-800/PS process is owing to free radical SO$_4^{\bullet-}$ and non-radical channels (electron transfer).

3.3. Physical Activation

In this process, different gases (air, steam, and CO$_2$) are used to activate the BC surface [119]. In general, physical activation consists of two stages: carbonization (occurs at lower temperatures of 427–877 °C) and activation (occurs at higher temperatures of 627–927 °C) [120]. BC activation can be carried out via steam or gas purging after BC is carbonized. Although this process is simple and cost-effective, it is less efficient than chemical activation approaches. BC physicochemical properties can be enhanced via physical activation by eliminating incomplete combustion by-products and the carbon surface oxidation to enhance BC surface area [121,122].

Pristine BC is activated by steam with improved polarity, surface area, and pore volume [123]. After preliminary pyrolysis, BC was partially vaporized by steam for 0.5 and 3 h [124]. Steam activation improves BC growth and the reachability of internal pores [124]. During steam activation, pore development is primarily associated with water-gas shift reaction, reduction of carbon [125], and elimination of trapped components (aldehydes, ketones, and some acids from biomass) [126].

In the presence of heated steam, BC is activated and displays high adsorption ability owing to improved surface area, oxygen-containing functional groups, and micropore volume. This allows for efficient removal of pollutants from wastewater [127]. The porosity of BC can be improved by physical activation at <250 °C in the presence of air or >750 °C in the presence of CO$_2$, steam, and blends. CO$_2$ and H$_2$O become reactive at high temperatures, allowing chemical reaction with BC [128]. CO$_2$ treatment increases the pore volume and surface area of BC by 1.5-fold and 5.9-fold, respectively [129]. At low temperatures ($\approx$275 °C), the activation process with air increased the sewage sludge-derived BC structure and texture [130]. Furthermore, Sewu et al. [131] observed an increase in BC SSA and porosity via steam activation and significantly improved the crystal violet adsorption capacity by 4.1-fold.

3.4. Biochar-Based Composites

BC-based composites allow for the combined advantages of BC and nanomaterials, resulting in composites with enhanced functional groups, pore size and volume, active surface sites, net surface charge, ease of recovery, and catalytic ability for degradation [132,133].

These composites are divided into three groups according to the synthesis process: (i) nano-metal oxide/hydroxide-BC composites, (ii) magnetic BC composites [80], (iii) clay mineral-based BC [134].

3.4.1. Nano-Metal Oxide/Hydroxide-Biochar Composites

These composites have been synthesized via numerous chemical additives such as H_3PO_4, NaCl, K_2CO_3, $ZnCl_2$, KOH, and $FeCl_3$, and can be incorporated prior or later biomass carbonization [89]. According to Zhao et al. [135], the synthesis of nano-metal BC composites can be divided into the following categories: (i) impregnation, categorized by the ease and enormous capacity of the attained composites, (ii) chemical coprecipitation, categorized by inexpensiveness, high purity, homogeneous nanoparticles, but both process causes chemical contamination, (iii) direct pyrolysis, considered as easy process and biomass supplemented with desired heavy metals; however, controlling the nano-metal oxide/BC ratio in the composites is difficult, and (iv) ball milling, simple, cost-effective, no generation of chemical pollutants, and efficient decrease of metal oxide particle size. However, during particle preparation via ball milling, the particles are easily dispersed in water, resulting in the migration of contaminants out of the polluted place, causing a possible hazard to groundwater [135]. Figure 5 represents various pathways for synthesis of nano-metal oxide/hydroxide-BC composites.

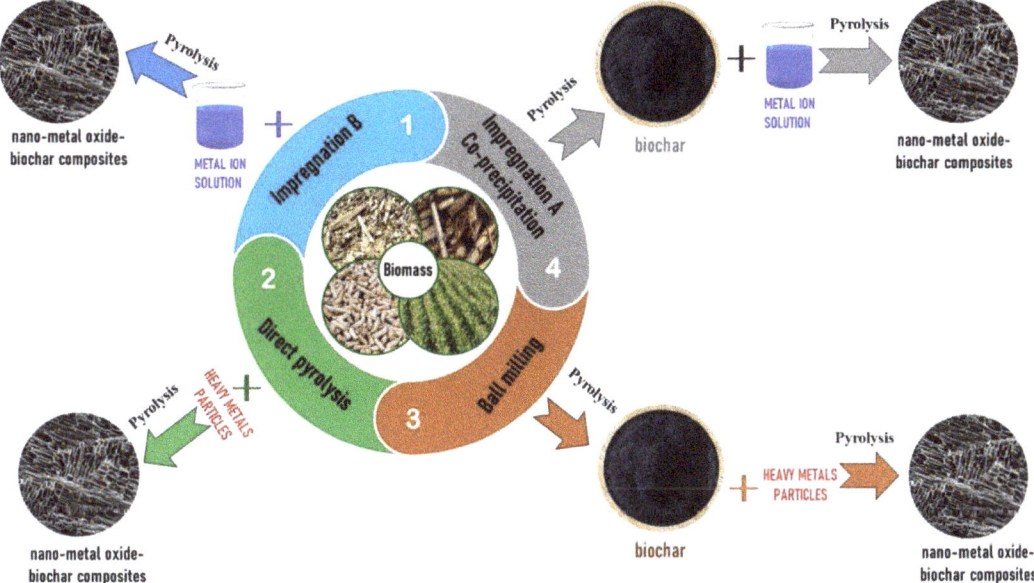

Figure 5. Various pathways for synthesis of nano-metal oxide/hydroxide-biochar composites. Reprinted with permission from [32].

Preparation of BC sample via ball milling surges SSA, possibly by revealing pores [136] and enhances carboxyl, lactone, and hydroxyl groups (oxygen-containing functional groups) [137]. Zheng et al. [138] synthesized a dual-function MgO/BC nanocomposite by ball milling to remove both cationic and anionic contaminants and showed an increased (8.4-fold) methylene blue adsorption, probably due to enhanced surface area and pore volume. Synthesis of BC by co-pyrolysis from firwood biomass utilizing non-magnetic goethite mineral proved to be a green method in comparison to the conventionally used $FeCl_3$ [139].

Sludge-derived BC was modified via Fe/Mn by a heterogeneous activator to attain enhanced PS activation to degrade reactive blue 19 (RB19) [140]. Experiments based on scavenger quenching and electron paramagnetic resonance displayed that radical and non-radical mechanisms are involved in the degradation of RB19 by Fe/MnBC combined PS. Sulfate radicals ($SO_4^{\bullet-}$), hydroxyl radicals ($OH^{\bullet}$), and singlet oxygen ($^1O_2$) were found to be involved in the process. It was also postulated that Fe(IV)/Mn(VII) (non-radical) was involved in the degradation process. These results provided a new understanding of the mechanism of PS activated by metal-BC composite. In addition, fixed-bed reactor studies revealed that Fe/MnBC has significant PS activation capability to remove dyes. Further, the authors analyzed degradation process by central composite design-response surface methodology (CCD-RSM) and ANN. Statistical analysis showed that the ANN model was better than the CCD-RSM model [140].

BC derived from pulp sludge was produced via $ZnCl_2$ modification to effectively remove methylene blue (MB) from aqueous solutions [141]. The maximum (590.20 mg/g) adsorption of MB was observed within 24 h at pH 8 when the initial amount of Zn2PT350-700 was 10 mg [141]. The primary mechanism involved in MB adsorption was electrostatic interaction between deprotonated functional groups and MB^+, cation exchange, and π-electron interactions followed by physical adsorption.

3.4.2. Magnetized Biochar Composites

Substantial studies have been conducted to enhance BC adsorption capacity. However, the biggest limitation of BC is its removal and reuse after wastewater treatment owing to its small size and low density [59]. Various studies of magnetic BC for successful production have been reported [142], displaying impregnation-pyrolysis, chemical co-precipitation, solvothermal, and reductive co-precipitation processes. It was observed that the composites have a significantly efficient adsorption, easy removal and recycling after application of an external magnetic field [143]. However, the conventional magnetic medium loading process surges the sorbent cost [144]. The primary common processes of magnetic material synthesis include pyrolysis, co-precipitation, and calcination [145]. Figure 6 shows a brief schematic of different magnetic BC preparation.

Further, traditional heating in electric furnaces [146] and microwave heating in modified furnaces have also been studied. Among them, the co-precipitation method is easy because mixing and heating processes are simple. It emphasizes the transition metal ion molar ratios used in mixing to improve magnetic BC characteristics concerning porosity and magnetism [145]. The co-precipitation process is used for BC surface coating with various iron oxides such as Fe_3O_4, c-Fe_2O_3, and $CoFe_2O_4$ particles to impart magnetic characteristics to the iron oxide active sites for the pollutant removal [147,148]. It was observed that the high decomposition ability of additives such as $AlCl_3$, $FeCl_2$, and $MgCl_2$ displays enhanced porosity, high catalytic and sorption activity, and creation of positively charged adsorption sites [89,149].

Fe-tanned collagen fibers were pyrolyzed to synthesize magnetic BC catalyst (Fe@BC) utilized as a persulfate activator to remove a refractory dye (methylene blue) [150]. Results have shown that within 20 min, methylene blue was completely degraded at a constant rate of 0.2246 min^{-1}, much higher than when using pure BC (0.0497 min^{-1}). The enhanced catalytic activity and exceptional recycling ability of the Fe@BC/PS system can be attributed to the homogeneously dispersed Fe ions, enhanced functional groups (oxygen containing), and deformed carbon matrix. $SO_4^{\bullet-}$ was found to be the primary free radical for the degradation of methylene blue. Table 2 enlists literature on surface-modified BC and its nanocomposites for dye removal from wastewater.

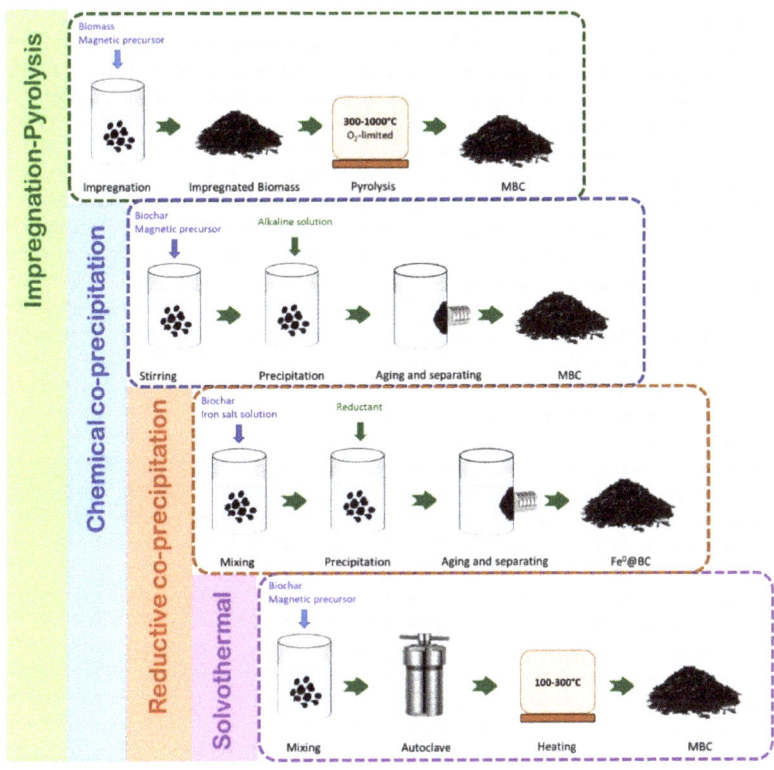

Figure 6. Brief schematic of different magnetic biochar preparation. Reprinted with permission from [32].

Table 2. Surface-modified biochar and its nanocomposites for dye removal from wastewater.

S. No.	Surface-Modified Biochar and Its Nanocomposites	Dye	Adsorption Model	Maximum Adsorption Capacity (mg/g)	Mechanism Involved	Reference
1.	TiO$_2$ supported BC	3,4-dimethylaniline	Toth adsorption models	285.71	-	[151]
2.	Layered double-oxide/BC composites	Congo red	Langmuir	344.83,	-	[152]
3.		Methyl orange	Langmuir	588.24	-	[152]
4.	CO$_2$ and H$_2$O activator Pine sawdust BC	Methylene blue	Langmuir	160	Hydrogen bonding, ion exchange, π–π interaction	[153]
5.	Phosphomolybdate-modified BC	Methylene blue	Langmuir	146.23	Hydrogen bonding, electrostatic interactions, and ion exchange	[154]
6.	Banana peel BC/iron oxide composite	Methylene blue	Langmuir	862	-	[54]
7.	Mixed municipal discarded material derived BC	Methylene blue	Langmuir	7.2	π–π interactions	[155]
8.	KOH modified lychee seed BC	Methylene blue	Langmuir	124.5	-	[156]
9.	Triethylenetetramine corncob BC	Sunset yellow	Langmuir	-	-	[74]
10.	Fe$_2$O$_3$/TiO$_2$ functionalized wasted tea leaves derived BC	Methylene blue	-	-	Reactive radical species	[157]
11.		Rhodamine B	-	-		
12.		Methyl orange	-	-		

Table 2. Cont.

S. No.	Surface-Modified Biochar and Its Nanocomposites	Dye	Adsorption Model	Maximum Adsorption Capacity (mg/g)	Mechanism Involved	Reference
13.	Manganese-modified lignin BC	Methylene blue	Langmuir	248.96	-	[158]
14.	KOH activated pine BC	Methylene blue	Freundlich	637.5	Primary polar and π–π interactions	[109]
15.	KMnO$_4$ activated pine BC	Methylene blue	Freundlich	439.5	Primary polar and π–π interactions	[109]
16.	Fe$_3$O$_4$-modified *Citrus bergamia* peel derived BC	Methylene blue	Langmuir	136.72	Electrostatic interaction	[26]
17.	Sulfuric acid modified BC from Pumpkin peel	Methylene Blue	Langmuir	208.3	-	[159]
18.	Acid activated Pine needle BC	Methylene blue	Langmuir	153.84	Hydrogen bonding, electrostatic interaction	[160]
19.	Laccase immobilized pine needle BC	Malachite green	-	-	Enzymatic degradation	[161]
20.	Ozonized saw dust BC	Methylene blue	Langmuir	200	Electrostatic interaction and hydrogen bonding	[162]
21.	Sonicated saw dust BC	Methylene blue	Langmuir	526		
22.	Nitric acid-treated Pterospermum acerifolium fruit waste BC	Methylene blue	Langmuir	-	-	[28]
23.	SDS-modified nitric acid-treated Pterospermum acerifolium fruit waste BC	Methylene blue	Langmuir	-	-	[28]
24.	Cetyl trimethyl ammonium bromide modified magnetic BC from pine nut shells	Acid chrome blue K	Langmuir	-	-	[27]

3.4.3. Modification via Clay Mineral

Pre-mixing of the feedstock prior to pyrolysis or post-mixing of the generated BC and clay minerals produces a clay-BC composite [95]. Arif et al. [134] studied the clay mineral-BC synthesis process and characteristics including pollutant interactions with the composite. This study displayed that clay-mineral amendment increases BC micropore area (>200%), whereas slight growth was detected in the mesopore area [163]. BC stability can also be enhanced by clay organo-mineral layers and avoid BC decomposition when applied to soil [164].

The interaction of the BC-clay/mineral composites with the pollutants involves various interactions such as ion exchange, precipitation, complexation, electrostatic interactions, hydrogen bonding, partitioning, electron–donor–acceptor interactions, hydrophobic interactions, and pore-filling [165]. Layered porous clay-BC (C-BC) was produced via layered bentonite clay intercalating properties comprising high cation exchange capacity with BC supporting redox-sensitive zero-valent iron (nZVI) nano-trident particles [166]. The synthesized C-BC-nZVI displayed a heterogeneous pore distribution on the C-BC-nZVI surface with dispersed un-oxidized nZVI particles (20–30 nm). Methylene blue was selectively adsorbed (52.1 mg/g) by nano-trident in the dye mixture. Further, pH and dissolved organic matter showed no influence on methylene blue sorption [166].

4. Application of Machine Learning and Artificial Neural Networks into Biochar-Facilitated Wastewater Remediation

The process initiates from BC production. Various techniques are then carried out to ensure the continued use of BC for wastewater treatment until it is landfilled or recycled. Precisely, the produced BCs are characterized as SSA, elemental composition and pH, used as input and aided in implementation [36]. Machine learning (ML) algorithms such as support vector regression, ANN, or random forests (RF) are applied at this stage. The data is processed via algorithm processes and the variables are transformed into files using weights [167,168]. The decision criterion is made from the index to choose BC for wastewater treatment. After BC selection, it is passed on for wastewater treatment, where it is otherwise activated or functionalized by chemical or physical approaches. BC is re-characterized after activation/functionalization, and ML algorithms reprocess the data to determine whether the functionalized BC fulfils the standards required for wastewater treatment [36,169]. Moosavi et al. [170] compared the model performance of activated carbon derived from agricultural waste to remove dyes using gradient boosting, decision tree, and RF models. Model accuracy can be evaluated as correlation coefficient (R), mean squared error (MSE), and root mean squared error ($RMSE$) according to the following equations:

$$R = \sqrt{1 - \frac{\sum_{i=1}^{N}(\check{y}_l - y_i)^2}{\sum_{i=1}^{N}(\check{y}_l - \bar{y})^2}}, \quad (1)$$

$$MSE = \frac{1}{N}\sum_{i=1}^{N}(\check{y}_l - \bar{y})^2, \quad (2)$$

$$RMSE = \sqrt{MSE}. \quad (3)$$

The linear dependences amid any two variables or each attribute and the target variable are calculated by the following equation.

$$r_{xy} = \frac{\sum_{i=1}^{n}(x_i - \bar{x})\sum_{i=1}^{n}(y_i - \bar{y})}{\sqrt{\sum_{i=1}^{n}(x_i - \bar{x})^2}\sqrt{\sum_{i=1}^{n}(y_i - \bar{y})^2}}, \quad (4)$$

where x or y is the mean of the factors x or y, respectively, and the data for each variable are normalized to a range of 0 and 1 using the equation below.

$$y = \frac{(x_i - x_{min})}{(x_{max} - x_{min})}, \quad (5)$$

where y is the normalized value of the initial x_i, and x_{max} and x_{min} are the maximum and minimum values of x_i, respectively.

Conventional modelling and optimization methods, i.e., RSM and statistical analysis, need a varied range of experiments, which are expensive and time consuming. Additionally, they do not precisely elucidate the association amid different factors affecting removal efficacy. ML is more robust and effective compared to statistical models when modelling complex data with possible nonlinearities or partial data [171]. ML offers stage for mapping associations amid input and output parameters during pollutant remediation. It is grounded on computer algorithms that are automatically developed and adjusted to specified conditions, with the ability to forecast the properties of concern via the knowledge and interpretation carried out by machines. It is used to study complex associations amid adsorbent properties (e.g., surface area, total pore volume, average pore diameter, and elemental composition), adsorption conditions, and adsorption performance without empirical assumptions [172]. Moreover, the foremost goal of system modelling and simulation is to obtain data on the progress of the system without conducting experiments. Thus, it is essential to develop a multidimensional nature model that can adapt to each process. Regarding the problem of overlooking variable relationships, ML models have demonstrated beneficial tools for method design, regression models, and response optimization [173].

Non-linear associations between independent and dependent variables can be created by ANN models grounded on a set of test results. Recently, the use of ANN models for modelling dye adsorption has become very common. Based on the ANN model type, dye adsorption can be categorized into the following groups: (i) multilayer feedforward neural networks, (ii) adaptive neuro-fuzzy inference systems, (iii) support vector regression, and (iv) hybrid models. Various ANN models were used to assess dye removal [174]. A three-layer feedforward backpropagation network (FFBPN) was used to study the removal of acid orange from an aqueous medium by powdered activated carbon [175]. In an input layer with three neurons (consisting of initial pH, dye concentration, and contact time) and an output layer with one neuron (dye concentration after time 't'), they established FFBPN. As a result of evaluating the model's performance using the mean relative error (MRE), it was found to be 5.81%. This MRE value signifies that the ANN model had good analytical performance and can be used in place of kinetic studies.

5. Conclusions and Future Perspectives

BC has been considered an efficient adsorbent to treat wastewater owing to the presence of abundant functional groups and large surface area. Few studies assessed expenses during the production of BC-based adsorbents. A cost-effective technique and the reuse of multiple cycles are possible. However, additional investigation of numerous outlooks is obligatory to confirm BC efficiency and inexpensiveness, mainly for pilot-scale implementation in subsequent fields.

(1) To attain optimum BC efficacy, it is crucial to study the relations amid various parameters, such as production process, modification/functionalization, and handling all in an eco-friendly way.
(2) Moreover, promising sorbents that are effectively suitable for pilot scale must be inexpensive and resources ought to be widely accessible in huge amounts in nature. Recycling of sorbents on a large scale can reduce costs and energy consumption to provide sustainable products.
(3) Involvement of software to optimize factors affecting pollutant removal by BC is an innovative and powerful tool in experimental design and analysis. Applications of ANN and ML can be used to predict and reduce explicit computer programming.

Modified clay/BC composites displayed significant benefits compared to modified clays alone, owing to their cost-effective, high adsorption ability, and anionic dye removal effect. However, few studies have been carried out on a pilot scale. The advancement of magnetic BC research with organic pollutant removal is a vital and new research area, and more research in this field should be conducted in the future. However, generation of noxious components during the synthesis of magnetic BC should be careful.

BC delivers a robust interaction between the BC surface and various pollutants existing in water and soil. However, as BC loaded with pollutants may pose serious hazard associated with disposal, a new challenge for researchers concerning the prospects for the transformed environmental movement of these pollutants is to change the context of pollutant hazard valuation in the milieu. Implementation of a sustainable management strategy for contaminant loaded-BC is life threatening. This approach should follow the concept of a circular economy and should allow for an eco-friendly and cost-effective recycling of these BCs in the forthcoming adsorption cycles. In addition, during laboratory scale studies, knowledge of the mechanisms of chemical and physical interaction of BC with pollutants is required. Promising technologies always have a secondary hazard, and the utilization of BC loaded with pollutant for bio-oil production in the future could be a fascinating area that must be explored.

ANN and ML predictive models reduce workload, budget, space requirements, and pollutant remediation time. For instance, ML can be used to develop models to improve pyrolysis process and forecast the yield of BC depending on raw material characteristics and process conditions, which can help develop a sustainable wastewater treatment system. Further, the commencement of more compacted reactors capable of accommodating BC-supported catalysts are required for implementation. Future research should be devoted to the application of BC-based catalysts in large-scale reactors, which can be implemented at the industrial level. It is urgent to further explore the potential of BC in real challenging systems.

Author Contributions: L.G.: writing—original draft preparation; A.K.: writing—review and editing; S.R.K.: writing—review and editing; B.-S.K.: writing—review and editing. All authors have read and agreed to the published version of the manuscript.

Funding: This research was supported by National Research Foundation of Korea (NRF-2019R1I1A-3A02058523).

Data Availability Statement: The study did not report any data.

Conflicts of Interest: There are no financial conflict of interest to disclose.

References

1. Bouckaert, S.; Pales, A.F.; McGlade, C.; Remme, U.; Wanner, B.; Varro, L.; Spencer, T. *Net Zero by 2050: A Roadmap for the Global Energy Sector*; International Energy Agency: Paris, France, 2021.
2. Sathe, S.S.; Goswami, L.; Mahanta, C. Arsenic reduction and mobilization cycle via microbial activities prevailing in the Holocene aquifers of Brahmaputra flood plain. *Groundw. Sustain. Dev.* **2021**, *13*, 100578. [CrossRef]
3. Kushwaha, A.; Goswami, S.; Hans, N.; Goswami, L.; Devi, G.; Deshavath, N.N.; Lall, A.M. An insight into biological and chemical technologies for micropollutant removal from wastewater. In *Fate and Transport of Subsurface Pollutants*; Springer: Singapore, 2021; pp. 199–226.
4. Mallakpour, S.; Lormahdiabadi, M. Removal of the Anionic Dye Congo Red from an Aqueous Solution Using a Crosslinked Poly (vinyl alcohol)-ZnO-Vitamin M Nanocomposite Film: A Study of the Recent Concerns about Nonlinear and Linear Forms of Isotherms and Kinetics. *Langmuir* **2022**, *38*, 4065–4076. [CrossRef] [PubMed]
5. Hussain, C.M.; Singh, S.; Goswami, L. *Emerging Trends to Approaching Zero Waste*; Elsevier: Amsterdam, The Netherlands, 2021.
6. Hussain, C.M.; Singh, S.; Goswami, L. (Eds.) *Waste-to-Energy Approaches towards Zero Waste: Interdisciplinary Methods of Controlling Waste*; Elsevier: Amsterdam, The Netherlands, 2021.
7. Choudhary, M.; Kumar, R.; Neogi, S. Activated biochar derived from Opuntia ficus-indica for the efficient adsorption of malachite green dye, Cu^{+2} and Ni^{+2} from water. *J. Hazard. Mater.* **2020**, *392*, 122441. [CrossRef]
8. Begum, W.; Goswami, L.; Sharma, B.B.; Kushwaha, A. Assessment of urban river pollution using the water quality index and macro-invertebrate community index. *Environ. Dev. Sustain.* **2022**, 1–26. [CrossRef]
9. Goswami, S.; Kushwaha, A.; Goswami, L.; Singh, N.; Bhan, U.; Daverey, A.; Hussain, C.M. Biological treatment, recovery, and recycling of metals from waste printed circuit boards. In *Environmental Management of Waste Electrical and Electronic Equipment*; Elsevier: Amsterdam, The Netherlands, 2021; pp. 163–184.
10. Lata, K.; Kushwaha, A.; Ramanathan, G. Bacterial enzymatic degradation and remediation of 2,4,6-trinitrotoluene. *Microb. Nat. Macromol.* **2021**, *23*, 623–659.
11. Vadivel, V.K.; Cikurel, H.; Mamane, H. Removal of Indigo Dye by $CaCO_3/Ca(OH)_2$ Composites and Resource Recovery. *Ind. Eng. Chem. Res.* **2021**, *60*, 10312–10318. [CrossRef]
12. Kumar, M.; Kushwaha, A.; Goswami, L.; Singh, A.K.; Sikandar, M. A review on advances and mechanism for the phycoremediation of cadmium contaminated wastewater. *Clean. Eng. Technol.* **2021**, *5*, 100288. [CrossRef]
13. Rizvi, S.; Singh, A.; Kushwaha, A.; Gupta, S.K. Recent advances in melanoidin removal from wastewater: Sources, properties, toxicity, and remediation strategies. In *Emerging Trends to Approaching Zero Waste*; Elsevier: Amsterdam, The Netherlands, 2022; pp. 361–386.
14. Patra, D.; Patra, B.R.; Pattnaik, F.; Hans, N.; Kushwaha, A. Recent evolution in green technologies for effective valorization of food and agricultural wastes. In *Emerging Trends to Approaching Zero Waste*; Elsevier: Amsterdam, The Netherlands, 2022; pp. 103–132.
15. Kushwaha, A.; Goswami, S.; Hans, N.; Singh, A.; Vishwakarma, H.S.; Devi, G.; Hussain, C.M. Sorption of pharmaceutical and personal care products from the wastewater by carbonaceous materials. In *Emerging Trends to Approaching Zero Waste*; Elsevier: Amsterdam, The Netherlands, 2021; pp. 175–196.
16. Saha, P.; Goswami, L.; Kim, B.S. Novel Biobased Non-Isocyanate Polyurethanes from Microbially Produced 7,10-Dihydroxy-8 (E)-Octadecenoic Acid for Potential Packaging and Coating Applications. *ACS Sustain. Chem. Eng.* **2022**, *10*, 4623–4633. [CrossRef]
17. Ojaswini, C.; Sathe, S.S.; Mahanta, C.; Kushwaha, A. Evaluation of iron coated natural sand for removal of dissolved arsenic from groundwater and develop sustainable filter media. *Environ. Nanotechnol. Monit. Manag.* **2022**, *18*, 100682. [CrossRef]
18. Gautam, A.; Kushwaha, A.; Rani, R. Microbial remediation of hexavalent chromium: An eco-friendly strategy for the remediation of chromium-contaminated wastewater. In *The Future of Effluent Treatment Plants*; Elsevier: Amsterdam, The Netherlands, 2021; pp. 361–384.
19. Gupt, C.B.; Kushwaha, A.; Prakash, A.; Chandra, A.; Goswami, L.; Sekharan, S. Mitigation of groundwater pollution: Heavy metal retention characteristics of fly ash based liner materials. In *Fate and Transport of Subsurface Pollutants*; Springer: Singapore, 2021; pp. 79–104.
20. Kushwaha, A.; Rani, R.; Patra, J.K. Adsorption kinetics and molecular interactions of lead [Pb (II)] with natural clay and humic acid. *Int. J. Environ. Sci. Technol.* **2020**, *17*, 1325–1336. [CrossRef]

21. Goswami, L.; Manikandan, N.A.; Taube, J.C.R.; Pakshirajan, K.; Pugazhenthi, G. Novel waste-derived biochar from biomass gasification effluent: Preparation, characterization, cost estimation, and application in polycyclic aromatic hydrocarbon biodegradation and lipid accumulation by *Rhodococcus opacus*. *Environ. Sci. Pollut. Res.* **2019**, *26*, 25154–25166. [CrossRef]
22. Goswami, L.; Kushwaha, A.; Singh, A.; Saha, P.; Choi, Y.; Maharana, M.; Kim, B.S. Nano-biochar as a sustainable catalyst for anaerobic digestion: A synergetic closed-loop approach. *Catalysts* **2022**, *12*, 186. [CrossRef]
23. Goswami, S.; Kushwaha, A.; Goswami, L.; Gupta, N.R.; Kumar, V.; Bhan, U.; Tripathi, K.M. Nanobiochar-a green catalyst for wastewater remediation. In *Bio-Based Nanomaterials*; Elsevier: Amsterdam, The Netherlands, 2022; pp. 109–132.
24. Goswami, L.; Pakshirajan, K.; Pugazhenthi, G. Biological treatment of biomass gasification wastewater using hydrocarbonoclastic bacterium *Rhodococcus opacus* in an up-flow packed bed bioreactor with a novel waste-derived nano-biochar based bio-support material. *J. Clean. Prod.* **2020**, *256*, 120253. [CrossRef]
25. Salleh, M.A.M.; Mahmoud, D.K.; Karim, W.A.W.A.; Idris, A. Cationic and anionic dye adsorption by agricultural solid wastes: A comprehensive review. *Desalination* **2011**, *280*, 1–13. [CrossRef]
26. Tripathy, S.; Sahu, S.; Patel, R.K.; Panda, R.B.; Kar, P.K. Novel Fe_3O_4-Modified Biochar Derived from Citrus Bergamia Peel: A Green Synthesis Approach for Adsorptive Removal of Methylene Blue. *ChemistrySelect* **2022**, *7*, e202103595. [CrossRef]
27. Wang, H.; Wang, S.; Gao, Y. Cetyl trimethyl ammonium bromide modified magnetic biochar from pine nut shells for efficient removal of acid chrome blue K. *Bioresour. Technol.* **2020**, *312*, 123564. [CrossRef]
28. Oraon, A.; Prajapati, A.K.; Ram, M.; Saxena, V.K.; Dutta, S. Synthesis, characterization, and application of microporous biochar prepared from *Pterospermum acerifolium* plant fruit shell waste for methylene blue dye adsorption: The role of surface modification by SDS surfactant. *Biomass Convers. Biorefin.* **2022**, 1–23. [CrossRef]
29. Al-Rumaihi, A.; Shahbaz, M.; Mckay, G.; Mackey, H.; Al-Ansari, T. A review of pyrolysis technologies and feedstock: A blending approach for plastic and biomass towards optimum biochar yield. *Renew. Sustain. Energy Rev.* **2022**, *167*, 112715. [CrossRef]
30. Zaman, C.Z.; Pal, K.; Yehye, W.A.; Sagadevan, S.; Shah, S.T.; Adebisi, G.A.; Johan, R.B. *Pyrolysis: A Sustainable Way to Generate Energy from Waste*; IntechOpen: Rijeka, Croatia, 2017; Volume 1, p. 316806.
31. Liu, W.J.; Jiang, H.; Yu, H.Q. Development of biochar-based functional materials: Toward a sustainable platform carbon material. *Chem. Rev.* **2015**, *115*, 12251–12285. [CrossRef] [PubMed]
32. Zeghioud, H.; Fryda, L.; Djelal, H.; Assadi, A.; Kane, A. A comprehensive review of biochar in removal of organic pollutants from wastewater: Characterization, toxicity, activation/functionalization and influencing treatment factors. *J. Water Process Eng.* **2022**, *47*, 102801. [CrossRef]
33. Tripathi, M.; Sahu, J.N.; Ganesan, P. Effect of process parameters on production of biochar from biomass waste through pyrolysis: A review. *Renew. Sustain. Energy Rev.* **2016**, *55*, 467–481. [CrossRef]
34. Yaashikaa, P.R.; Kumar, P.S.; Varjani, S.; Saravanan, A. A critical review on the biochar production techniques, characterization, stability and applications for circular bioeconomy. *Biotechnol. Rep.* **2020**, *28*, e00570. [CrossRef]
35. Liu, Y.; Chen, Y.; Li, Y.; Chen, L.; Jiang, H.; Li, H.; Hou, S. Fabrication, application, and mechanism of metal and heteroatom co-doped biochar composites (MHBCs) for the removal of contaminants in water: A review. *J. Hazard. Mater.* **2022**, *431*, 128584. [CrossRef]
36. Da Silva Medeiros, D.C.C.; Nzediegwu, C.; Benally, C.; Messele, S.A.; Kwak, J.H.; Naeth, M.A.; El-Din, M.G. Pristine and engineered biochar for the removal of contaminants co-existing in several types of industrial wastewaters: A critical review. *Sci. Total Environ.* **2021**, *809*, 151120. [CrossRef]
37. Das, O.; Bhattacharyya, D.; Hui, D.; Lau, K.T. Mechanical and flammability characterisations of biochar/polypropylene biocomposites. *Compos. Part B Eng.* **2016**, *106*, 120–128. [CrossRef]
38. Zou, H.; Zhao, J.; He, F.; Zhong, Z.; Huang, J.; Zheng, Y.; Gao, B. Ball milling biochar iron oxide composites for the removal of chromium (Cr (VI)) from water: Performance and mechanisms. *J. Hazard. Mater.* **2021**, *413*, 125252. [CrossRef]
39. Li, S.; Harris, S.; Anandhi, A.; Chen, G. Predicting biochar properties and functions based on feedstock and pyrolysis temperature: A review and data syntheses. *J. Clean. Prod.* **2019**, *215*, 890–902. [CrossRef]
40. Agrafioti, E.; Bouras, G.; Kalderis, D.; Diamadopoulos, E. Biochar production by sewage sludge pyrolysis. *J. Anal. Appl. Pyrolysis* **2013**, *101*, 72–78. [CrossRef]
41. Amonette, J.E.; Joseph, S. Characteristics of biochar: Microchemical properties. In *Biochar for Environmental Management*; Routledge: London, UK, 2012; pp. 65–84.
42. Huang, Q.; Song, S.; Chen, Z.; Hu, B.; Chen, J.; Wang, X. Biochar-based materials and their applications in removal of organic contaminants from wastewater: State-of-the-art review. *Biochar* **2019**, *1*, 45–73. [CrossRef]
43. Tan, X.; Liu, Y.; Zeng, G.; Wang, X.; Hu, X.; Gu, Y.; Yang, Z. Application of biochar for the removal of pollutants from aqueous solutions. *Chemosphere* **2015**, *125*, 70–85. [CrossRef] [PubMed]
44. Ambaye, T.G.; Vaccari, M.; van Hullebusch, E.D.; Amrane, A.; Rtimi, S. Mechanisms and adsorption capacities of biochar for the removal of organic and inorganic pollutants from industrial wastewater. *Int. J. Environ. Sci. Technol.* **2021**, *18*, 3273–3294. [CrossRef]
45. Wu, J.; Yang, J.; Feng, P.; Huang, G.; Xu, C.; Lin, B. High-efficiency removal of dyes from wastewater by fully recycling litchi peel biochar. *Chemosphere* **2020**, *246*, 125734. [CrossRef] [PubMed]
46. Qambrani, N.A.; Rahman, M.M.; Won, S.; Shim, S.; Ra, C. Biochar properties and eco-friendly applications for climate change mitigation, waste management, and wastewater treatment: A review. *Renew. Sustain. Energy Rev.* **2017**, *79*, 255–273. [CrossRef]

47. Kushwaha, A.; Rani, R.; Kumar, S.; Thomas, T.; David, A.A.; Ahmed, M. A new insight to adsorption and accumulation of high lead concentration by exopolymer and whole cells of lead-resistant bacterium *Acinetobacter junii* L. Pb1 isolated from coal mine dump. *Environ. Sci. Pollut. Res.* **2017**, *24*, 10652–10661. [CrossRef] [PubMed]
48. Enaime, G.; Baçaoui, A.; Yaacoubi, A.; Lübken, M. Biochar for wastewater treatment-conversion technologies and applications. *Appl. Sci.* **2020**, *10*, 3492. [CrossRef]
49. Gautam, R.K.; Goswami, M.; Mishra, R.K.; Chaturvedi, P.; Awashthi, M.K.; Singh, R.S.; Pandey, A. Biochar for remediation of agrochemicals and synthetic organic dyes from environmental samples: A review. *Chemosphere* **2021**, *272*, 129917. [CrossRef]
50. Qiu, Y.; Zheng, Z.; Zhou, Z.; Sheng, G.D. Effectiveness and mechanisms of dye adsorption on a straw-based biochar. *Bioresour. Technol.* **2009**, *100*, 5348–5351. [CrossRef] [PubMed]
51. Kaya, N.; Yıldız, Z.; Ceylan, S. Preparation and characterisation of biochar from hazelnut shell and its adsorption properties for methylene blue dye. *Politek. Derg.* **2018**, *21*, 765–776. [CrossRef]
52. Lin, Y.C.; Ho, S.H.; Zhou, Y.; Ren, N.Q. Highly efficient adsorption of dyes by biochar derived from pigments-extracted macroalgae pyrolyzed at different temperature. *Bioresour. Technol.* **2018**, *259*, 104–110.
53. Nautiyal, P.; Subramanian, K.A.; Dastidar, M.G. Adsorptive removal of dye using biochar derived from residual algae after in-situ transesterification: Alternate use of waste of biodiesel industry. *J. Environ. Manag.* **2016**, *182*, 187–197. [CrossRef] [PubMed]
54. Zhang, P.; O'Connor, D.; Wang, Y.; Jiang, L.; Xia, T.; Wang, L.; Hou, D. A green biochar/iron oxide composite for methylene blue removal. *J. Hazard. Mater.* **2020**, *384*, 121286. [CrossRef] [PubMed]
55. Chowdhury, S.; Chakraborty, S.; Saha, P. Biosorption of Basic Green 4 from aqueous solution by Ananas comosus (pineapple) leaf powder. *Colloids Surf. B Biointerfaces* **2011**, *84*, 520–527. [CrossRef]
56. Dawood, S.; Sen, T.K.; Phan, C. Adsorption removal of Methylene Blue (MB) dye from aqueous solution by bio-char prepared from Eucalyptus sheathiana bark: Kinetic, equilibrium, mechanism, thermodynamic and process design. *Desalination Water Treat.* **2016**, *57*, 28964–28980. [CrossRef]
57. Biswas, S.; Mohapatra, S.S.; Kumari, U.; Meikap, B.C.; Sen, T.K. Batch and continuous closed circuit semi-fluidized bed operation: Removal of MB dye using sugarcane bagasse biochar and alginate composite adsorbents. *J. Environ. Chem. Eng.* **2020**, *8*, 103637. [CrossRef]
58. Xu, Y.; Chen, B. Investigation of thermodynamic parameters in the pyrolysis conversion of biomass and manure to biochars using thermogravimetric analysis. *Bioresour. Technol.* **2013**, *146*, 485–493. [CrossRef]
59. Li, C.; Hayashi, J.I.; Sun, Y.; Zhang, L.; Zhang, S.; Wang, S.; Hu, X. Impact of heating rates on the evolution of function groups of the biochar from lignin pyrolysis. *J. Anal. Appl. Pyrolysis* **2021**, *155*, 105031. [CrossRef]
60. Yorgun, S.; Yıldız, D.; Şimşek, Y.E. Activated carbon from paulownia wood: Yields of chemical activation stages. *Energy Sources Part A Recovery Util. Environ. Eff.* **2016**, *38*, 2035–2042. [CrossRef]
61. Harun, N.Y.; Afzal, M.T. Effect of particle size on mechanical properties of pellets made from biomass blends. *Procedia Eng.* **2016**, *148*, 93–99. [CrossRef]
62. Bourke, J.; Manley-Harris, M.; Fushimi, C.; Dowaki, K.; Nunoura, T.; Antal, M.J. Do all carbonized charcoals have the same chemical structure? 2. A model of the chemical structure of carbonized charcoal. *Ind. Eng. Chem. Res.* **2007**, *46*, 5954–5967. [CrossRef]
63. Aliasa, N.; Zaini, M.A.A.; Kamaruddin, M.J. Roles of impregnation ratio of K_2CO_3 and NaOH in chemical activation of palm kernel shell. *J. Appl. Sci. Process Eng.* **2017**, *4*, 195–204. [CrossRef]
64. Inyang, M.I.; Gao, B.; Yao, Y.; Xue, Y.; Zimmerman, A.; Mosa, A.; Cao, X. A review of biochar as a low-cost adsorbent for aqueous heavy metal removal. *Crit. Rev. Environ. Sci. Technol.* **2016**, *46*, 406–433. [CrossRef]
65. Ma, J.; Zhou, B.; Zhang, H.; Zhang, W. Fe/S modified sludge-based biochar for tetracycline removal from water. *Powder Technol.* **2020**, *364*, 889–900. [CrossRef]
66. Wu, M.; Wang, Y.; Lu, B.; Xiao, B.; Chen, R.; Liu, H. Efficient activation of peroxymonosulfate and degradation of Orange G in iron phosphide prepared by pickling waste liquor. *Chemosphere* **2021**, *269*, 129398. [CrossRef]
67. Vigneshwaran, S.; Sirajudheen, P.; Karthikeyan, P.; Meenakshi, S. Fabrication of sulfur-doped biochar derived from tapioca peel waste with superior adsorption performance for the removal of Malachite green and Rhodamine B dyes. *Surf. Interfaces* **2021**, *23*, 100920. [CrossRef]
68. Wang, K.; Peng, N.; Sun, J.; Lu, G.; Chen, M.; Deng, F.; Zhong, Y. Synthesis of silica-composited biochars from alkali-fused fly ash and agricultural wastes for enhanced adsorption of methylene blue. *Sci. Total Environ.* **2020**, *729*, 139055. [CrossRef]
69. Cojocaru, C.; Samoila, P.; Pascariu, P. Chitosan-based magnetic adsorbent for removal of water-soluble anionic dye: Artificial neural network modeling and molecular docking insights. *Int. J. Biol. Macromol.* **2019**, *123*, 587–599. [CrossRef]
70. Siddiqui, S.I.; Manzoor, O.; Mohsin, M.; Chaudhry, S.A. Nigella sativa seed-based nanocomposite-MnO_2/BC: An antibacterial material for photocatalytic degradation, and adsorptive removal of Methylene blue from water. *Environ. Res.* **2019**, *171*, 328–340. [CrossRef]
71. Pirbazari, A.E.; Saberikhah, E.; Kozani, S.H. Fe_3O_4–wheat straw: Preparation, characterization and its application for methylene blue adsorption. *Water Resour. Ind.* **2014**, *7*, 23–37. [CrossRef]
72. Zheng, Y.; Cheng, B.; You, W.; Yu, J.; Ho, W. 3D hierarchical graphene oxide-NiFe LDH composite with enhanced adsorption affinity to Congo red, methyl orange and Cr (VI) ions. *J. Hazard. Mater.* **2019**, *369*, 214–225. [CrossRef] [PubMed]

73. Li, H.; Cao, X.; Zhang, C.; Yu, Q.; Zhao, Z.; Niu, X.; Li, Z. Enhanced adsorptive removal of anionic and cationic dyes from single or mixed dye solutions using MOF PCN-222. *RSC Adv.* **2017**, *7*, 16273–16281. [CrossRef]
74. Mahmoud, M.E.; Abdelfattah, A.M.; Tharwat, R.M.; Nabil, G.M. Adsorption of negatively charged food tartrazine and sunset yellow dyes onto positively charged triethylenetetramine biochar: Optimization, kinetics and thermodynamic study. *J. Mol. Liq.* **2020**, *318*, 114297. [CrossRef]
75. Ho, S.H.; Zhu, S.; Chang, J.S. Recent advances in nanoscale-metal assisted biochar derived from waste biomass used for heavy metals removal. *Bioresour. Technol.* **2017**, *246*, 123–134. [CrossRef]
76. Shen, K.; Gondal, M.A. Removal of hazardous Rhodamine dye from water by adsorption onto exhausted coffee ground. *J. Saudi Chem. Soc.* **2017**, *21*, S120–S127. [CrossRef]
77. Dutta, S.; Gupta, B.; Srivastava, S.K.; Gupta, A.K. Recent advances on the removal of dyes from wastewater using various adsorbents: A critical review. *Mater. Adv.* **2021**, *2*, 4497–4531. [CrossRef]
78. Vijayaraghavan, K. The importance of mineral ingredients in biochar production, properties and applications. *Crit. Rev. Environ. Sci. Technol.* **2021**, *51*, 113–139. [CrossRef]
79. Yang, X.; Wan, Y.; Zheng, Y.; He, F.; Yu, Z.; Huang, J.; Gao, B. Surface functional groups of carbon-based adsorbents and their roles in the removal of heavy metals from aqueous solutions: A critical review. *Chem. Eng. J.* **2019**, *366*, 608–621. [CrossRef]
80. Tan, X.F.; Liu, Y.G.; Gu, Y.L.; Xu, Y.; Zeng, G.M.; Hu, X.J.; Li, J. Biochar-based nano-composites for the decontamination of wastewater: A review. *Bioresour. Technol.* **2016**, *212*, 318–333. [CrossRef]
81. Dai, L.; Lu, Q.; Zhou, H.; Shen, F.; Liu, Z.; Zhu, W.; Huang, H. Tuning oxygenated functional groups on biochar for water pollution control: A critical review. *J. Hazard. Mater.* **2021**, *420*, 126547. [CrossRef]
82. Yek, P.N.Y.; Peng, W.; Wong, C.C.; Liew, R.K.; Ho, Y.L.; Mahari, W.A.W.; Lam, S.S. Engineered biochar via microwave CO_2 and steam pyrolysis to treat carcinogenic Congo red dye. *J. Hazard. Mater.* **2020**, *395*, 122636. [CrossRef] [PubMed]
83. Park, J.H.; Wang, J.J.; Meng, Y.; Wei, Z.; DeLaune, R.D.; Seo, D.C. Adsorption/desorption behavior of cationic and anionic dyes by biochars prepared at normal and high pyrolysis temperatures. *Colloids Surf. A Physicochem. Eng. Asp.* **2019**, *572*, 274–282. [CrossRef]
84. Sterenzon, E.; Vadivel, V.K.; Gerchman, Y.; Luxbacher, T.; Narayanan, R.; Mamane, H. Effective Removal of Acid Dye in Synthetic and Silk Dyeing Effluent: Isotherm and Kinetic Studies. *ACS Omega* **2021**, *7*, 118–128. [CrossRef] [PubMed]
85. Chowdhury, M.F.; Khandaker, S.; Sarker, F.; Islam, A.; Rahman, M.T.; Awual, M.R. Current treatment technologies and mechanisms for removal of indigo carmine dyes from wastewater: A review. *J. Mol. Liq.* **2020**, *318*, 114061. [CrossRef]
86. Eltaweil, A.S.; Mohamed, H.A.; Abd El-Monaem, E.M.; El-Subruiti, G.M. Mesoporous magnetic biochar composite for enhanced adsorption of malachite green dye: Characterization, adsorption kinetics, thermodynamics and isotherms. *Adv. Powder Technol.* **2020**, *31*, 1253–1263. [CrossRef]
87. Yang, S.S.; Kang, J.H.; Xie, T.R.; He, L.; Xing, D.F.; Ren, N.Q.; Wu, W.M. Generation of high-efficient biochar for dye adsorption using frass of yellow mealworms (larvae of Tenebrio molitor Linnaeus) fed with wheat straw for insect biomass production. *J. Clean. Prod.* **2019**, *227*, 33–47. [CrossRef]
88. Yao, X.; Ji, L.; Guo, J.; Ge, S.; Lu, W.; Chen, Y.; Song, W. An abundant porous biochar material derived from wakame (Undaria pinnatifida) with high adsorption performance for three organic dyes. *Bioresour. Technol.* **2020**, *318*, 124082. [CrossRef]
89. Li, R.; Wang, J.J.; Gaston, L.A.; Zhou, B.; Li, M.; Xiao, R.; Zhang, X. An overview of carbothermal synthesis of metal–biochar composites for the removal of oxyanion contaminants from aqueous solution. *Carbon* **2018**, *129*, 674–687. [CrossRef]
90. Liu, P.; Liu, W.J.; Jiang, H.; Chen, J.J.; Li, W.W.; Yu, H.Q. Modification of bio-char derived from fast pyrolysis of biomass and its application in removal of tetracycline from aqueous solution. *Bioresour. Technol.* **2012**, *121*, 235–240. [CrossRef]
91. Shen, W.; Li, Z.; Liu, Y. Surface chemical functional groups modification of porous carbon. *Recent Pat. Chem. Eng.* **2008**, *1*, 27–40. [CrossRef]
92. Ahmaruzzaman, M. Biochar based nanocomposites for photocatalytic degradation of emerging organic pollutants from water and wastewater. *Mater. Res. Bull.* **2021**, *140*, 111262. [CrossRef]
93. Sun, J.; Lin, X.; Xie, J.; Zhang, Y.; Wang, Q.; Ying, Z. Facile synthesis of novel ternary g-C_3N_4/ferrite/biochar hybrid photocatalyst for efficient degradation of methylene blue under visible-light irradiation. *Colloids Surf. A Physicochem. Eng. Asp.* **2020**, *606*, 125556. [CrossRef]
94. Yu, F.; Tian, F.; Zou, H.; Ye, Z.; Peng, C.; Huang, J.; Gao, B. ZnO/biochar nanocomposites via solvent free ball milling for enhanced adsorption and photocatalytic degradation of methylene blue. *J. Hazard. Mater.* **2021**, *415*, 125511. [CrossRef] [PubMed]
95. Cheng, N.; Wang, B.; Wu, P.; Lee, X.; Xing, Y.; Chen, M.; Gao, B. Adsorption of emerging contaminants from water and wastewater by modified biochar: A review. *Environ. Pollut.* **2021**, *273*, 116448. [CrossRef]
96. Benis, K.Z.; Damuchali, A.M.; Soltan, J.; McPhedran, K.N. Treatment of aqueous arsenic–A review of biochar modification methods. *Sci. Total Environ.* **2020**, *739*, 139750. [CrossRef] [PubMed]
97. Barquilha, C.E.; Braga, M.C. Adsorption of organic and inorganic pollutants onto biochars: Challenges, operating conditions, and mechanisms. *Bioresour. Technol. Rep.* **2021**, *15*, 100728. [CrossRef]
98. Cho, H.H.; Wepasnick, K.; Smith, B.A.; Bangash, F.K.; Fairbrother, D.H.; Ball, W.P. Sorption of aqueous Zn [II] and Cd [II] by multiwall carbon nanotubes: The relative roles of oxygen-containing functional groups and graphenic carbon. *Langmuir* **2010**, *26*, 967–981. [CrossRef]

99. Wang, Y.; Zhong, B.; Shafi, M.; Ma, J.; Guo, J.; Wu, J.; Jin, H. Effects of biochar on growth, and heavy metals accumulation of moso bamboo (Phyllostachy pubescens), soil physical properties, and heavy metals solubility in soil. *Chemosphere* **2019**, *219*, 510–516. [CrossRef] [PubMed]
100. Li, Y.; Shao, J.; Wang, X.; Deng, Y.; Yang, H.; Chen, H. Characterization of modified biochars derived from bamboo pyrolysis and their utilization for target component (furfural) adsorption. *Energy Fuels* **2014**, *28*, 5119–5127. [CrossRef]
101. Peng, P.; Lang, Y.H.; Wang, X.M. Adsorption behavior and mechanism of pentachlorophenol on reed biochars: pH effect, pyrolysis temperature, hydrochloric acid treatment and isotherms. *Ecol. Eng.* **2016**, *90*, 225–233. [CrossRef]
102. Jin, Y.; Zhang, M.; Jin, Z.; Wang, G.; Li, R.; Zhang, X.; Wang, H. Characterization of biochars derived from various spent mushroom substrates and evaluation of their adsorption performance of Cu (II) ions from aqueous solution. *Environ. Res.* **2021**, *196*, 110323. [CrossRef] [PubMed]
103. Wei, D.; Li, B.; Huang, H.; Luo, L.; Zhang, J.; Yang, Y.; Zhou, Y. Biochar-based functional materials in the purification of agricultural wastewater: Fabrication, application and future research needs. *Chemosphere* **2018**, *197*, 165–180. [CrossRef]
104. Li, B.; Yang, L.; Wang, C.Q.; Zhang, Q.P.; Liu, Q.C.; Li, Y.D.; Xiao, R. Adsorption of Cd (II) from aqueous solutions by rape straw biochar derived from different modification processes. *Chemosphere* **2017**, *175*, 332–340. [CrossRef]
105. Sizmur, T.; Fresno, T.; Akgül, G.; Frost, H.; Moreno-Jiménez, E. Biochar modification to enhance sorption of inorganics from water. *Bioresour. Technol.* **2017**, *246*, 34–47. [CrossRef] [PubMed]
106. Cazetta, A.L.; Vargas, A.M.; Nogami, E.M.; Kunita, M.H.; Guilherme, M.R.; Martins, A.C.; Almeida, V.C. NaOH-activated carbon of high surface area produced from coconut shell: Kinetics and equilibrium studies from the methylene blue adsorption. *Chem. Eng. J.* **2011**, *174*, 117–125. [CrossRef]
107. Shen, Y.; Zhou, Y.; Fu, Y.; Zhang, N. Activated carbons synthesized from unaltered and pelletized biomass wastes for bio-tar adsorption in different phases. *Renew. Energy* **2020**, *146*, 1700–1709. [CrossRef]
108. Zheng, Y.; Wang, J.; Li, D.; Liu, C.; Lu, Y.; Lin, X.; Zheng, Z. Insight into the KOH/KMnO$_4$ activation mechanism of oxygen-enriched hierarchical porous biochar derived from biomass waste by in-situ pyrolysis for methylene blue enhanced adsorption. *J. Anal. Appl. Pyrolysis* **2021**, *158*, 105269. [CrossRef]
109. Dong, H.; Zhang, C.; Deng, J.; Jiang, Z.; Zhang, L.; Cheng, Y.; Zeng, G. Factors influencing degradation of trichloroethylene by sulfide-modified nanoscale zero-valent iron in aqueous solution. *Water Res.* **2018**, *135*, 1–10. [CrossRef] [PubMed]
110. Yang, X.; Xu, G.; Yu, H.; Zhang, Z. Preparation of ferric-activated sludge-based adsorbent from biological sludge for tetracycline removal. *Bioresour. Technol.* **2016**, *211*, 566–573. [CrossRef] [PubMed]
111. Navarathna, C.M.; Bombuwala Dewage, N.; Keeton, C.; Pennisson, J.; Henderson, R.; Lashley, B.; Mlsna, T. Biochar adsorbents with enhanced hydrophobicity for oil spill removal. *ACS Appl. Mater. Interfaces* **2020**, *12*, 9248–9260. [CrossRef] [PubMed]
112. Zulbadli, N.; Zaini, N.; Soon, L.Y.; Kang, T.Y.; Al-Harf, A.A.A.; Nasri, N.S.; Kamarudin, K.S.N. Acid-modified adsorbents from sustainable green-based materials for crude oil removal. *J. Adv. Res. Fluid Mech. Therm. Sci.* **2018**, *46*, 11–20.
113. Chen, L.; Jiang, X.; Xie, R.; Zhang, Y.; Jin, Y.; Jiang, W. A novel porous biochar-supported Fe-Mn composite as a persulfate activator for the removal of acid red 88. *Sep. Purif. Technol.* **2020**, *250*, 117232. [CrossRef]
114. Zhao, C.; Wang, B.; Theng, B.K.; Wu, P.; Liu, F.; Wang, S.; Zhang, X. Formation and mechanisms of nano-metal oxide-biochar composites for pollutants removal: A review. *Sci. Total Environ.* **2021**, *767*, 145305. [CrossRef]
115. Liu, J.; Huang, S.; Wang, T.; Mei, M.; Chen, S.; Li, J. Peroxydisulfate activation by digestate-derived biochar for azo dye degradation: Mechanism and performance. *Sep. Purif. Technol.* **2021**, *279*, 119687. [CrossRef]
116. Huang, Q.; Chen, C.; Zhao, X.; Bu, X.; Liao, X.; Fan, H.; Huang, Z. Malachite green degradation by persulfate activation with CuFe$_2$O$_4$@ biochar composite: Efficiency, stability and mechanism. *J. Environ. Chem. Eng.* **2021**, *9*, 105800. [CrossRef]
117. Liu, H.; Hu, M.; Zhang, H.; Wei, J. Biosynthesis of Stalk Biochar-nZVI and its Catalytic Reactivity in Degradation of Dyes by Persulfate. *Surf. Interfaces* **2022**, *31*, 102098. [CrossRef]
118. Zhao, Y.; Dai, H.; Ji, J.; Yuan, X.; Li, X.; Jiang, L.; Wang, H. Resource utilization of luffa sponge to produce biochar for effective degradation of organic contaminants through persulfate activation. *Sep. Purif. Technol.* **2022**, *288*, 120650. [CrossRef]
119. Gaspard, S.; Ncibi, M.C. (Eds.) *Activated Carbon from Biomass for Water Treatment. Biomass for Sustainable Applications: Pollution Remediation and Energy*; Royal Society of Chemistry: London, UK, 2013.
120. Kim, J.R.; Kan, E. Heterogeneous photocatalytic degradation of sulfamethoxazole in water using a biochar-supported TiO$_2$ photocatalyst. *J. Environ. Manag.* **2016**, *180*, 94–101. [CrossRef] [PubMed]
121. Ahmed, M.B.; Zhou, J.L.; Ngo, H.H.; Guo, W.; Chen, M. Progress in the preparation and application of modified biochar for improved contaminant removal from water and wastewater. *Bioresour. Technol.* **2016**, *214*, 836–851. [CrossRef] [PubMed]
122. Cha, J.S.; Park, S.H.; Jung, S.C.; Ryu, C.; Jeon, J.K.; Shin, M.C.; Park, Y.K. Production and utilization of biochar: A review. *J. Ind. Eng. Chem.* **2016**, *40*, 1–15. [CrossRef]
123. Panwar, N.L.; Pawar, A. Influence of activation conditions on the physicochemical properties of activated biochar: A review. *Biomass Convers. Biorefin.* **2020**, *12*, 925–947. [CrossRef]
124. Chia, C.H.; Downie, A.; Munroe, P. Characteristics of biochar: Physical and structural properties. In *Biochar for Environmental Management*; Routledge: London, UK, 2015; pp. 89–109.
125. Rajapaksha, A.U.; Vithanage, M.; Ahmad, M.; Seo, D.C.; Cho, J.S.; Lee, S.E.; Ok, Y.S. Enhanced sulfamethazine removal by steam-activated invasive plant-derived biochar. *J. Hazard. Mater.* **2015**, *290*, 43–50. [CrossRef] [PubMed]

126. Santos, R.M.; Santos, A.O.; Sussuchi, E.M.; Nascimento, J.S.; Lima, Á.S.; Freitas, L.S. Pyrolysis of mangaba seed: Production and characterization of bio-oil. *Bioresour. Technol.* **2015**, *196*, 43–48. [CrossRef] [PubMed]
127. Uchimiya, M.; Lima, I.M.; Klasson, K.T.; Wartelle, L.H. Contaminant immobilization and nutrient release by biochar soil amendment: Roles of natural organic matter. *Chemosphere* **2010**, *80*, 935–940. [CrossRef] [PubMed]
128. Hagemann, N.; Spokas, K.; Schmidt, H.P.; Kägi, R.; Böhler, M.; Bucheli, T. Activated Carbon, Biochar and Charcoal: Linkages and Synergies across Pyrogenic Carbon's ABCs. *Water* **2018**, *10*, 182. [CrossRef]
129. Jellali, S.; Khiari, B.; Usman, M.; Hamdi, H.; Charabi, Y.; Jeguirim, M. Sludge-derived biochars: A review on the influence of synthesis conditions on pollutants removal efficiency from wastewaters. *Renew. Sustain. Energy Rev.* **2021**, *144*, 111068. [CrossRef]
130. Méndez, A.; Gascó, G.; Freitas, M.M.A.; Siebielec, G.; Stuczynski, T.; Figueiredo, J.L. Preparation of carbon-based adsorbents from pyrolysis and air activation of sewage sludges. *Chem. Eng. J.* **2005**, *108*, 169–177. [CrossRef]
131. Sewu, D.D.; Jung, H.; Kim, S.S.; Lee, D.S.; Woo, S.H. Decolorization of cationic and anionic dye-laden wastewater by steam-activated biochar produced at an industrial-scale from spent mushroom substrate. *Bioresour. Technol.* **2019**, *277*, 77–86. [CrossRef]
132. Jin, H.; Capareda, S.; Chang, Z.; Gao, J.; Xu, Y.; Zhang, J. Biochar pyrolytically produced from municipal solid wastes for aqueous As (V) removal: Adsorption property and its improvement with KOH activation. *Bioresour. Technol.* **2014**, *169*, 622–629. [CrossRef] [PubMed]
133. Zhang, Q.; Wang, J.; Lyu, H.; Zhao, Q.; Jiang, L.; Liu, L. Ball-milled biochar for galaxolide removal: Sorption performance and governing mechanisms. *Sci. Total Environ.* **2019**, *659*, 1537–1545. [CrossRef] [PubMed]
134. Arif, M.; Liu, G.; Yousaf, B.; Ahmed, R.; Irshad, S.; Ashraf, A.; Rashid, M.S. Synthesis, characteristics and mechanistic insight into the clays and clay minerals-biochar surface interactions for contaminants removal—A review. *J. Clean. Prod.* **2021**, *310*, 127548. [CrossRef]
135. Zhao, Y.; Yuan, X.; Li, X.; Jiang, L.; Wang, H. Burgeoning prospects of biochar and its composite in persulfate-advanced oxidation process. *J. Hazard. Mater.* **2021**, *409*, 124893. [CrossRef] [PubMed]
136. Peterson, S.C.; Jackson, M.A.; Kim, S.; Palmquist, D.E. Increasing biochar surface area: Optimization of ball milling parameters. *Powder Technol.* **2012**, *228*, 115–120. [CrossRef]
137. Lyu, H.; Gao, B.; He, F.; Zimmerman, A.R.; Ding, C.; Huang, H.; Tang, J. Effects of ball milling on the physicochemical and sorptive properties of biochar: Experimental observations and governing mechanisms. *Environ. Pollut.* **2018**, *233*, 54–63. [CrossRef]
138. Zheng, Y.; Wan, Y.; Chen, J.; Chen, H.; Gao, B. MgO modified biochar produced through ball milling: A dual-functional adsorbent for removal of different contaminants. *Chemosphere* **2020**, *243*, 125344. [CrossRef]
139. Sewu, D.D.; Tran, H.N.; Ohemeng-Boahen, G.; Woo, S.H. Facile magnetic biochar production route with new goethite nanoparticle precursor. *Sci. Total Environ.* **2020**, *717*, 137091. [CrossRef]
140. Qiu, Y.; Zhang, Q.; Wang, Z.; Gao, B.; Fan, Z.; Li, M.; Zhong, M. Degradation of anthraquinone dye reactive blue 19 using persulfate activated with Fe/Mn modified biochar: Radical/non-radical mechanisms and fixed-bed reactor study. *Sci. Total Environ.* **2021**, *758*, 143584. [CrossRef]
141. Zhao, F.; Shan, R.; Li, W.; Zhang, Y.; Yuan, H.; Chen, Y. Synthesis, characterization, and dye removal of $ZnCl_2$-modified biochar derived from pulp and paper sludge. *ACS Omega* **2021**, *6*, 34712–34723. [CrossRef]
142. Feng, Z.; Yuan, R.; Wang, F.; Chen, Z.; Zhou, B.; Chen, H. Preparation of magnetic biochar and its application in catalytic degradation of organic pollutants: A review. *Sci. Total Environ.* **2021**, *765*, 142673. [CrossRef]
143. Reddy, D.H.K.; Lee, S.M. Magnetic biochar composite: Facile synthesis, characterization, and application for heavy metal removal. *Colloids Surf. A Physicochem. Eng. Asp.* **2014**, *454*, 96–103. [CrossRef]
144. Qu, S.; Huang, F.; Yu, S.; Chen, G.; Kong, J. Magnetic removal of dyes from aqueous solution using multi-walled carbon nanotubes filled with Fe_2O_3 particles. *J. Hazard. Mater.* **2008**, *160*, 643–647. [CrossRef]
145. Thines, K.R.; Abdullah, E.C.; Mubarak, N.M.; Ruthiraan, M. Synthesis of magnetic biochar from agricultural waste biomass to enhancing route for waste water and polymer application: A review. *Renew. Sustain. Energy Rev.* **2017**, *67*, 257–276. [CrossRef]
146. Theydan, S.K.; Ahmed, M.J. Adsorption of methylene blue onto biomass-based activated carbon by $FeCl_3$ activation: Equilibrium, kinetics, and thermodynamic studies. *J. Anal. Appl. Pyrolysis* **2012**, *97*, 116–122. [CrossRef]
147. Wang, S.; Gao, B.; Zimmerman, A.R.; Li, Y.; Ma, L.; Harris, W.G.; Migliaccio, K.W. Removal of arsenic by magnetic biochar prepared from pinewood and natural hematite. *Bioresour. Technol.* **2015**, *175*, 391–395. [CrossRef]
148. Baig, S.A.; Zhu, J.; Muhammad, N.; Sheng, T.; Xu, X. Effect of synthesis methods on magnetic Kans grass biochar for enhanced As (III, V) adsorption from aqueous solutions. *Biomass Bioenergy* **2014**, *71*, 299–310. [CrossRef]
149. Han, Y.; Cao, X.; Ouyang, X.; Sohi, S.P.; Chen, J. Adsorption kinetics of magnetic biochar derived from peanut hull on removal of Cr (VI) from aqueous solution: Effects of production conditions and particle size. *Chemosphere* **2016**, *145*, 336–341. [CrossRef] [PubMed]
150. Guo, L.; Zhao, L.; Tang, Y.; Zhou, J.; Shi, B. An iron–based biochar for persulfate activation with highly efficient and durable removal of refractory dyes. *J. Environ. Chem. Eng.* **2022**, *10*, 106979. [CrossRef]
151. Abodif, A.M.; Meng, L.; Ma, S.; Ahmed, A.S.; Belvett, N.; Wei, Z.Z.; Ning, D. Mechanisms and models of adsorption: TiO_2-supported biochar for removal of 3,4-dimethylaniline. *ACS Omega* **2020**, *5*, 13630–13640. [CrossRef]
152. Deng, H.; Li, A.; Ye, C.; Sheng, L.; Li, Z.; Jiang, Y. Green Removal of Various Pollutants by Microsphere Adsorption: Material Characterization and Adsorption Behavior. *Energy Fuels* **2020**, *34*, 16330–16340. [CrossRef]

153. Liu, S.; Shen, C.; Wang, Y.; Huang, Y.; Hu, X.; Li, B.; Zhang, H. Development of CO_2/H_2O activated biochar derived from pine pyrolysis: Application in methylene blue adsorption. *J. Chem. Technol. Biotechnol.* **2022**, *97*, 885–893. [CrossRef]
154. Liu, S.; Li, J.; Xu, S.; Wang, M.; Zhang, Y.; Xue, X. A modified method for enhancing adsorption capability of banana pseudostem biochar towards methylene blue at low temperature. *Bioresour. Technol.* **2019**, *282*, 48–55. [CrossRef]
155. Hoslett, J.; Ghazal, H.; Mohamad, N.; Jouhara, H. Removal of methylene blue from aqueous solutions by biochar prepared from the pyrolysis of mixed municipal discarded material. *Sci. Total Environ.* **2020**, *714*, 136832. [CrossRef]
156. Sahu, S.; Pahi, S.; Tripathy, S.; Singh, S.K.; Behera, A.; Sahu, U.K.; Patel, R.K. Adsorption of methylene blue on chemically modified lychee seed biochar: Dynamic, equilibrium, and thermodynamic study. *J. Mol. Liq.* **2020**, *315*, 113743. [CrossRef]
157. Chen, X.L.; Li, F.; Chen, H.; Wang, H.; Li, G. Fe_2O_3/TiO_2 functionalized biochar as a heterogeneous catalyst for dyes degradation in water under Fenton processes. *J. Environ. Chem. Eng.* **2020**, *8*, 103905. [CrossRef]
158. Liu, X.J.; Li, M.F.; Singh, S.K. Manganese-modified lignin biochar as adsorbent for removal of methylene blue. *J. Mater. Res. Technol.* **2021**, *12*, 1434–1445. [CrossRef]
159. Bal, D.; Özer, Ç.; İmamoğlu, M. Green and ecofriendly biochar preparation from pumpkin peel and its usage as an adsorbent for methylene blue removal from aqueous solutions. *Water Air Soil Pollut.* **2021**, *232*, 457. [CrossRef]
160. Pandey, D.; Daverey, A.; Dutta, K.; Yata, V.K.; Arunachalam, K. Valorization of waste pine needle biomass into biosorbents for the removal of methylene blue dye from water: Kinetics, equilibrium and thermodynamics study. *Environ. Technol. Innov.* **2022**, *25*, 102200. [CrossRef]
161. Pandey, D.; Daverey, A.; Dutta, K.; Arunachalam, K. Bioremoval of toxic malachite green from water through simultaneous decolorization and degradation using laccase immobilized biochar. *Chemosphere* **2022**, *297*, 134126. [CrossRef]
162. Eldeeb, T.M.; Aigbe, U.O.; Ukhurebor, K.E.; Onyancha, R.B.; El-Nemr, M.A.; Hassaan, M.A.; El Nemr, A. Adsorption of methylene blue (MB) dye on ozone, purified and sonicated sawdust biochars. *Biomass Convers. Biorefin.* **2022**, 1–23. [CrossRef]
163. Zhao, Z.; Zhou, W. Insight into interaction between biochar and soil minerals in changing biochar properties and adsorption capacities for sulfamethoxazole. *Environ. Pollut.* **2019**, *245*, 208–217. [CrossRef] [PubMed]
164. Lu, J.; Yang, Y.; Liu, P.; Li, Y.; Huang, F.; Zeng, L.; Hou, B. Iron-montmorillonite treated corn straw biochar: Interfacial chemical behavior and stability. *Sci. Total Environ.* **2020**, *708*, 134773. [CrossRef]
165. Han, H.; Rafiq, M.K.; Zhou, T.; Xu, R.; Mašek, O.; Li, X. A critical review of clay-based composites with enhanced adsorption performance for metal and organic pollutants. *J. Hazard. Mater.* **2019**, *369*, 780–796. [CrossRef]
166. Khandelwal, N.; Tiwari, E.; Singh, N.; Darbha, G.K. Heterogeneously Porous Multiadsorbent Clay–Biochar Surface to Support Redox-Sensitive Nanoparticles: Applications of Novel Clay–Biochar–Nanoscale Zerovalent Iron Nanotrident (C-BC-nZVI) in Continuous Water Filtration. *ACS EST Water* **2021**, *1*, 641–652. [CrossRef]
167. Lakshmi, D.; Akhil, D.; Kartik, A.; Gopinath, K.P.; Arun, J.; Bhatnagar, A.; Muthusamy, G. Artificial intelligence (AI) applications in adsorption of heavy metals using modified biochar. *Sci. Total Environ.* **2021**, *801*, 149623. [CrossRef]
168. Mishra, P.; Gupta, C. Cookies in a Cross-site scripting: Type, Utilization, Detection, Protection and Remediation. In Proceedings of the 2020 8th International Conference on Reliability, Infocom Technologies and Optimization (Trends and Future Directions) (ICRITO), Noida, India, 4–5 June 2020; pp. 1056–1059. [CrossRef]
169. Babaei, A.A.; Khataee, A.; Ahmadpour, E.; Sheydaei, M.; Kakavandi, B.; Alaee, Z. Optimization of cationic dye adsorption on activated spent tea: Equilibrium, kinetics, thermodynamic and artificial neural network modeling. *Korean J. Chem. Eng.* **2016**, *33*, 1352–1361. [CrossRef]
170. Moosavi, S.; Manta, O.; El-Badry, Y.A.; Hussein, E.E.; El-Bahy, Z.M.; Mohd Fawzi, N.F.B.; Moosavi, S.M.H. A study on machine learning methods' application for dye adsorption prediction onto agricultural waste activated carbon. *Nanomaterials* **2021**, *11*, 2734. [CrossRef]
171. Alam, G.; Ihsanullah, I.; Naushad, M.; Sillanpää, M. Applications of artificial intelligence in water treatment for optimization and automation of adsorption processes: Recent advances and prospects. *Chem. Eng. J.* **2022**, *427*, 130011. [CrossRef]
172. Taoufik, N.; Boumya, W.; Achak, M.; Chennouk, H.; Dewil, R.; Barka, N. The state of art on the prediction of efficiency and modeling of the processes of pollutants removal based on machine learning. *Sci. Total Environ.* **2022**, *807*, 150554. [CrossRef]
173. Beck, D.A.; Carothers, J.M.; Subramanian, V.R.; Pfaendtner, J. Data science: Accelerating innovation and discovery in chemical engineering. *AIChE J.* **2016**, *62*, 1402–1416. [CrossRef]
174. Ghaedi, A.M.; Vafaei, A. Applications of artificial neural networks for adsorption removal of dyes from aqueous solution: A review. *Adv. Colloid Interface Sci.* **2017**, *245*, 20–39. [CrossRef]
175. Aber, S.; Daneshvar, N.; Soroureddin, S.M.; Chabok, A.; Asadpour-Zeynali, K. Study of acid orange 7 removal from aqueous solutions by powdered activated carbon and modeling of experimental results by artificial neural network. *Desalination* **2007**, *211*, 87–95. [CrossRef]

Article

Biochar from Lemon Stalks: A Highly Active and Selective Carbocatalyst for the Oxidation of Sulfamethoxazole with Persulfate

Spyridon Giannakopoulos [1], John Vakros [1,2,*], Zacharias Frontistis [3], Ioannis D. Manariotis [4], Danae Venieri [5], Stavros G. Poulopoulos [6] and Dionissios Mantzavinos [1]

1. Department of Chemical Engineering, University of Patras, Caratheodory 1, University Campus, GR-26504 Patras, Greece
2. School of Sciences and Engineering, University of Nicosia, Nicosia 2417, Cyprus
3. Department of Chemical Engineering, University of Western Macedonia, GR-50132 Kozani, Greece
4. Environmental Engineering Laboratory, Department of Civil Engineering, University of Patras, University Campus, GR-26504 Patras, Greece
5. School of Chemical & Environmental Engineering, Technical University of Crete, GR-73100 Chania, Greece
6. Department of Chemical and Materials Engineering, School of Engineering and Digital Sciences, Nazarbayev University, Astana 010000, Kazakhstan
* Correspondence: vakros@chemistry.upatras.gr

Abstract: Pyrolysis of lemon stalks at 850 °C under a limited oxygen atmosphere yields a highly active and selective biochar for the activation of persulfate ion and the oxidation of sulfamethoxazole (SMX). The biochar mainly consists of C and O atoms, with Ca and K being the most abundant minerals. It has a moderate specific surface area of 154 m^2 g^{-1} and carbonate species, probably in the form of calcium carbonate. Complete degradation of 0.5 mg L^{-1} SMX can be achieved within 20 min using 500 mg L^{-1} sodium persulfate (SPS) and 100 mg L^{-1} biochar in ultrapure water (UPW). The acidic environment positively influences the degradation and adsorption processes, while the complexity of the water matrices usually has a negative impact on the degradation. The presence of chloride accelerates the oxidation of SMX, whose mechanism follows radical and non-radical pathways. Hydroxyl radicals seem to have the dominant contribution, while the electron transfer pathway was proven with electrochemical characterization. The biochar is stable for at least five cycles, and this makes it a good candidate for a sustainable, metal-free catalyst.

Keywords: biochar; sulfamethoxazole; persulfate; electron transfer

1. Introduction

Wastes from wastewater treatment plants (WWTPs), hospitals, agriculture, and farms are likely to contain various antibiotics, sometimes at high concentrations, since these pharmaceuticals are widely used in various activities; such wastes may end up in the environment such as rivers, surface waters, and the sea. In fact, many antibiotics are resistant to degradation and mineralization in WWTPs due to their complex chemical structures. It is important to note that in areas with high levels of contamination, these chemicals can remain in the environment due to their physicochemical properties, making certain bacteria resistant to antibiotics. This can lead to the natural environment becoming a pool of antibiotic-resistant pathogens, which under certain conditions can enter the food chain and create antibiotic-resistant microbes that pose a threat to human health [1].

Sulfonamides is one of the most common families of antibiotics with sulfamethoxazole (SMX) being a typical representative. They are typically found in terrestrial and aqueous matrices [2–4], mainly due to their high resistance to soil and use in excessive quantities [5]. For instance, detection of sulfonamides has been reported in 80% of 139 rivers and streams sampled in 1999–2000 in 30 states in the US [6]. From that period onwards, sulfonamides are found in livestock-related rivers, agro-industrial streams [7–9], and wetlands and surface waters [10]. Although antibiotics are present in surface waters at concentrations of ng L^{-1} [11,12], maximum concentration values in rivers in the order of μg L^{-1} have also been reported [13,14]. More specifically, SMX was detected in all samples from river Seine, France at a maximum concentration of 544 ng L^{-1} [15]. The occurrence of SMX at concentrations up to 1 μg L^{-1} was reported in 20% of groundwater samples during a monitoring campaign in the US [16].

Therefore, there is a need to develop efficient technologies to adequately remove persistent pollutants, such as SMX, which are not fully degraded in conventional WWTPs [17,18]. In recent decades, advanced oxidation processes (AOPs) have evolved as promising treatment technologies based on the production of reactive oxygen species (ROS) that completely oxidize organic pollutants into end products such as water, carbon dioxide, and minerals. ROS mainly involve radicals that are essentially derived from an ion, molecule or atom having at least one unpaired electron in the outer electron layer. Due to such an electronic configuration, a free radical becomes very active as it is very unstable and can react rapidly with several types of organic pollutants [19]. Hydroxyl radicals, which perhaps are the most common ones, can be produced from the reaction of hydrogen peroxide with iron ions in acidic conditions [20]. In the past few years, sulfate radicals have become a good alternative to hydroxyl radicals [21] since they are more selective for the oxidation of organic molecules by electron transfer [22]. Persulfate ions, $S_2O_8^{2-}$, have an O–O bond with length proportional to that in H_2O_2 [23,24] but exhibit higher water solubility and stability [25] than H_2O_2. The redox potential of $S_2O_8^{2-}$ is 2.01 V, but it requires an activator to produce sulfate radicals:

$$S_2O_8^{2-} + 2e^- \rightarrow 2SO_4^{\bullet -} \tag{1}$$

Activation can be performed with heat [26], ultrasound [27], solar or UV irradiation [28], and in the presence of a transition metal [29–31], i.e., Equations (2)–(4):

$$S_2O_8^{2-} + M^{n+} \rightarrow M^{(n+1)+} + SO_4^{\bullet -} + SO_4^{2-} \tag{2}$$

$$SO_4^{\bullet -} + M^{n+} \rightarrow M^{(n+1)+} + SO_4^{2-} \tag{3}$$

$$S_2O_8^{2-} + 2M^{n+} \rightarrow 2M^{(n+1)+} + 2SO_4^{2-} \tag{4}$$

In the last decade, metal-free catalysis has gained considerable attention. Carbonaceous materials, such as biochar (BC), can be used for the activation of persulfate. Pyrolysis of waste biomass yields BC with moderate to high specific surface area (SSA), hierarchical pore structure, and plenty of surface functional groups. Biochar find significant applications as supercapacitors [32], adsorbents [33], catalysts in transesterification reactions [34], and persulfate activators to oxidize emerging contaminants. They can be prepared from any kind of raw biomass and their properties; for instance, aromaticity and the polarity index depend on the nature and type of oxygenated functional groups formed on the surface [35–37], which is a function of the pyrolysis conditions and the characteristics of raw biomass.

A type of biomass that is found in abundance worldwide is the trimmings of citrus and fruit trees, containing high levels of lignin and cellulose but low humidity. All over the world, there are large areas with orchards where fruit trees are grown such as orange, lemon, pear, apple, and so on. Due to the need for regular care of these arable lands through annual pruning of the trees, very large quantities of unwanted wood (trunks, branches, stalks) and leaves are obtained [38], which are most often burned. For instance, the annual production of lemons for the period 2019–2020 was estimated at 7.55 million metric tons, with Mexico holding first place with 2.2 million metric tons, followed by the European Union and Argentina with 1.42 and 1.4 million tons, respectively [39].

In this work, biochar prepared from lemon stalk was used to activate persulfate for the destruction of SMX. To our knowledge, this is the first time that this kind of biomass was employed for the production of BC, which was then applied for SMX oxidation. Previous studies have dealt with SMX oxidation through the carbocatalytic activation of persulfate with various biochar from spent coffee grounds [40], olive stones [41], spent malt rootlets [42], and rice husks [43]. In general, agro-industrial BC is an excellent persulfate activator for SPS and the subsequent degradation of antibiotics. In this perspective, a new BC was synthesized, characterized with different physicochemical techniques, and tested for the degradation of SMX. Electrochemical characterization was applied to elucidate the degradation mechanism.

2. Results and Discussion

2.1. Biochar Characterization

The SEM image (Figure 1) shows that there are some crystallites formed on the surface of carbon phase. These crystallites have relatively long dimensions and are rich in Ca, as was shown from the EDX analysis.

Figure 1. Representative SEM image of the biochar showing deposits of minerals.

The formation of carbonate species may either be due to pyrolysis or, more likely, the exposure of BC in the air and the subsequent reaction of CaO with CO_2. Based on EDX analysis, the percent elemental composition was as follows: 86.5% C, 12.2% O, 0.5% Ca, 0.3% K, 0.2% Mg, and 0.1% from each of P, Si, and Cl. Minerals were in the form of oxides and carbonates, since only small amounts of Cl were detected.

The XRD pattern (Figure 2) exhibits two broad peaks centered at 2θ about 23° and 43.6°, which are typical of the graphitization process during pyrolysis. The broad peak at 23° refers to graphitic carbon and is due to the (002) plane. The second peak describes the (100) plane of the sp^2 hybridization of the carbon atoms [44,45]. There are some other sharp peaks centered at 26.5° and 29.4° due to carbonate minerals such as calcite, in accordance with the EDX analysis. Their content is rather limited, and the sharp peak indicates the absence of significant interactions with the carbon phase.

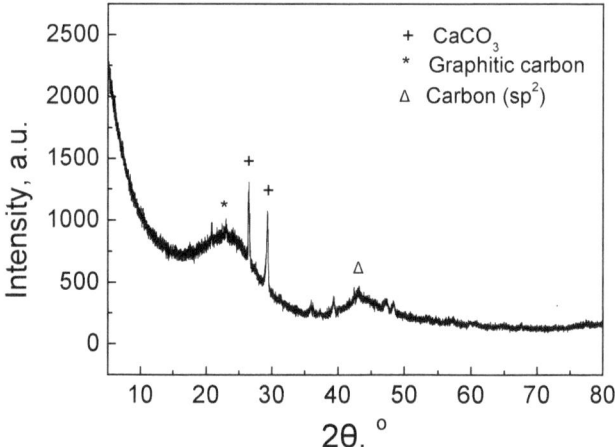

Figure 2. XRD pattern of the biochar.

As can be seen in the FTIR spectrum (Figure 3), there is a broad peak centered at 3430 cm^{-1} and a peak at 1105 cm^{-1}. These peaks are due to –OH and C–O(H) bonds and refer to the surface hydroxyl species and/or adsorbed water molecules. The peak at 1432 cm^{-1} is due to the carbonate species, probably closely related with Ca (or K) ions, while the 1631 cm^{-1} describes the C=C bonds. There also are two peaks at 2917 and 2845 cm^{-1}, which are due to C–H aliphatic bonds. The absence of any peak above 3000 cm^{-1} denotes that the aromatic phase is poor in H, since the C–H bonds in aromatic or unsaturated bonds generate peaks above 3000 cm^{-1} [41].

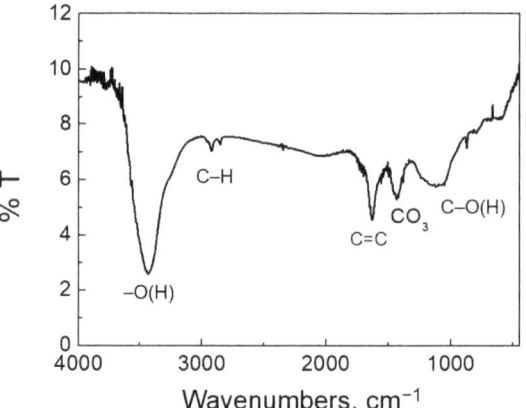

Figure 3. FTIR spectrum of biochar.

The TGA curve (Figure 4) clearly shows that the organic phase is highly homogeneous. There is a sharp decrease in mass starting at 450 °C, while the differential curve reveals that there is a sharp peak at 505 °C. Interestingly, the mass left after the TGA run is only 7%; therefore, the minerals content is limited in the biochar.

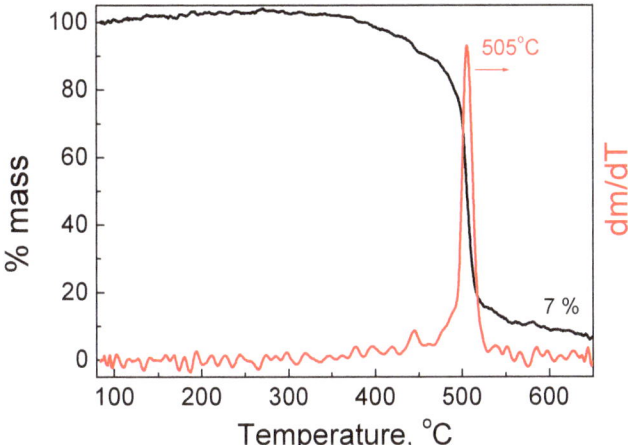

Figure 4. TGA (black line) and differential curve, dm/dT, (red line) of biochar. TGA was performed under air atmosphere at 10 °C min^{-1} heating rate.

The SSA of the biochar was measured equal to 154 m^2 g^{-1}. The micropores' SSA was 105 m^2 g^{-1}, while the total pore volume was 0.08 mL g^{-1}. The pore width was about 1.9 nm, while some macropores were also present, as can be seen in Figure 5.

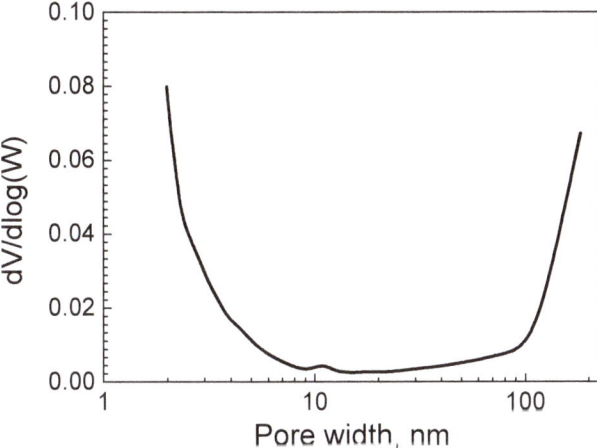

Figure 5. Pore width distribution of biochar.

The point of zero charge was determined using the potentiometric mass titration method [46] and it was found equal to 9.2. This value confirms the basic character of the biochar surface, which can be attributed to CaCO$_3$.

2.2. Biochar Activity

2.2.1. Preliminary Screening Experiments

Figure 6 shows the extent of SMX removal by adsorption and oxidation during preliminary experiments with biochar produced from various lemon tree parts, i.e., leaves pyrolyzed at 400 or 850 °C (LL400 or LL850), stalks (LST), and branches (LBR). SMX removal is mainly due to the oxidation process (Figure 6b), although adsorption contributes considerably in certain occasions (Figure 6a).

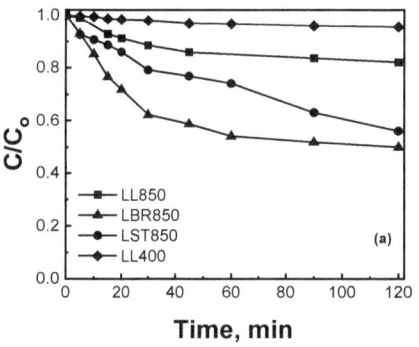 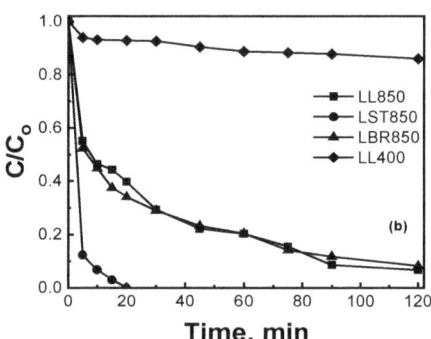

Figure 6. Removal of 0.5 mg L^{-1} SMX 100 mg L^{-1} BC under (**a**) adsorption and (**b**) oxidation with 500 mg L^{-1} SPS in UPW and inherent pH.

Adsorption is either due to interactions between opposite-charged species in the solution and surface functional groups (electrostatic adsorption) or π–π interactions between the organic contaminant and the graphitic phase [47]. As can be seen in Figure 6a, biochar prepared from biomass with higher lignin content exhibits greater adsorption capacity. Moreover, the pyrolysis temperature positively affects the degree of adsorption, probably due to higher graphitization levels occurring at higher temperatures. A similar trend regarding pyrolysis temperature occurs for the oxidation process (Figure 6b) and is consistent with previous studies [43]. SMX decomposition is complete within 20 min of the reaction with the biochar from stalks, while 90% can be achieved after 120 min with biochar from leaves or brunch.

Table 1 summarizes the conditions needed (i.e., biochar and persulfate concentrations and reaction time) to achieve maximum SMX degradation with various biochar tested in previous studies of our group. Evidently, the LST850 sample prepared in this work showed higher activity than previous materials in terms of reaction time and/or chemicals concentrations. In this respect, all subsequent experiments were performed with LST850. Figure 7 shows the effect of SPS, biochar, or initial SMX concentration on its degradation.

Table 1. Removal of SMX with biochar from different biomasses pyrolyzed at 850 °C. Values with asterisks refer to an experiment with trimethoprim as the target compound.

Biomass	BC (mg L^{-1})	SPS (mg L^{-1})	Removal (%)	Time (min)	Reference
Coffee grounds	200	1000	97	75	[40]
Malt rootlets	90	250	94	90	[42]
		500 *	74 *	120 *	[48]
Olive stones	200	1000	65	75	[41]
Rice husks	100	500	96	120	[43]

* The values are refer to the degradation of trimethoprim instead of SMX.

The beneficial effect of increasing SPS concentration in the range of 0–500 mg L^{-1} on SMX degradation is presented in Figure 7a. Assuming that degradation follows a pseudo-first-order rate expression with regards to SMX concentration, apparent kinetic constants can be computed from the data of Figure 7a; the respective values are 0.0052 min^{-1}, 0.0581 min^{-1}, 0.0738 min^{-1}, 0.0938 min^{-1}, 0.1227 min^{-1}, and 0.2574 min^{-1} at 0, 25, 50, 125, 250, and 500 mg L^{-1} SPS. Evidently, more radicals are produced at higher oxidant source concentrations, and this may possibly offset the competition between SMX and SPS for adsorption on the biochar's active sites. It should be noted here that excessive oxidant concentrations may be detrimental to the process due to radical scavenging effects and/or the conversion of sulfate and hydroxyl radicals to less powerful species such as $S_2O_8^{\bullet-}$

and O_2 [49]; this effect was not observed at the maximum SPS concentration employed in this work.

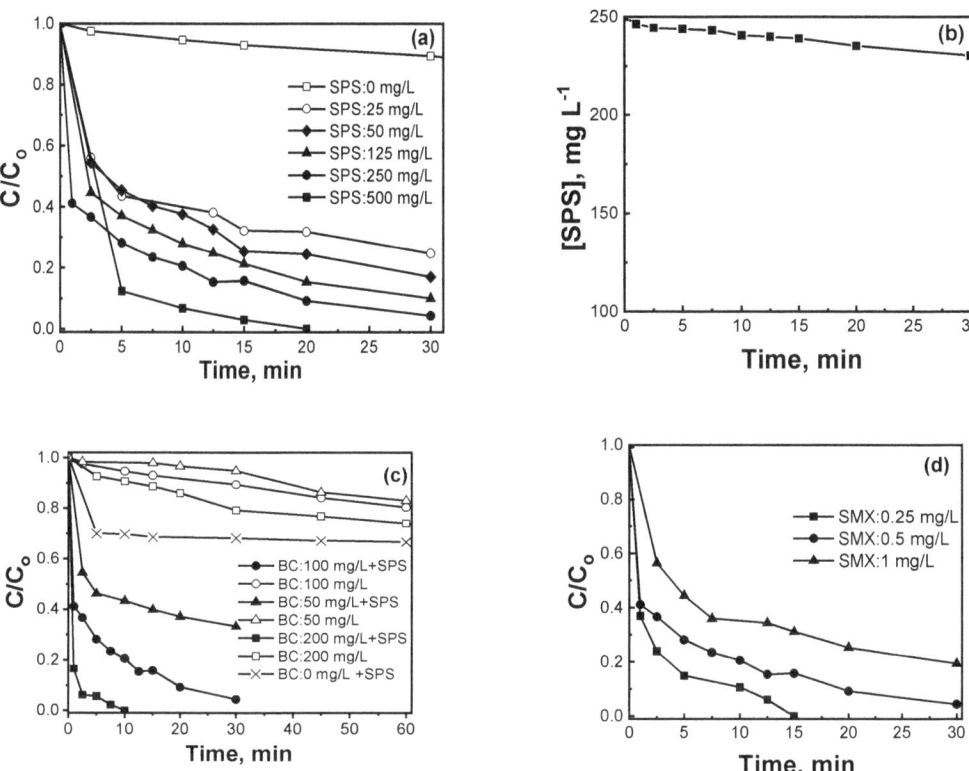

Figure 7. Removal of SMX in UPW and inherent pH as a function of (**a**) SPS concentration (0.5 mg L^{-1} SMX and 100 mg L^{-1} BC), (**b**) shows SPS consumption with 100 mg L^{-1} BC, (**c**) BC concentration (0.5 mg L^{-1} SMX with and without 250 mg L^{-1} SPS), and (**d**) SMX concentration (100 mg L^{-1} BC and 250 mg L^{-1} SPS).

Subsequent experiments were performed at 250 mg L^{-1} SPS in order to (i) keep the final concentration of sulfate ions in the treated effluent as low as possible and (ii) deliberately prolong the reaction time to make discrepancies in activity more readily observable. Interestingly, SPS was only partially converted to sulfate radicals and ions, as can be seen in Figure 7b; however, the concentration of radicals related to the 30-min SPS conversion of 8% sufficed to degrade SMX.

The positive effect of increasing biochar concentration in the range 0–200 mg L^{-1} on the adsorption and oxidation of 0.5 mg L^{-1} SMX is depicted in Figure 7c. Notably, SPS alone in the absence of biochar resulted in 30% SMX degradation in 60 min; this is because SPS is a mild oxidant on its own. However, the coexistence of persulfate and biochar is clearly beneficial since persulfate is activated, producing reactive oxygen species on the biochar surface and/or at the interfacial region or accelerating the transfer of electrons from organic molecules through the biochar's organic lattice. The apparent rate constants took values of 0.046 min^{-1}, 0.1227 min^{-1}, and 0.546 min^{-1} at 50, 100, and 200 mg L^{-1} of biochar, respectively.

The effect of SMX concentration in the range 0.25–1 mg L^{-1} is depicted in Figure 7d; conversion decreases with increasing SMX concentration with the apparent rate constants being equal to 0.0682 min^{-1}, 0.1227 min^{-1}, and 0.2483 min^{-1} at 1, 0.5, and 0.25 mg L^{-1} SMX, respectively. Although data fitting to a first-order rate expression is good and allows the computation of rate constants, the reaction is not true first order since the rate constant depends on the initial concentration.

2.2.2. Effect of pH

The solution pH strongly influences the surface charge of biochar, the form of SMX, and the activity of radicals. In this view, experiments were performed at initial pH values varying between 3 and 9 to study the effect of pH on adsorption and oxidation, and the results are shown in Figure 8. SMX adsorption was maximized at pH = 3 reaching 84% removal in 30 min, while it was considerably lower at pH = 7 (29%) and 9 (19%). The biochar whose pzc is 9.2 had its surface positively charged in a wide pH range. On the other hand, SMX has two pKa values of 1.4 and 5.7, while its isoelectric point is 4.5 [40,50]; this means that SMX is negatively charged at pH > 5.7 and neutral at 1.4 < pH < 5.7.

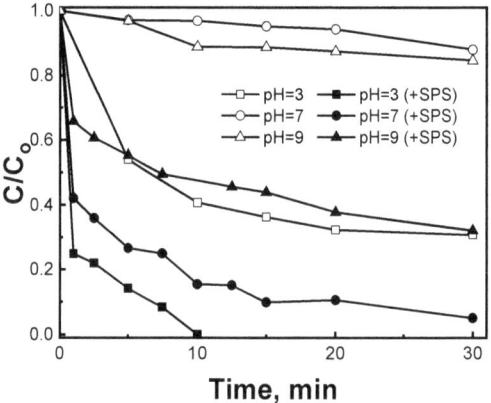

Figure 8. Effect of initial solution pH on the removal of 0.5 mg L^{-1} SMX in UPW with 100 mg L^{-1} biochar and in the presence or absence of 250 mg L^{-1} SPS.

This explains the high level of SMX adsorption at pH = 3, where the biochar exhibited the higher positive charge, and the adsorption values at pH = 7 and 9 where the surface was less acidic and close to neutral. On the addition of SPS, its activation is accompanied by the formation of HSO_4^-, a moderate acid, thus reducing the solution pH. SMX oxidation is much faster than adsorption, thus proving the catalytic action of biochar. Interestingly, complete SMX degradation can be achieved in just 10 min at pH = 3.

2.2.3. Effect of the Water Matrix

All experiments so far were performed in UPW, which is an unrealistic aqueous matrix, free of impurities, particles, and organic and inorganic compounds. Therefore, it was decided to test the process in various environmental matrices to assess the effect of their complexity on SMX degradation. Figure 9 shows experiments in secondary treated wastewater (WW), river water (RW), bottled water (BW), and seawater (SW); with the exception of BW, where the rate of degradation is comparable to UPW, all other matrices resulted in reduced degradation. The occurrence of various non-target inorganic and organic species in environmental matrices gives competitive reactions with SMX for the generated radicals, as well as the surface active sites available for adsorption [51]. Reactions in WW may also be decreased by its alkaline pH value of 8.4, as is demonstrated in Figure 8. Reduced rates in SW may partially be attributed to the high salinity and the presence of high amounts of different inorganic ions and carbon. On the contrary, the presence in BW of cations like Ca^{2+}, which interact positively with the catalyst surface and especially carbonate species, may explain the similar reactivities recorded in UPW and BW.

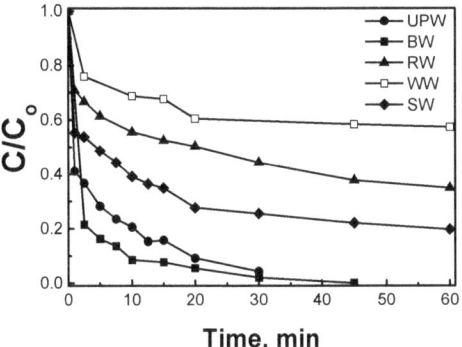

Figure 9. Effect of water matrix on the degradation of 0.5 mg L^{-1} SMX with 100 mg L^{-1} biochar and 250 mg L^{-1} SPS at inherent pH.

To shed more light on the influence of different matrix species, experiments were performed in UPW spiked with bicarbonate, chloride, or humic acid (HA), and the results are shown in Figure 10. Bicarbonate at 250 mg L^{-1} (i.e., concentration that matches that in BW and WW) has practically little effect, leading to 90% SMX degradation in 30 min. Although bicarbonate may scavenge sulfate and hydroxyl radicals, this is partially compensated by the formation of carbonate radicals, i.e., Equations (5) and (6):

$$HCO_3^- + SO_4^{\bullet -} \leftrightarrow SO_4^{2-} + CO_3^{\bullet} + H^+ \tag{5}$$

$$HCO_3^- + HO^{\bullet} \leftrightarrow CO_3^{\bullet} + H_2O \tag{6}$$

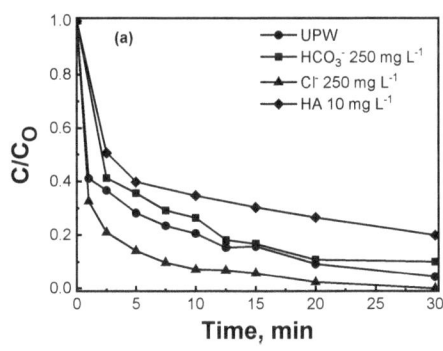
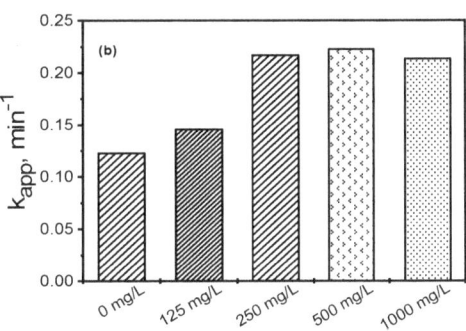

Figure 10. Degradation of 0.5 mg L^{-1} SMX with 100 mg L^{-1} biochar and 250 mg L^{-1} SPS in UPW and inherent pH as a function of (**a**) spiked chloride, bicarbonate, or humic acid; (**b**) Chloride spiked at various concentrations.

Interestingly, chloride at 250 mg L^{-1} favors SMX degradation; the computed apparent rate constant increased from 0.1227 min^{-1} in UPW to 0.2168 min^{-1} at 250 mg L^{-1} chloride and remained unchanged at concentrations up to 1000 mg L^{-1} (Figure 10b). This effect is ascribed to the formation of Cl$_2$/HOCl/Cl$^{\bullet}$ [52], which stimulate the reaction rate as described in Equations (7)–(9):

$$2Cl^- + HSO_5^- + H^+ \rightarrow SO_4^{2-} + Cl_2 + H_2O \tag{7}$$

$$Cl^- + HSO_5^- + H^+ \rightarrow SO_4^{2-} + HOCl \tag{8}$$

$$SO_4^{\bullet -} + Cl^- \leftrightarrow SO_4^{2-} + Cl^{\bullet} \tag{9}$$

Although a large excess of Cl$^-$ could intercept/interrupt the production of Cl$^\bullet$ and facilitate the formation of the less reactive Cl$_2^{\bullet-}$, thus causing reduced rates [53], this has not been observed in this study for the range of chloride concentrations spiked in UPW; however, this effect may be associated with the reduced rate recorded in SW. Other studies have reported the formation of HOCl/Cl$_2$ through a mechanism of two-stage electron transfer [54] leading to the fast abatement of SMX and ciprofloxacin [55]. Also of interest is the reported reactivity of active chlorine species against organic pollutants having excess electron molecular moieties [56,57]. Moreover, enhanced reaction rates of pharmaceuticals have been related to the increased ionic strength induced by chloride [58]. Finally, the pair HO$^\bullet$/HOCl$^{\bullet-}$ (Equation (10)) may contribute to rate enhancement [59]:

$$HOCl^{\bullet-} \leftrightarrow {}^\bullet OH + Cl^- \tag{10}$$

Figure 10a also shows the detrimental effect of HA on SMX degradation; HA was spiked in UPW to simulate the natural organic matter typically found in waters, while its concentration was selected to match the organic carbon content of WW. Humic acid competes with SMX for either the surface active sites for adsorption and/or the precious oxidants, and this presumably decreases SMX degradation.

2.2.4. Effect of the Type of Pharmaceutical

To assess process applicability to treat other pharmaceuticals, the decomposition of losartan (LOS), valsartan (VAL), and dexamethasone (DEX) at initial concentrations of 0.5 mg L^{-1} was examined. LOS, primarily employed to treat high blood pressure, belongs to angiotensin receptor blockers. These compounds are partly metabolized and have been detected in wastewater treatment plants [60–62], hospital discharges [63], rivers [64], and seawater [65] at concentration levels from ng L^{-1} to μg L^{-1}. VAL, which is also used to regulate blood pressure [66], accumulates in the environment through feces and urine and has been detected in the natural ecosystem at concentration levels from ng L^{-1} to μg L^{-1} [67,68]. DEX is a corticosteroid drug that is used to treat severe allergies, asthma, rheumatic problems, and skin diseases, amongst other diseases, and is mainly detected in the sewage of hospitals and pharmaceutical centers [69].

Figure 11 shows the relative reactivity of the four drugs, which decreases in the order: SMX > LOS > VAL > DEX. The ionization state of each compound plays an important role for the adsorption and oxidation, as has already been discussed in Section 2.2.2. The pKa values of LOS, VAL, and DEX are 4, 3.6, and 1.2, respectively [66,70], which are lower than the upper value of SMX. This trend was followed in the degradation experiments, pointing out that the physicochemical properties of SMX favor its interactions with the biochar surface. Depending on the compound, the biochar can be more or less effective for its degradation. The selectivity of the biochar can be influenced by the speciation of the antibiotic. The acidity of the surface, and thus the solution pH, is one of the crucial parameters for the selectivity of the biochar in the oxidation of different antibiotics. This is the first evidence that oxidation is surface-sensitive and mainly occurs on the surface and interface of the biochar. In this respect, further experiments were performed to investigate possible mechanisms.

2.3. Degradation Mechanism

2.3.1. Electrochemical Characterization

In general, degradation of pollutants by AOPs can occur via different mechanisms including radical and non-radical pathways. To investigate the possible contribution of an electron transfer mechanism to SMX degradation, electrochemical measurements were performed [71], where the biochar was employed as an anodic electrode.

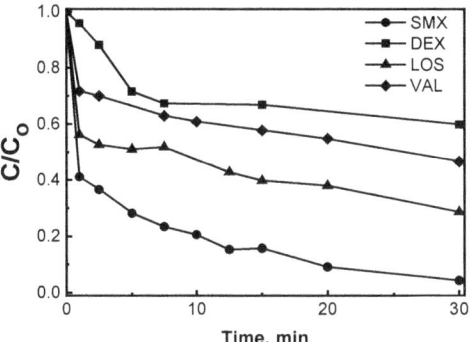

Figure 11. Degradation of 0.5 mg L^{-1} of various pharmaceuticals with 100 mg L^{-1} biochar and 250 mg L^{-1} SPS in UPW and inherent pH.

As depicted from the linear sweep voltammetry (LSV) curves (Figure 12a), the addition of 250 mg L^{-1} SPS to the solution (BC/SPS system) caused an increase in current density, indicating an interaction of SPS with the electrode surface due to the creation of metastable reactive complexes. A similar trend was recorded in the presence of 0.5 mg L^{-1} SMX (BC/SMX system), revealing charge flow between the organic molecule, which acts as an electron donor, and the electrode, and this possibly enhances degradation. A noteworthy current-density intensification was observed with the coexistence of the oxidant and the organic molecule (BC/SPS/SMX system), verifying the formation of an electron transfer pathway based on the establishment of a ternary interfacial complex between SMX, SPS, and BC surface groups. This increment in the charge transfer rate may be due to the easier and faster electron loss from the organic molecule towards the anode due to the redox potential change of the electrode in the presence of SPS, facilitating the direct oxidation of SMX by the biochar.

Then, an amperometry measurement was carried out, applying a voltage of 0 V (vs. Ag/AgCl) in the presence of 0.1 M Na$_2$SO$_4$ electrolyte in order to accurately determine the electron transfer between the catalyst, SPS, and SMX, as well as the direction of flow. According to Figure 12b, the addition of 250 mg L^{-1} SPS caused a dramatic current augmentation (from -0.25 mA to -1.4 mA), which is equivalent to a strong electron flow between the biochar and the oxidant and the function of SPS as an electron acceptor. In contrast, the following addition of 0.5 mg L^{-1} SMX resulted in a slight current decrease, thus revealing the electron displacement from the organic molecule to the BC/SPS complex, either to the carbonaceous surface, directly to the adsorbed SPS, or to the adsorbed SPS via the carbonaceous surface acting as an electron bridge. In this case, the possibility of the formation of a ternary complex, which favors electron transfer, can be supported [72].

In general, heterogeneous electron transfer is facilitated when the catalyst is characterized by high electrical conductivity. In this light, the charge transfer resistance (R_{CT}) of the material was examined using electrochemical impedance spectroscopy (EIS). The Nyquist plots for various conditions are given in Figure 12c, where the imaginary part of the impedance is represented as a function of its real part for the frequency range 0.1–10^5 Hz by imposing a voltage of 0.2 V, which is sinusoidally perturbed with an amplitude of 0.01 V. In each curve, there is an incomplete semicircle whose diameter expresses the R_{CT}. Electrical conductivity is inversely proportional to the length of the diameter. When 250 mg L^{-1} SPS or 0.5 mg L^{-1} SMX are added, the diameter diminishes, equating to a greater mobility of charge carriers. In fact, when the oxidant and the organic molecule coexist, the resistance decreases further, confirming the charge flow in the BC/SPS/SMX system.

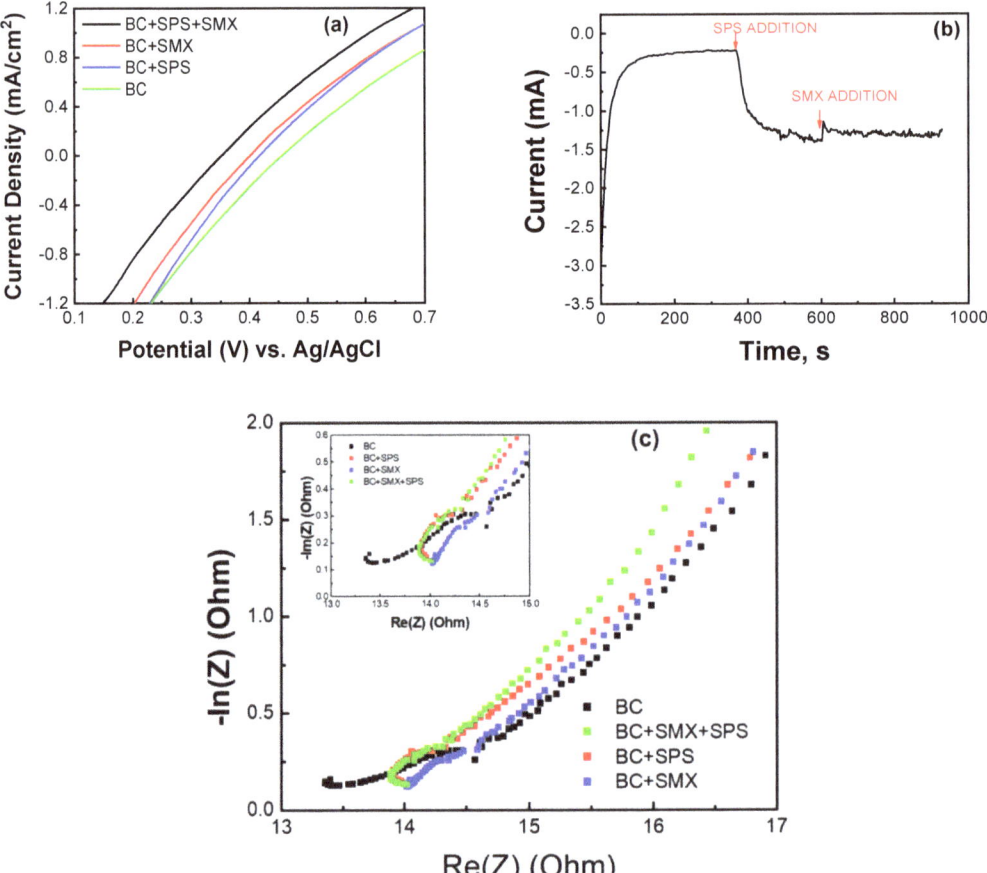

Figure 12. (a) LSV curves obtained under different conditions (SMX = 0.5 mg L^{-1}, SPS = 250 mg L^{-1}, Na$_2$SO$_4$ = 0.1 M); (b) I–t curves obtained at 0 V vs. Ag/AgCl (0.1 M Na$_2$SO$_4$); (c) EIS analysis of BC, BC/SPS, BC/SMX, and BC/SPS/SMX systems (SMX = 0.5 mg L^{-1}, SPS = 250 mg L^{-1}, Na$_2$SO$_4$ = 0.1 M).

In summary and based on the above analysis, the following electron transfer mechanism can be proposed in the BC/SPS system for the SMX decomposition:

$$BC + S_2O_8^{2-} \rightarrow \left[BC - S_2O_8^{2-}\right] \quad (11)$$

$$\left[BC - S_2O_8^{2-}\right] \xrightarrow{e^- \text{ from BC to PS}} BC_{ox} + 2SO_4^{2-} \quad (12)$$

$$SMX + BC_{ox} \xrightarrow{e^- \text{ from SMX to BC}} BC + SMX_{ox} \quad (13)$$

$$SMX + \left[BC - S_2O_8^{2-}\right] \xrightarrow[\text{through BC}]{e^- \text{ from SMX to PS}} SMX_{ox} + BC + 2SO_4^{2-} \quad (14)$$

where [BC–S$_2$O$_8^{2-}$] symbolizes the BC–PS complex, BC$_{OX}$ is the oxidation state of the biochar, and SMX$_{OX}$ is the oxidized form of SMX and intermediates. The regions of the carbon surface characterized by high electron density contribute to the creation of active centers, consequently initiating the interaction between biochar and SPS and finally resulting in the formation of surface-bound complexes (Equation (11)). This is followed by internal electron transfer within the complex (Equation (12)) and electron uptake into the SPS by SMX through the carbon structure of biochar (Equations (13) and (14)). Therefore, SMX oxidation is achieved, to a certain extent, regardless of the existence of active oxidizing species such as $HO^{\bullet}$, $SO_4^{\bullet-}$.

2.3.2. Effect of Scavengers

To estimate the radical pathway and the participation of singlet oxygen, $^1O_2$, [73], degradation experiments were conducted in the presence of 10 g L^{-1} methanol and t-butanol in large excess (i.e., alcohol to SMX ratio of 20,000) and 100 mg L^{-1} sodium azide [74,75]. In particular, methanol reacts with sulfate and hydroxyl radicals at a similar rate, while t-butanol reacts with hydroxyl radicals at a rate that is three times greater than with sulfate radicals [76,77]. To inhibit charge transfer via $^1O_2$, sodium azide was used as a quenching agent at a concentration equal to that of the biochar. Figure 13 shows that methanol does not practically affect degradation. On the contrary, SMX degradation was inhibited in the presence of t-butanol, implying that hydroxyl radicals were the dominant ones in the solution, while singlet oxygen may also participate in the oxidation process, since NaN_3 seems to slow degradation.

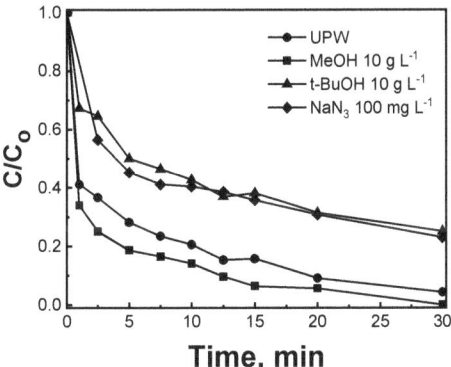

Figure 13. Effect of scavengers on the degradation of 0.5 mg L^{-1} SMX with 100 mg L^{-1} biochar and 250 mg L^{-1} SPS in UPW and inherent pH.

2.4. Biochar Reuse

To assess the stability of biochar upon repeated use, five consecutive runs were performed as follows: fresh biochar at 250 mg L^{-1} was mixed with 300 mg L^{-1} SPS and 0.5 mg L^{-1} SMX and the mixture was left to react for 30 min. The used biochar was removed by filtration, washed with UPW and reused in the following experiment and this cycle was repeated four times. The 30-min SMX conversion partly decreased from 100% in the first run to 85% in the fifth run, thus implying that the biochar can retain most of its activity although SPS may oxidize its surface according to the electron transfer mechanism discussed in Section 2.3.1. These results are in good agreement with previous studies reporting good stability of alike biochars [78].

3. Materials and Methods

3.1. Biochar Preparation

Branches of lemon tree were collected from the area of Patras, W. Greece, and lemon stalks were separated and dried in an oven at 105 °C for 2 d to remove moisture. The dried biomass was fragmented, weighed, and pyrolyzed at 850 °C using a cylindrical ceramic vessel (LH 60/12, Nabertherm GmbH, Lilienthal, Germany). The vessel was sealed with a ceramic lid to reduce oxygen by up to 20% of the quantity required for the complete burning of biomass. The biochar mass obtained after pyrolysis was 26% of the starting biomass. Finally, the biochar was sieved, and the fraction with particles < 75 μm in diameter was used to conduct experiments. For preliminary runs, other parts of the lemon tree were also used to prepare biochar, e.g., leaves (pyrolyzed at 850 °C and 400 °C) and branches (pyrolyzed at 850 °C).

3.2. Biochar Characterization

The N_2 adsorption-desorption isotherms at liquid N_2 temperature were used for the determination of specific surface area (SSA) and pore size distribution (Tristar 3000 porosimeter, Micromeritics, Norcross, GA, USA). The point of zero charge was measured with the Potentiometric Mass titration technique (Tim Talk 8 Radiometer, Copenhagen, Denmark). A scanning electron microscope (SEM) (FEI Quanta 250 FEG, Hillsboro, OR, USA) worked under various pressures (10–4000 Pa). X-ray diffraction (XRD) patterns were recorded in a Bruker D8 (Billerica, MA, USA). An advance diffractometer was equipped with a nickel-filtered CuKa (1.5418 Å) radiation source. Fourier transform infrared (FTIR) spectroscopy was performed using a Perkin Elmer Spectrum RX FTIR system (Waltham, MA, USA). The measurement range was 4000–400 cm^{-1}. More details about the characterization can be found in [79].

3.3. Electrode Preparation

Anodic materials for electrochemical measurements were prepared mixing 270 mg of biochar with 30 mg of carbon black; the latter was employed to achieve increased electrode stability and electrical conductivity. The mixture also had 100 mg PTFE added to guarantee the binding of the deposition material to the electrode surface. Ten mL of isopropanol was then added to the solid mixture under magnetic stirring, and this was subject to ultrasonic treatment for 30 min to dissolve PTFE. An electric mixer was operated for 10 min at 7000 rpm to establish a stable, homogenous suspension. Due to the mixture's insolubility in the solvent, the material was deposited on the carbon cloth using an electronic pipette, with the active electrode area being 9 cm^2 (3 cm × 3 cm). To ensure successful deposition of the carbonaceous mixture on the organic substrate, the electrode was first placed into an oven at 105 °C for 15 min and then into a furnace at 340 °C for 20 min. The electrochemical characterization was performed with an Autolab potentiostat PGSTAT128N (Utrecht, The Netherlands). When needed, an Ag/AgCl electrode was used as a reference electrode.

3.4. Chemicals and Aqueous Matrices

Sulfamethoxazole (SMX, $C_{10}H_{11}N_3O_3S$, analytical standard, C.A.S. number: 723–46–6), Valsartan (VAL, $C_{24}H_{29}N_5O_3$ analytical standard, C.A.S. number: 137862-53-4), Losartan (LOS, $C_{22}H_{23}ClN_6O$ analytical standard, C.A.S. number: 114798-26-4), Dexamethasone (DEX, $C_{22}H_{29}FO_5$ analytical standard, C.A.S. number: 50-02-2), sodium persulfate (SPS, $Na_2S_2O_8$, 99 +%, C.A.S. number: 7775-27-1), sulfuric acid (H_2SO_4, 95 wt%), sodium hydroxide (NaOH, 98 wt%), tert-butanol (t–BuOH, C.A.S. number: 75-65-0), methanol (MeOH, 99.9%, C.A.S. number: 67-56-1), sodium azide (NaN_3, C.A.S. number: 26628-22-8), humic acid (HA, C.A.S. number: 1415-93-6), sodium chloride (NaCl, C.A.S. number: 7647-14-5), and sodium bicarbonate ($NaHCO_3$, C.A.S. number: 144-55-8) were purchased from Sigma–Aldrich (St. Louis, MO, USA).

Most of the experiments were performed in ultrapure water (UPW, pH = 5.9, conductivity = 18.2 $M\Omega^{-1}$ cm^{-1}), while other matrices included: (i) commercially available bottled water (brand name: ZAGORI) (BW, pH = 7.6, conductivity = 359 $\mu S\ cm^{-1}$, TDS = 260 mg L^{-1}, total hardness ($CaCO_3$) = 219 mg L^{-1}, Cl^- = 4.3 mg L^{-1}, SO_4^{2-} = 9.2 mg L^{-1}, HCO_3^- = 244 mg L^{-1}, NO_3^- = 1.93 mg L^{-1}, Mg^{2+} = 3.1 mg L^{-1}, Ca^{2+} = 83 mg L^{-1}, K^+ = 1 mg L^{-1} and Na^+ = 2.9 mg L^{-1}); (ii) seawater (SW) collected from a coastal area in Patras city, W. Greece; (iii) surface water collected from river Glafkos, Patras city, W. Greece (RW, pH = 8.1, conductivity = 283 $\mu S\ cm^{-1}$, TSS = 17 mg L^{-1}, VSS = 7 mg L^{-1}, COD = 3.8 mg L^{-1}, SO_4^{2-} = 8 mg L^{-1}, NO_3^- = 0.9 mg L^{-1}, Cl^- = 1 mg L^{-1}, NH_3 < 1 mg L^{-1});

and (iv) secondary effluent taken from the WWTP of University of Patras (WW, COD = 48.5 mg L^{-1}, TOC = 2.5 mg L^{-1}, TSS = 22 mg L^{-1}, HCO$_3^-$ = 278 mg L^{-1}, Cl$^-$ = 262.4 mg L^{-1}, PO$_4^{3-}$ = 15 mg L^{-1}, Br$^-$ = 165.6 mg L^{-1}, Ca^{2+} = 112 mg L^{-1}).

3.5. Experimental Procedure

A 50 mg L^{-1} stock solution of SMX in UPW was used to prepare the working solutions at the desired concentration and matrix. Experiments were performed in a 250 mL beaker made of borosilicate glass, and the working volume was 120 mL under continuous magnetic stirring. Liquid samples of about 1.2 mL were withdrawn at regular intervals, then 0.3 mL of methanol was added to quench the reaction, and it was finally filtered with a 0.22 μm pore diameter filter (PVDF, Whatman). For those experiments performed at pH values other than the inherent, adjustment was done adding the appropriate volume of acid or alkali. A similar approach was followed with pharmaceuticals other than SMX.

Apparent rate constants were computed assuming that the oxidation follows a pseudo-first-order kinetic model:

$$\ln C = \ln C_o - k_{app} t \quad (15)$$

where C and C$_o$ is SMX concentration at t = t and t = 0, respectively, and k$_{app}$ is the apparent rate constant.

3.6. Analytical Methods

SMX concentration was monitored with high-performance liquid chromatography using a Waters Alliance 2695 system equipped with (i) a gradient pump (Waters 2695, Milford, PA, USA), (ii) a photodiode array detector (Waters 2996, Milford, PA, USA), (iii) a Kinetex C18 100A column (150 × 3 mm; 2.6 μm particle size) (Phenomenex, Torrance, CA, USA) where separation was achieved, and (iv) a 0.5 μm inline filter (KrudKatcher Ultra, Phenomenex, Torrance, CA, USA). The elution of the kinetic phase was done isocratically at a flowrate of 0.25 mL min^{-1} at 45 °C and consisted of 75% UPW (0.1% phosphoric acid) and 25% acetonitrile. The injection volume was 100 μL.

4. Conclusions

The valorization of citrus residues can produce carbonaceous materials, such as biochar, which can act as catalysts to activate persulfate and break down emerging micropollutants, such as antibiotics. In this work, a highly active and stable biochar was produced from lemon stalks through pyrolysis at 850 °C and used for the removal of sulfamethoxazole by adsorption/oxidation. The main conclusions are summarized as follows:

(1) Biochar characterization by various techniques gives useful information regarding its composition, as well as its structural and electronic properties; such information can be related to its activity and associated mechanisms (i.e., radical and electron transfer) for persulfate activation and subsequent SMX degradation. Specifically, the biochar has a significant amount of calcium carbonate, moderate SSA, and a pzc value 9.2. A total of 100% degradation of SMX can be achieved under pH = 3, 100 mg L^{-1} biochar, and 250 g L^{-1} SPS.

(2) Process efficiency is dictated by the interactions between the properties of the biochar, the organic pollutant, the oxidant source, and the water matrix. This interplay eventually determines the dominant reaction pathways and kinetics.

(3) Although several operating variables may determine efficiency to various degrees, particular emphasis must be given to the water matrix effect; this usually is underestimated since studies are mainly performed at conditions that are not environmentally realistic.

(4) The biochar can be applied in more realistic applications, although higher amounts of biochar may be needed.

Author Contributions: S.G., investigation, data curation; J.V., investigation, data curation, writing—original draft preparation; Z.F., validation, writing—review and editing; I.D.M., investigation; D.V., writing—review and editing.; S.G.P., writing—review and editing; D.M., writing—review and editing. All authors have read and agreed to the published version of the manuscript.

Funding: This research was funded by the Hellenic Foundation for Research and Innovation (H.F.R.I.) under the "First Call for H.F.R.I. Research Projects to support Faculty members and Researchers and the procurement of high–cost research equipment grant" (Project Number: 81080).

Data Availability Statement: The data presented in this study are available on request from the corresponding author.

Acknowledgments: S.G., J.V., D.V., and D.M. acknowledge support by the H.F.R.I.

Conflicts of Interest: The authors declare no conflict of interest.

References

1. Alduina, R. Antibiotics and Environment. *Antibiotics* **2020**, *9*, 202. [CrossRef]
2. Baran, W.; Sochacka, J.; Wardas, W. Toxicity and Biodegradability of Sulfonamides and Products of Their Photocatalytic Degradation in Aqueous Solutions. *Chemosphere* **2006**, *65*, 1295–1299. [CrossRef]
3. Deng, Y.; Li, B.; Zhang, T. Bacteria That Make a Meal of Sulfonamide Antibiotics: Blind Spots and Emerging Opportunities. *Environ. Sci. Technol.* **2018**, *52*, 3854–3868. [CrossRef]
4. Manzetti, S.; Ghisi, R. The Environmental Release and Fate of Antibiotics. *Mar. Pollut. Bull.* **2014**, *79*, 7–15. [CrossRef]
5. Thiele-Bruhn, D.S. *Environmental Risks from Mixtures of Antibiotic Pharmaceuticals in Soils—A Literature Review*; Umweltbundesamt: Dessau-Roßlau, Germany, 2019.
6. Kolpin, D.W.; Furlong, E.T.; Meyer, M.T.; Thurman, E.M.; Zaugg, S.D.; Barber, L.B.; Buxton, H.T. Pharmaceuticals, Hormones, and Other Organic Wastewater Contaminants in U.S. Streams, 1999−2000: A National Reconnaissance. *Environ. Sci. Technol.* **2002**, *36*, 1202–1211. [CrossRef]
7. Bernot, M.J.; Smith, L.; Frey, J. Human and Veterinary Pharmaceutical Abundance and Transport in a Rural Central Indiana Stream Influenced by Confined Animal Feeding Operations (CAFOs). *Sci. Total Environ.* **2013**, *445–446*, 219–230. [CrossRef]
8. Iglesias, A.; Nebot, C.; Vázquez, B.I.; Miranda, J.M.; Abuín, C.M.F.; Cepeda, A. Detection of Veterinary Drug Residues in Surface Waters Collected Nearby Farming Areas in Galicia, North of Spain. *Environ. Sci. Pollut. Res.* **2014**, *21*, 2367–2377. [CrossRef]
9. Veach, A.M.; Bernot, M.J. Temporal Variation of Pharmaceuticals in an Urban and Agriculturally Influenced Stream. *Sci. Total Environ.* **2011**, *409*, 4553–4563. [CrossRef]
10. Hruska, K.; Franek, M. Sulfonamides in the Environment: A Review and a Case Report. *Veterinární Medicína* **2012**, *57*, 1–35. [CrossRef]
11. Grenni, P.; Ancona, V.; Barra Caracciolo, A. Ecological Effects of Antibiotics on Natural Ecosystems: A Review. *Microchem. J.* **2018**, *136*, 25–39. [CrossRef]
12. Matongo, S.; Birungi, G.; Moodley, B.; Ndungu, P. Occurrence of Selected Pharmaceuticals in Water and Sediment of Umgeni River, KwaZulu-Natal, South Africa. *Environ. Sci. Pollut. Res.* **2015**, *22*, 10298–10308. [CrossRef]
13. Peng, X.; Tan, J.; Tang, C.; Yu, Y.; Wang, Z. Multiresidue determination of fluoroquinolone, sulfonamide, trimethoprim, and chloramphenicol antibiotics in urban waters in china. *Environ. Toxicol. Chem.* **2008**, *27*, 73. [CrossRef]
14. Yargeau, V.; Lopata, A.; Metcalfe, C. Pharmaceuticals in the Yamaska River, Quebec, Canada. *Water Qual. Res. J.* **2007**, *42*, 231–239. [CrossRef]
15. Tamtam, F.; Mercier, F.; Le Bot, B.; Eurin, J.; Tuc Dinh, Q.; Clément, M.; Chevreuil, M. Occurrence and Fate of Antibiotics in the Seine River in Various Hydrological Conditions. *Sci. Total Environ.* **2008**, *393*, 84–95. [CrossRef]
16. Barnes, K.K.; Kolpin, D.W.; Furlong, E.T.; Zaugg, S.D.; Meyer, M.T.; Barber, L.B. A National Reconnaissance of Pharmaceuticals and Other Organic Wastewater Contaminants in the United States—I) Groundwater. *Sci. Total Environ.* **2008**, *402*, 192–200. [CrossRef]
17. Qu, J.; Fan, M. The Current State of Water Quality and Technology Development for Water Pollution Control in China. *Crit. Rev. Environ. Sci. Technol.* **2010**, *40*, 519–560. [CrossRef]
18. Zheng, X.; Zhang, B.-T.; Teng, Y. Distribution of Phthalate Acid Esters in Lakes of Beijing and Its Relationship with Anthropogenic Activities. *Sci. Total Environ.* **2014**, *476–477*, 107–113. [CrossRef]
19. Zhang, B.-T.; Zhao, L.-X.; Lin, J.-M. Study on Superoxide and Hydroxyl Radicals Generated in Indirect Electrochemical Oxidation by Chemiluminescence and UV-Visible Spectra. *J. Environ. Sci.* **2008**, *20*, 1006–1011. [CrossRef]
20. Shukla, P.; Sun, H.; Wang, S.; Ang, H.M.; Tadé, M.O. Co-SBA-15 for Heterogeneous Oxidation of Phenol with Sulfate Radical for Wastewater Treatment. *Catal. Today* **2011**, *175*, 380–385. [CrossRef]
21. Anipsitakis, G.P.; Dionysiou, D.D. Radical Generation by the Interaction of Transition Metals with Common Oxidants. *Environ. Sci. Technol.* **2004**, *38*, 3705–3712. [CrossRef]
22. Anipsitakis, G.P.; Dionysiou, D.D. Degradation of Organic Contaminants in Water with Sulfate Radicals Generated by the Conjunction of Peroxymonosulfate with Cobalt. *Environ. Sci. Technol.* **2003**, *37*, 4790–4797. [CrossRef]

23. Zhang, B.-T.; Zhao, L.; Lin, J.-M. Determination of Folic Acid by Chemiluminescence Based on Peroxomonosulfate-Cobalt(II) System. *Talanta* **2008**, *74*, 1154–1159. [CrossRef] [PubMed]
24. Zhang, B.-T.; Lin, J.-M. Chemiluminescence and Energy Transfer Mechanism of Lanthanide Ions in Different Media Based on Peroxomonosulfate System. *Luminescence* **2010**, *25*, 322–327. [CrossRef] [PubMed]
25. Liang, C.J.; Bruell, C.J.; Marley, M.C.; Sperry, K.L. Thermally Activated Persulfate Oxidation of Trichloroethylene (T.C.E.) and 1,1,1-Trichloroethane (T.C.A.) in Aqueous Systems and Soil Slurries. *Soil Sediment Contam. Int. J.* **2003**, *12*, 207–228. [CrossRef]
26. Ioannidi, A.; Arvaniti, O.S.; Nika, M.-C.; Aalizadeh, R.; Thomaidis, N.S.; Mantzavinos, D.; Frontistis, Z. Removal of Drug Losartan in Environmental Aquatic Matrices by Heat-Activated Persulfate: Kinetics, Transformation Products and Synergistic Effects. *Chemosphere* **2022**, *287*, 131952. [CrossRef]
27. Darsinou, B.; Frontistis, Z.; Antonopoulou, M.; Konstantinou, I.; Mantzavinos, D. Sono-Activated Persulfate Oxidation of Bisphenol A: Kinetics, Pathways and the Controversial Role of Temperature. *Chem. Eng. J.* **2015**, *280*, 623–633. [CrossRef]
28. Giannakopoulos, S.; Frontistis, Z.; Vakros, J.; Poulopoulos, S.G.; Manariotis, I.D.; Mantzavinos, D. Combined Activation of Persulfate by Biochars and Artificial Light for the Degradation of Sulfamethoxazole in Aqueous Matrices. *J. Taiwan Inst. Chem. Eng.* **2022**, *136*, 104440. [CrossRef]
29. Xiang, Y.; Yang, K.; Zhai, Z.; Zhao, T.; Yuan, D.; Jiao, T.; Zhang, Q.; Tang, S. Molybdenum co-catalytic promotion for Fe^{3+}/peroxydisulfate process: Performance, mechanism, and immobilization. *Chem. Eng. J.* **2022**, *438*, 135656. [CrossRef]
30. Pan, S.; Zhai, Z.; Yang, K.; Xiang, Y.; Tang, S.; Zhang, Y.; Jiao, T.; Zhang, Q.; Yuan, D. β-Lactoglobulin amyloid fibrils supported Fe(III) to activate peroxydisulfate for organic pollutants elimination. *Sep. Purif. Technol.* **2022**, *289*, 120806. [CrossRef]
31. Xiang, Y.; Liu, H.; Zhu, E.; Yang, K.; Yuan, D.; Jiao, T.; Zhang, Q.; Tang, S. Application of inorganic materials as heterogeneous cocatalyst in Fenton/Fenton-like processes for wastewater treatment. *Sep. Purif. Technol.* **2022**, *295*, 121293. [CrossRef]
32. Giannakopoulos, S.; Vakros, J.; Dracopoulos, V.; Manariotis, I.D.; Mantzavinos, D.; Lianos, P. Enhancement of the Photoelectrocatalytic Degradation Rate of a Pollutant in the Presence of a Supercapacitor. *J. Clean. Prod.* **2022**, *377*, 134456. [CrossRef]
33. Srivatsav, P.; Bhargav, B.S.; Shanmugasundaram, V.; Arun, J.; Gopinath, K.P.; Bhatnagar, A. Biochar as an Eco-Friendly and Economical Adsorbent for the Removal of Colorants (Dyes) from Aqueous Environment: A Review. *Water* **2020**, *12*, 3561. [CrossRef]
34. Ntaflou, M.; Vakros, J. Transesterification Activity of Modified Biochars from Spent Malt Rootlets Using Triacetin. *J. Clean. Prod.* **2020**, *259*, 120931. [CrossRef]
35. Azargohar, R.; Dalai, A.K. Steam and K.O.H. Activation of Biochar: Experimental and Modeling Studies. *Microporous Mesoporous Mater.* **2008**, *110*, 413–421. [CrossRef]
36. Kong, S.-H.; Loh, S.-K.; Bachmann, R.T.; Rahim, S.A.; Salimon, J. Biochar from Oil Palm Biomass: A Review of Its Potential and Challenges. *Renew. Sustain. Energy Rev.* **2014**, *39*, 729–739. [CrossRef]
37. Sharma, A.; Pareek, V.; Zhang, D. Biomass Pyrolysis—A Review of Modelling, Process Parameters and Catalytic Studies. *Renew. Sustain. Energy Rev.* **2015**, *50*, 1081–1096. [CrossRef]
38. *Renewable Energy Sources: Engineering, Technology, Innovation: I.C.O.R.E.S. 2017*; Mudryk, K.; Werle, S. (Eds.) Springer Proceedings in Energy; Springer International Publishing: Cham, Switzerland, 2018; ISBN 978-3-319-72370-9.
39. Fresh Lemons and Limes: Leading Producers Worldwide 2021. Statista. Available online: https://www.statista.com/statistics/1045016/world-lemons-and-limes-major-producers/ (accessed on 24 August 2021).
40. Lykoudi, A.; Frontistis, Z.; Vakros, J.; Manariotis, I.D.; Mantzavinos, D. Degradation of Sulfamethoxazole with Persulfate Using Spent Coffee Grounds Biochar as Activator. *J. Environ. Manage.* **2020**, *271*, 111022. [CrossRef]
41. Magioglou, E.; Frontistis, Z.; Vakros, J.; Manariotis, I.; Mantzavinos, D. Activation of Persulfate by Biochars from Valorized Olive Stones for the Degradation of Sulfamethoxazole. *Catalysts* **2019**, *9*, 419. [CrossRef]
42. Kemmou, L.; Frontistis, Z.; Vakros, J.; Manariotis, I.D.; Mantzavinos, D. Degradation of Antibiotic Sulfamethoxazole by Biochar-Activated Persulfate: Factors Affecting the Activation and Degradation Processes. *Catal. Today* **2018**, *313*, 128–133. [CrossRef]
43. Avramiotis, E.; Frontistis, Z.; Manariotis, I.D.; Vakros, J.; Mantzavinos, D. Oxidation of Sulfamethoxazole by Rice Husk Biochar-Activated Persulfate. *Catalysts* **2021**, *11*, 850. [CrossRef]
44. Andrade, T.S.; Vakros, J.; Mantzavinos, D.; Lianos, P. Biochar Obtained by Carbonization of Spent Coffee Grounds and Its Application in the Construction of an Energy Storage Device. *Chem. Eng. J. Adv.* **2020**, *4*, 100061. [CrossRef]
45. Gao, G.; Cheong, L.-Z.; Wang, D.; Shen, C. Pyrolytic Carbon Derived from Spent Coffee Grounds as Anode for Sodium-Ion Batteries. *Carbon Resour. Convers.* **2018**, *1*, 104–108. [CrossRef]
46. Bourikas, K.; Vakros, J.; Kordulis, C.; Lycourghiotis, A. Potentiometric Mass Titrations: Experimental and Theoretical Establishment of a New Technique for Determining the Point of Zero Charge (P.Z.C.) of Metal (Hydr)Oxides. *J. Phys. Chem. B* **2003**, *107*, 9441–9451. [CrossRef]
47. Mrozik, W.; Minofar, B.; Thongsamer, T.; Wiriyaphong, N.; Khawkomol, S.; Plaimart, J.; Vakros, J.; Karapanagioti, H.; Vinitnantharat, S.; Werner, D. Valorisation of agricultural waste derived biochars in aquaculture to remove organic micropollutants from water—Experimental study and molecular dynamics simulations. *J. Environ. Manag.* **2021**, *300*, 113717. [CrossRef] [PubMed]
48. Grilla, E.; Vakros, J.; Konstantinou, I.; Manariotis, I.D.; Mantzavinos, D. Activation of persulfate by biochar from spent malt rootlets for the degradation of trimethoprim in the presence of inorganic ions. *J. Chem. Technol. Biotechnol.* **2020**, *95*, 2348–2358. [CrossRef]

49. Fang, G.; Wu, W.; Liu, C.; Dionysiou, D.D.; Deng, Y.; Zhou, D. Activation of Persulfate with Vanadium Species for P.C.B.s Degradation: A Mechanistic Study. *Appl. Catal. B Environ.* **2017**, *202*, 1–11. [CrossRef]
50. Avisar, D.; Primor, O.; Gozlan, I.; Mamane, H. Sorption of Sulfonamides and Tetracyclines to Montmorillonite Clay. *Water. Air. Soil Pollut.* **2010**, *209*, 439–450. [CrossRef]
51. Ribeiro, R.S.; Frontistis, Z.; Mantzavinos, D.; Silva, A.M.T.; Faria, J.L.; Gomes, H.T. Screening of Heterogeneous Catalysts for the Activated Persulfate Oxidation of Sulfamethoxazole in Aqueous Matrices. Does the Matrix Affect the Selection of Catalyst? *J. Chem. Technol. Biotechnol.* **2019**, *8*, 2425–2432. [CrossRef]
52. Wang, Y.; Cao, D.; Zhao, X. Heterogeneous Degradation of Refractory Pollutants by Peroxymonosulfate Activated by CoOx-Doped Ordered Mesoporous Carbon. *Chem. Eng. J.* **2017**, *328*, 1112–1121. [CrossRef]
53. Chen, L.; Zuo, X.; Yang, S.; Cai, T.; Ding, D. Rational Design and Synthesis of Hollow Co3O4@Fe2O3 Core-Shell Nanostructure for the Catalytic Degradation of Norfloxacin by Coupling with Peroxymonosulfate. *Chem. Eng. J.* **2019**, *359*, 373–384. [CrossRef]
54. Lai, L.; Yan, J.; Li, J.; Lai, B. Co/Al2O3-EPM as Peroxymonosulfate Activator for Sulfamethoxazole Removal: Performance, Biotoxicity, Degradation Pathways and Mechanism. *Chem. Eng. J.* **2018**, *343*, 676–688. [CrossRef]
55. Serna-Galvis, E.A.; Jojoa-Sierra, S.D.; Berrio-Perlaza, K.E.; Ferraro, F.; Torres-Palma, R.A. Structure-Reactivity Relationship in the Degradation of Three Representative Fluoroquinolone Antibiotics in Water by Electrogenerated Active Chlorine. *Chem. Eng. J.* **2017**, *315*, 552–561. [CrossRef]
56. Ao, X.; Liu, W.; Sun, W.; Cai, M.; Ye, Z.; Yang, C.; Lu, Z.; Li, C. Medium Pressure UV-Activated Peroxymonosulfate for Ciprofloxacin Degradation: Kinetics, Mechanism, and Genotoxicity. *Chem. Eng. J.* **2018**, *345*, 87–97. [CrossRef]
57. Sichel, C.; Garcia, C.; Andre, K. Feasibility Studies: U.V./Chlorine Advanced Oxidation Treatment for the Removal of Emerging Contaminants. *Water Res.* **2011**, *45*, 6371–6380. [CrossRef]
58. Li, A.; Wu, Z.; Wang, T.; Hou, S.; Huang, B.; Kong, X.; Li, X.; Guan, Y.; Qiu, R.; Fang, J. Kinetics and Mechanisms of the Degradation of P.P.C.P.s by Zero-Valent Iron (Fe°) Activated Peroxydisulfate (P.D.S.) System in Groundwater. *J. Hazard. Mater.* **2018**, *357*, 207–216. [CrossRef]
59. Xiao, S.; Cheng, M.; Zhong, H.; Liu, Z.; Liu, Y.; Yang, X.; Liang, Q. Iron-Mediated Activation of Persulfate and Peroxymonosulfate in Both Homogeneous and Heterogeneous Ways: A Review. *Chem. Eng. J.* **2020**, *384*, 123265. [CrossRef]
60. Ashfaq, M.; Li, Y.; Wang, Y.; Chen, W.; Wang, H.; Chen, X.; Wu, W.; Huang, Z.; Yu, C.-P.; Sun, Q. Occurrence, Fate, and Mass Balance of Different Classes of Pharmaceuticals and Personal Care Products in an Anaerobic-Anoxic-Oxic Wastewater Treatment Plant in Xiamen, China. *Water Res.* **2017**, *123*, 655–667. [CrossRef]
61. Botero-Coy, A.M.; Martínez-Pachón, D.; Boix, C.; Rincón, R.J.; Castillo, N.; Arias-Marín, L.P.; Manrique-Losada, L.; Torres-Palma, R.; Moncayo-Lasso, A.; Hernández, F. An Investigation into the Occurrence and Removal of Pharmaceuticals in Colombian Wastewater. *Sci. Total Environ.* **2018**, *642*, 842–853. [CrossRef]
62. Casado, J.; Rodríguez, I.; Ramil, M.; Cela, R. Selective Determination of Antimycotic Drugs in Environmental Water Samples by Mixed-Mode Solid-Phase Extraction and Liquid Chromatography Quadrupole Time-of-Flight Mass Spectrometry. *J. Chromatogr. A* **2014**, *1339*, 42–49. [CrossRef]
63. Azuma, T.; Otomo, K.; Kunitou, M.; Shimizu, M.; Hosomaru, K.; Mikata, S.; Ishida, M.; Hisamatsu, K.; Yunoki, A.; Mino, Y.; et al. Environmental Fate of Pharmaceutical Compounds and Antimicrobial-Resistant Bacteria in Hospital Effluents, and Contributions to Pollutant Loads in the Surface Waters in Japan. *Sci. Total Environ.* **2019**, *657*, 476–484. [CrossRef]
64. Mandaric, L.; Diamantini, E.; Stella, E.; Cano-Paoli, K.; Valle-Sistac, J.; Molins-Delgado, D.; Bellin, A.; Chiogna, G.; Majone, B.; Diaz-Cruz, M.S.; et al. Contamination Sources and Distribution Patterns of Pharmaceuticals and Personal Care Products in Alpine Rivers Strongly Affected by Tourism. *Sci. Total Environ.* **2017**, *590–591*, 484–494. [CrossRef]
65. Cortez, F.S.; Souza, L.D.S.; Guimarães, L.L.; Almeida, J.E.; Pusceddu, F.H.; Maranho, L.A.; Mota, L.G.; Nobre, C.R.; Moreno, B.B.; Abessa, D.M.D.S. et al. Ecotoxicological Effects of Losartan on the Brown Mussel Perna Perna and Its Occurrence in Seawater from Santos Bay (Brazil). *Sci. Total Environ.* **2018**, *637–638*, 1363–1371. [CrossRef] [PubMed]
66. Arvaniti, O.S.; Bairamis, F.; Konstantinou, I.; Mantzavinos, D.; Frontistis, Z. Degradation of Antihypertensive Drug Valsartan in Water Matrices by Heat and Heat/Ultrasound Activated Persulfate: Kinetics, Synergy Effect and Transformation Products. *Chem. Eng. J. Adv.* **2020**, *4*, 100062. [CrossRef]
67. Martínez-Pachón, D.; Ibáñez, M.; Hernández, F.; Torres-Palma, R.A.; Moncayo-Lasso, A. Photo-Electro-Fenton Process Applied to the Degradation of Valsartan: Effect of Parameters, Identification of Degradation Routes and Mineralization in Combination with a Biological System. *J. Environ. Chem. Eng.* **2018**, *6*, 7302–7311. [CrossRef]
68. Dasenaki, M.E.; Thomaidis, N.S. Multianalyte Method for the Determination of Pharmaceuticals in Wastewater Samples Using Solid-Phase Extraction and Liquid Chromatography-Tandem Mass Spectrometry. *Anal. Bioanal. Chem.* **2015**, *407*, 4229–4245. [CrossRef]
69. Mohseni, S.N.; Amooey, A.A.; Tashakkorian, H.; Amouei, A.I. Removal of Dexamethasone from Aqueous Solutions Using Modified Clinoptilolite Zeolite (Equilibrium and Kinetic). *Int. J. Environ. Sci. Technol.* **2016**, *13*, 2261–2268. [CrossRef]
70. Ioannidi, A.A.; Vakros, J.; Frontistis, Z.; Mantzavinos, D. Tailoring the Biochar Physicochemical Properties Using a Friendly Eco-Method and Its Application on the Oxidation of the Drug Losartan through Persulfate Activation. *Catalysts* **2022**, *12*, 1245. [CrossRef]

71. Li, M.; Bi, G.Y.; Xiang, L.; Chen, X.T.; Qin, Y.J.; Mo, C.H.; Zhou, S.Q. Improved cathodic oxygen reduction and bioelectricity generation of electrochemical reactor based on reduced graphene oxide decorated with titanium-based composites. *Bioresour. Techn.* **2020**, *296*, 122319. [CrossRef]
72. Yang, Z.; Wang, Z.; Liang, G.; Zhang, X.; Xie, X. Catalyst Bridging-Mediated Electron Transfer for Nonradical Degradation of Bisphenol A via Natural Manganese Ore-Cornstalk Biochar Composite Activated Peroxymonosulfate. *Chem. Eng. J.* **2021**, *426*, 131777. [CrossRef]
73. Yun, E.-T.; Yoo, H.-Y.; Bae, H.; Kim, H.-I.; Lee, J. Exploring the Role of Persulfate in the Activation Process: Radical Precursor Versus Electron Acceptor. *Environ. Sci. Technol.* **2017**, *51*, 10090–10099. [CrossRef]
74. Li, M.; Li, Z.; Yu, X.; Wu, Y.; Mo, C.; Luo, M.; Li, L.; Zhou, S.; Liu, Q.; Wang, N.; et al. FeN$_4$-doped carbon nanotubes derived from metal organic frameworks for effective degradation of organic dyes by peroxymonosulfate: Impacts of FeN$_4$ spin states. *Chem. Eng. J.* **2022**, *431*, 133339. [CrossRef]
75. Li, M.; Mo, C.H.; Luo, X.; He, K.Y.; Yan, J.F.; Wu, Q.; Yu, P.F.; Han, W.; Feng, N.X.; Yeung, K.L.; et al. Exploring key reaction sites and deep degradation mechanism of perfluorooctane sulfonate via peroxymonosulfate activation under electrocoagulation process. *Water Res.* **2021**, *207*, 117849. [CrossRef] [PubMed]
76. Yin, R.; Guo, W.; Wang, H.; Du, J.; Wu, Q.; Chang, J.-S.; Ren, N. Singlet Oxygen-Dominated Peroxydisulfate Activation by Sludge-Derived Biochar for Sulfamethoxazole Degradation through a Nonradical Oxidation Pathway: Performance and Mechanism. *Chem. Eng. J.* **2019**, *357*, 589–599. [CrossRef]
77. Yu, J.; Tang, L.; Pang, Y.; Zeng, G.; Wang, J.; Deng, Y.; Liu, Y.; Feng, H.; Chen, S.; Ren, X. Magnetic Nitrogen-Doped Sludge-Derived Biochar Catalysts for Persulfate Activation: Internal Electron Transfer Mechanism. *Chem. Eng. J.* **2019**, *364*, 146–159. [CrossRef]
78. Ntzoufra, P.; Vakros, J.; Frontistis, Z.; Tsatsos, S.; Kyriakou, G.; Kennou, S.; Manariotis, I.D.; Mantzavinos, D. Effect of Sodium Persulfate Treatment on the Physicochemical Properties and Catalytic Activity of Biochar Prepared from Spent Malt Rootlets. *J. Environ. Chem. Eng.* **2021**, *9*, 105071. [CrossRef]
79. Avramiotis, E.; Frontistis, Z.; Manariotis, I.D.; Vakros, J.; Mantzavinos, D. On the Performance of a Sustainable Rice Husk Biochar for the Activation of Persulfate and the Degradation of Antibiotics. *Catalysts* **2021**, *11*, 1303. [CrossRef]

Disclaimer/Publisher's Note: The statements, opinions and data contained in all publications are solely those of the individual author(s) and contributor(s) and not of MDPI and/or the editor(s). MDPI and/or the editor(s) disclaim responsibility for any injury to people or property resulting from any ideas, methods, instructions or products referred to in the content.

Article

Degradation of Sulfamethoxazole Using a Hybrid CuO$_x$–BiVO$_4$/SPS/Solar System

Konstantinos Kouvelis [1], Adamantia A. Kampioti [2], Athanasia Petala [2] and Zacharias Frontistis [3,*]

[1] Department of Chemical Engineering, University of Patras, GR-26504 Patras, Greece
[2] Department of Environment, Ionian University, GR-29100 Zakynthos, Greece
[3] Department of Chemical Engineering, University of Western Macedonia, GR-50132 Kozani, Greece
* Correspondence: zfronistis@uowm.gr

Citation: Kouvelis, K.; Kampioti, A.A.; Petala, A.; Frontistis, Z. Degradation of Sulfamethoxazole Using a Hybrid CuO$_x$–BiVO$_4$/SPS/Solar System. *Catalysts* **2022**, *12*, 882. https://doi.org/10.3390/catal12080882

Academic Editor: Consuelo Alvarez-Galvan

Received: 11 July 2022
Accepted: 7 August 2022
Published: 11 August 2022

Publisher's Note: MDPI stays neutral with regard to jurisdictional claims in published maps and institutional affiliations.

Copyright: © 2022 by the authors. Licensee MDPI, Basel, Switzerland. This article is an open access article distributed under the terms and conditions of the Creative Commons Attribution (CC BY) license (https://creativecommons.org/licenses/by/4.0/).

Abstract: In recent years, advanced oxidation processes (AOPs) demonstrated great efficiency in eliminating emerging contaminants in aqueous media. However, a majority of scientists believe that one of the main reasons hindering their industrial application is the low efficiencies recorded. This can be partially attributed to reactive oxygen species (ROS) scavenging from real water matrix constituents. A promising strategy to cost-effectively increase efficiency is the simultaneous use of different AOPs. Herein, photocatalysis and sodium persulfate activation (SPS) were used simultaneously to decompose the antibiotic sulfamethoxazole (SMX) in ultrapure water (UPW) and real water matrices, such as bottled water (BW) and wastewater (WW). Specifically, copper-promoted BiVO$_4$ photocatalysts with variable CuO$_x$ (0.75–10% wt.) content were synthesized in powder form and characterized using BET, XRD, DRS, SEM, and HRTEM. Results showed that under simulated solar light irradiation alone, 0.75 Cu.BVO leads to 0.5 mg/L SMX destruction in UPW in a very short treatment time, whereas higher amounts of copper loading decreased SMX degradation. In contrast, the efficiency of all photocatalytic materials dropped significantly in BW and WW. This phenomenon was surpassed using persulfate in the proposed system resulting in synergistic effects, thus significantly improving the efficiency of the combined process. Specifically, when 0.75 Cu.BVO was added in BW, only 40% SMX degradation took place in 120 min under simulated solar irradiation alone, whereas in the solar/SPS/Cu.BVO system, complete elimination was achieved after 60 min. Moreover, ~37%, 45%, and 66% synergy degrees were recorded in WW using 0.75 Cu, 3.0 Cu, and 10.0 Cu.BVO, respectively. Interestingly, experimental results highlight that catalyst screening or process/system examination must be performed in a wide window of operating parameters to avoid erroneous conclusions regarding optimal materials or process combinations for a specific application.

Keywords: photocatalysis; sodium persulfate; antibiotics; water treatment; hybrid system; water matrix

1. Introduction

In the last decade, scientists have pointed out the urgent need to develop and apply new wastewater treatment technologies able to eliminate emerging contaminants such as pharmaceuticals and endocrine-disrupting compounds (EDCs) detected at trace levels in environmental systems [1–4]. To this end, the European Commission is currently reviewing urban wastewater treatment and is considering including, where possible, requirements for elimination of micropollutants from domestic wastewater. Specifically, in early 2022, the Commission stated that drinking water across the EU should be monitored more closely for the possible presence of two endocrine disruptors (β-estradiol and nonylphenol) throughout the water supply chain, with more chemical substances expected to be added to this "watch-list" soon.

To address this challenge, the application of advanced oxidation processes (AOPs) involving the in situ production of reactive species, which degrade organic matter in efficient and usually non-selective ways, has attracted researchers' interest. AOPs constitute a family of similar (but not identical) processes including heterogeneous photocatalysis, electrochemical oxidation, (photo) Fenton, UV/oxidants, ozonation/H_2O_2, and ultrasound [1,5].

Among AOPs, heterogeneous photocatalysis is probably considered the greenest technology, as it has the potential to use sunlight to initiate reactions for micropollutant degradation. Moreover, it eliminates the requirement for a constant supply of precursor chemicals, making it particularly attractive for applications in remote places [6,7]. Sulfamethoxazole (SMX), a well-known antibiotic identified in surface water, has been the subject of many studies regarding the efficiency of photocatalytic systems [8,9]. Parabens, another class of emerging contaminants found in various environmental matrices, have been photocatalytically treated with promising results [10,11].

However, photocatalytic industrial applications in wastewater treatment are still minimal. The already reported low quantum yields are mainly attributed to the small portion of reactive oxygen species (ROS) generated in a photocatalytic system that eventually participates in the degradation reactions. This observation is strongly related to scavenging phenomena due to the presence of a significant variety of organic and inorganic species in real water matrices not considered in most studies carried out in UPW [2,6]. The majority of photocatalytic systems exhibit significantly lower performance in water matrices of environmental concern than in UPW; therefore, the effect of the water matrix is a crucial criterion that should be taken into account to make photocatalysis more attractive for commercial applications. Another factor that suppresses photocatalytic efficiency is the high recombination rate of photogenerated species that characterizes most visible light active semiconductors used as photocatalysts. In attempts to deal with this, many new materials configurations, such as heterojunctions, doped samples, and so on, were introduced as a promising approach to enhance photocatalytic quantum yield [12–15]. However, the photocatalytic materials with improved properties were not necessarily accompanied by higher performance regarding pollutant decomposition.

To overcome these limitations, some researchers suggested that photocatalysis could be applied together with other AOPs, such as persulfate oxidation [3]. In persulfate oxidation, sulfate radicals ($SO_4^{\bullet-}$) are produced after persulfate activation. Typically, heat or UV irradiation is used to break O-O bonds in the persulfate structure and the formation of $SO_4^{\bullet-}$ [16]. However, to decrease the energy requirements, an alternative activation method involving reactions with transition metals has been proposed in recent years [17]. The benefits of that process are multiplied when the system transition metal/persulfate works heterogeneously, as it avoids secondary pollution due to metal leaching. At first, the most promising results were from using copper or cobalt-based oxides as heterogeneous activators [18]. In recent studies, other configurations, such as phosphides [19] or perovskites [20], have shown high catalytic performance in micropollutant degradation systems. For example, diclofenac degradation in water was studied in a peroxymonosulfate/$LaFeO_3$ heterogeneous system [21], and lanthanum cobaltite perovskite was proposed as a promising peroxymonosulfate activator for carbamazepine degradation [22].

The addition of persulfate in a photocatalytic system has a potential dual benefit. Persulfate can react with photogenerated electrons, thus suppressing the recombination rate and enhancing photocatalytic reaction rate. At the same time, it can be activated by light irradiation and the appropriate photocatalytic material, producing highly reactive sulfate radicals that can participate in micropollutant degradation. However, the maximum benefit is gained only if the interaction between the two systems results in a synergistic, rather than cumulative, effect.

Taking the aforementioned into consideration, in the present study, SMX degradation was tested in a hybrid solar/SPS/CuO_x/$BiVO_4$ system in real water matrices. Copper-promoted $BiVO_4$ samples were selected as photocatalysts based on our previous studies that showed, on the one hand, their high photocatalytic efficiency for micropollutant degradation in ultrapure water (UPW). However, the oxidation of pollutants was inhibited in secondary effluent (WW) [23]. On the other hand, previous studies showed the demonstrated activity of copper configurations as efficient heterogeneous persulfate activators [19,24]. Specifically, excellent results have been reported for a CuO/persulfate system for 2,4-dichlorophenol degradation and a magnetic CuO-Fe_3O_4/persulfate system for phenol degradation [25,26]. Furthermore, apart from CuO_x formulations, an up-to-date work proved the applicability of Cu_3P as a heterogeneous sodium persulfate (SPS) activator for the degradation of the antibiotic agent sulfamethoxazole (SMX) [19]. Based on these findings, and considering the low cost and toxicity of copper, CuO_x was adopted in the present system. Monoclinic bismuth vanadate (m$BiVO_4$) has been recognized as one of the most promising photocatalysts. This is due to its narrow band gap of 2.3–2.5 eV, low synthesis cost and low toxicity [23,27]; m$BiVO_4$ has been used in various photocatalytic applications such as decomposition of micropollutants, production of hydrogen through water splitting, and elimination of pathogenic microorganisms [28,29]. However, the photocatalytic efficiency of m$BiVO_4$ is very low, due to the high recombination rate of photogenerated electrons and holes [30]. Kanigaridou et al. [23] mentioned the beneficial role of copper oxide coupling with $BiVO_4$ in the photocatalytic destruction of the endocrine disruptor bisphenol A (BPA). They reported that the apparent rate constant (k_{app}) in 0.75 wt% CuO_x/$BiVO_4$ was almost twice the k_{app} of pure $BiVO_4$ [23].

The main goal of the present study is to add important information regarding the efficiency of the hybrid photocatalytic/persulfate system for effective micropollutant degradation in water to the literature. As far as we know, this is the first time that SMX degradation has been studied in real water matrices in the proposed hybrid system. The novelty of the present study also lies in the fact that photocatalysts were tested in all water matrices, thus avoiding misleading conclusions arising from the fact that in most reported studies the screening of photocatalytic materials takes place only in ultrapure water.

Copper-promoted $BiVO_4$ samples were synthesized using a polyol-reduction method and characterized utilizing BET, XRD, DRS, TEM/HRTEM, and SEM/EDS. The effectiveness of the hybrid solar/SPS/CuO_x/$BiVO_4$ system for SMX degradation was studied in ultrapure water (UPW), bottled water (BW), and secondary treated wastewater (WW).

2. Results and Discussion

According to XRD analysis, all photocatalysts used in the present study are characterized by the scheelite-monoclinic phase of $BiVO_4$ (JCPDS No. 14-0688) with no additional peaks due to copper species being discerned in copper-promoted samples. Characteristic XRD patterns of pure $BiVO_4$ and 3.0 Cu.BVO are shown in Figure 1.

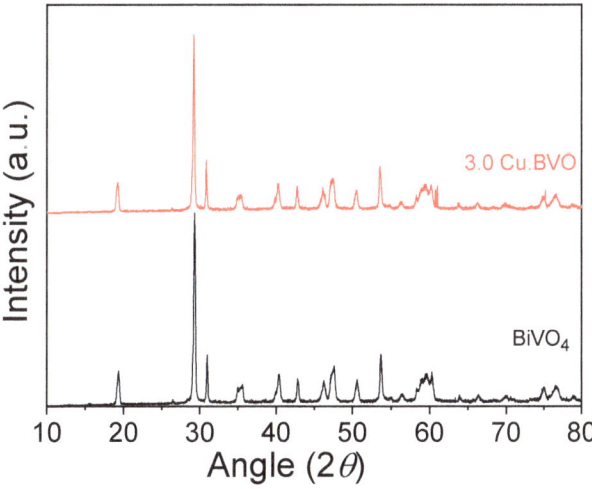

Figure 1. XRD patterns of pure $BiVO_4$ and 3.0 Cu.BVO.

Their specific surface area, as determined utilizing the BET method, was found to be <2 m^2/g in accordance with previous studies [31]. The samples' optical properties were examined employing UV-Vis DRS, as shown in Figure 2; it is observed that pure BiVO$_4$ can absorb light at wavelengths < 550 nm, corresponding to an optical band gap of 2.25–2.30 eV. The addition of copper does not significantly alter the position of the absorption edge, but it leads to increased absorption at longer wavelengths, which increases as Cu content increases. Their morphology was further examined through SEM/EDS and TEM/HRTEM analysis, as shown in Figure 3. SEM images of the 0.75 Cu.BVO and 3.0 Cu.BVO catalysts with EDS mapping of Cu (Figure 3A–D) show that copper is homogeneously distributed on the surface of BiVO$_4$. Similar results were obtained for all copper-promoted samples (not shown for brevity). The HR-TEM image of pure BiVO$_4$ (Figure 3F) shows the interplanar spacings of 0.31 nm and 0.47 nm, which correspond to the (121) and (110) planes of monoclinic BiVO$_4$, respectively. The presence of the CuO phase with a crystal size of less than 5 nm was confirmed for the 3.0 Cu.BVO sample. As shown in Figure 3E, CuO nanoparticles are uniformly dispersed on the BiVO$_4$ surface and are in intimate contact with the BiVO$_4$ nanocrystals. Qualitatively similar results were obtained for xCu.BVO samples with lower Cu loading. It should be noted, however, that the presence of Cu$_2$O nanoparticles should not be excluded. This is because of the low crystallinity of the observed CuO$_x$ nanoparticles and the quite similar (0.25 nm and 0.24 nm) d-spacing values of the (111) planes of the CuO and Cu$_2$O phases. Details concerning their physicochemical characteristics can be found elsewhere [23,32].

In order to assess the photocatalytic efficiency of the samples, a set of experiments dealing with 0.5 mg/L SMX photodecomposition under solar irradiation in UPW was undertaken. As shown in Figure 4A, 0.75 Cu.BVO is the most photocatalytically active copper-promoted sample; its k_{app} is equal to 0.0991 min^{-1}. Further addition of copper species did not favor SMX degradation, resulting in lower k_{app}s equal to 0.0062 and 0.0079 min^{-1} in 3.0 Cu.BVO and 10.0 Cu.BVO, respectively. The observed lowering of photocatalytic activity at higher copper oxide loadings can be attributed to the formation of bulk agglomerates which act as recombination centers and hinder the irradiation of BiVO$_4$, thus lowering the number of photogenerated species [33–35].

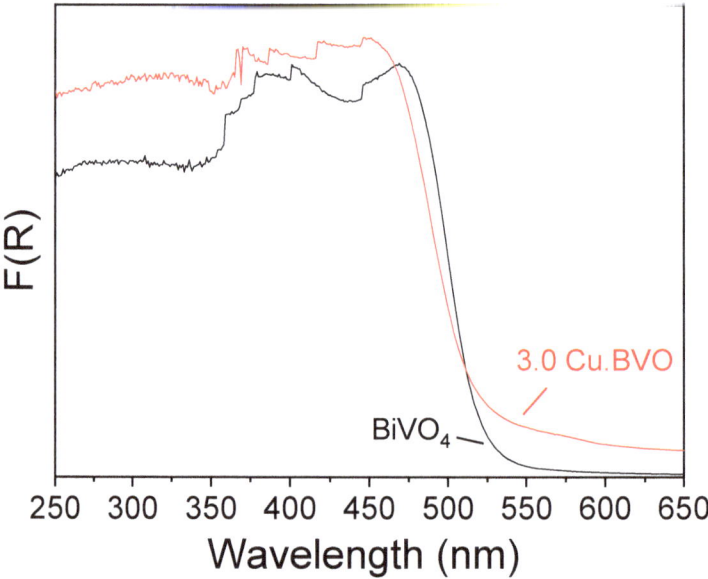

Figure 2. UV-vis diffuse reflectance spectra of BiVO$_4$ and 3.0 Cu.BVO.

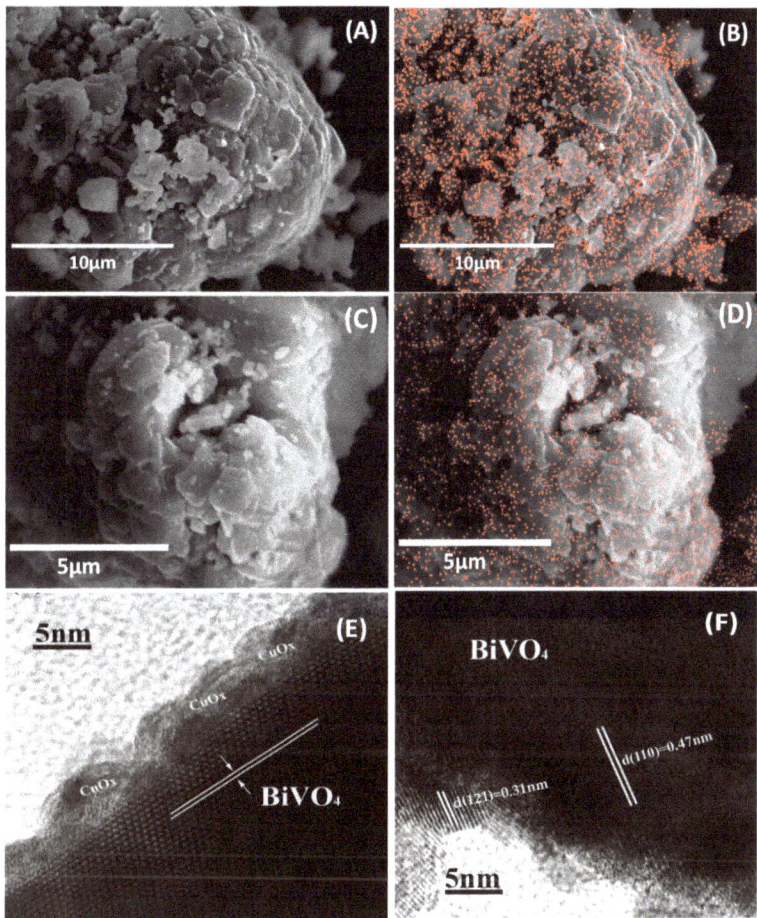

Figure 3. SEM images with EDS mapping of Cu of (**A,B**) 3.0 Cu.BVO, and (**C,D**) 0.75 Cu.BVO. HRTEM images of (**E**) 3.0 Cu.BVO and (**F**) pure BiVO$_4$.

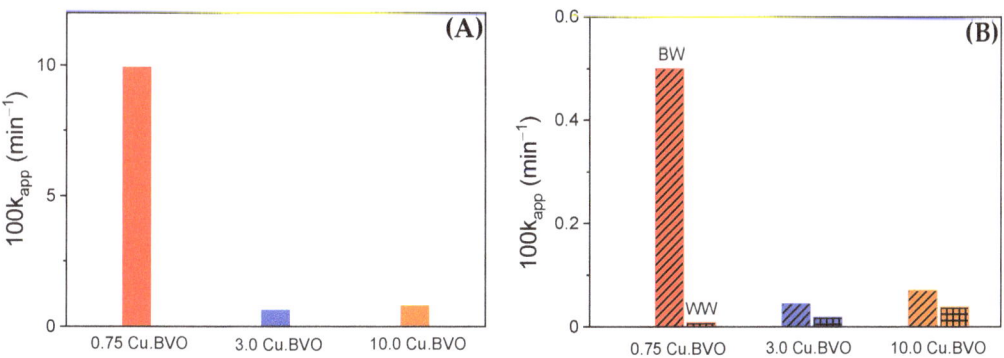

Figure 4. Apparent kinetic constants for Cu.BVO photocatalysts in (**A**) UPW, (**B**) BW, and WW. Experimental conditions: 0.5 mg/L SMX, 100 mg/L SPS, and 500 mg/L catalyst.

On the contrary, as shown in Figure 4B, apparent kinetic constants for all copper loadings tested were some orders of magnitude lower in real water matrices due to the presence of co-existing inorganic and organic ions competing with SMX for ROS [36] and catalyst surface. Specifically, in BW, 0.75 Cu.BVO was also the best performing photocatalytic material; however, its k_{app} was found to equal 0.0433 min^{-1}. Interestingly, its k_{app} decreased to 0.0191 min^{-1} for 3.0 Cu.BVO, whereas it was found to equal 0.0227 min^{-1} for 10.0 Cu.BVO. This observation implies that the optimum loading of a co-catalyst, such as metal oxides, is closely connected with the experimental parameters, such as water matrix, in agreement with other studies [37]. This behavior was further confirmed by experiments carried out in WW. As shown in Figure 4B, SMX degradation was faster when a copper-promoted sample with higher copper loading, 10.0 Cu.BVO, was used. In addition, a 0.75 Cu.BVO sample that showed the best photocatalytic results in UPW and BW was now characterized by the lowest k_{app}. In general, retarding phenomena were more intense in WW than in BW, making SMX degradation practically non-viable.

In order to moderate hindering phenomena in real water matrices, SPS was added to the photocatalytic system, and the degree of synergy was quantified. At first, the efficiency of the hybrid system solar/SPS/catalyst was studied in UPW; results are shown in Figure 5.

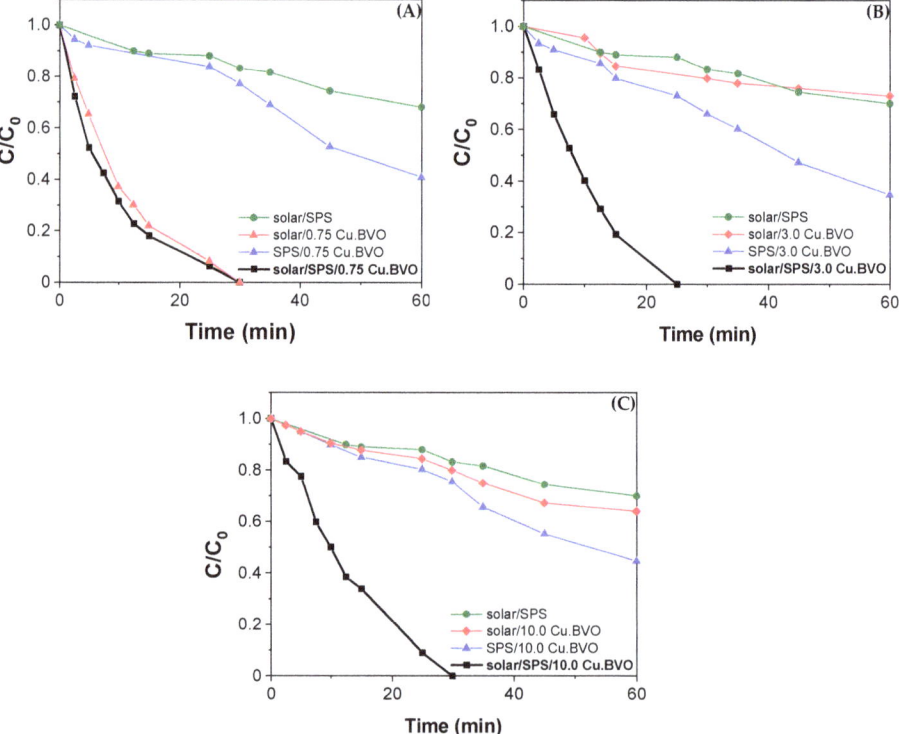

Figure 5. Removal of SMX using solar/SPS, solar/catalyst, SPS/catalyst, and solar/SPS/catalyst for (**A**) 0.75 Cu.BVO, (**B**) 3.0 Cu.BVO, and (**C**) 10.0 Cu.BVO in UPW. Experimental conditions: 0.5 mg/L SMX, 100 mg/L SPS, and 500 mg/L catalyst.

Figure 5A illustrates that 0.75 Cu.BVO is characterized by high photocatalytic activity which resulted in complete 0.5 mg/L SMX decomposition after 30 min of irradiation (solar/0.75 Cu.BVO system). Considering its activity toward SPS activation (SPS/0.75 Cu.BVO), SMX degradation did not exceed 60% in 60 min. SPS activation was more restricted using solar irradiation alone (solar/SPS). When solar irradiation and

SPS were applied together, complete SMX removal was achieved in 30 min, resulting in a zero synergy degree. In contrast, Figure 5B and Table 1 show that for 3.0 Cu.BVO the combined process resulted in 71% synergy.

Table 1. Apparent kinetic constants and synergy degree for all CuO_x loadings examined in UPW, BW, and WW.

	Matrix	k_{total}, 10^{-2} min^{-1}	S (%)	R^2
0.75 Cu.BVO	UPW	11.33	0	0.998
	BW	4.33	75.5	0.966
	WW	0.67	37	0.984
3.0 Cu.BVO	UPW	9.95	71.6	0.989
	BW	2.27	61.1	0.968
	WW	0.92	45	0.997
10.0 Cu.BVO	UPW	8.45	68.2	0.975
	BW	1.91	70.4	0.932
	WW	0.85	66.5	0.951

Considering the individual processes, 3.0 Cu.BVO was less photocatalytically active and moderately activated SPS. Qualitatively similar results were obtained for 10.0 Cu.BVO (Figure 5C). Moreover, the amount of copper leached in the liquid phase in the solar/CuO_x/SPS system was determined to be very low at the end of the run, corresponding to only 0.2% of the Cu contained in the 500 mg/L of the sample with the higher copper content (10.0 Cu.BVO).

Results for the combined processes are summarized in Figure 6; complete SMX degradation occurred in 30 min of irradiation for all catalysts tested. This is attributed to the high photocatalytic activity of 0.75 Cu.BVO and the existence of powerful synergistic effects in samples with higher copper loading.

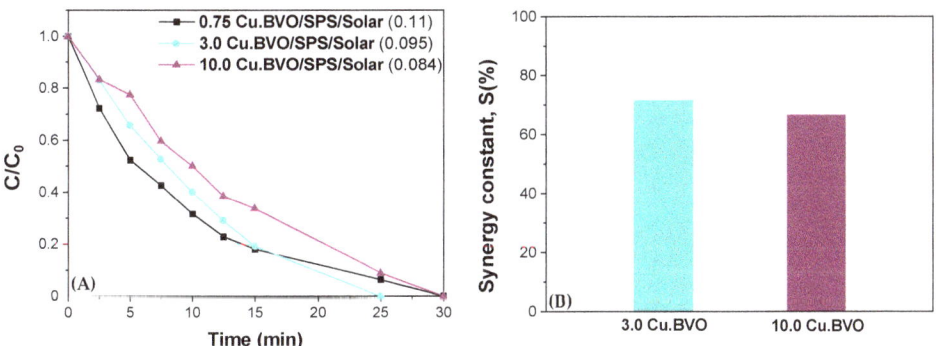

Figure 6. Effect of copper loading on (**A**) SMX removal and (**B**) synergy constant S in solar/SPS/catalyst systems in UPW.

In order to gain insight into the efficacy of the present system towards SMX degradation, Table 2 summarizes the main characteristics of other solar/SPS/photocatalyst systems demonstrating SMX degradation. Kemmou et al. [38], who studied the degradation of 0.5 mg/L SMX using -activated persulfate, reported longer time periods required for complete SMX degradation than those reported in Figure 5. Alexopoulou et al. [19] used Cu_3P as a heterogeneous persulfate activator. They found that SMX degrades quickly in UPW; the degradation rate depends on parameters such as SMX, SPS, and catalyst concentrations. Their results are of the same order of magnitude as the present study. Furthermore, Yin et al. [39] found that SMX was rapidly eliminated in the vis/PDS/MIF-100(Fe) system, and Song et al. [40] investigated the degradation of SMX in a solar/PDS/g-C_3N_4 system.

Table 2. SMX degradation in different solar/SPS/photocatalyst systems.

	SMX Concentration (Mg/L)	Catalyst Concentration (Mg/L)	Persulfate Concentration (Mg/L)	k (Min^{-1})	Time Period for Complete Degradation (Min)	Ref.
Solar/SPS/Biochar (BC)	0.25	90	250	0.065	90	[38]
Solar/SPS/Cu$_3$P	0.5	40	100	0.114	20	[19]
vis/PDS/MIF-100(Fe)	10	500	1000	0.012	180	[39]
solar/PDS/g-C$_3$N$_4$	1	500	120	0.068	60	[40]
Solar/SPS/3.0 Cu.BVO	0.5	500	100	0.099	25	Present study

SMX degradation profiles in BW under solar/SPS, solar/0.75 Cu.BVO, SPS/0.75 Cu.BVO, and solar/SPS/0.75 Cu.BVO systems are shown in Figure 7A. In the presence of only 100 mg/L SPS under simulated solar irradiation, 40% SMX degradation was observed after 120 min, showing the limited activation of SPS by sunlight. Similar SMX removal was recorded in the solar/0.75 Cu.BVO system. Furthermore, 0.75 Cu.BVO showed restricted activity as a SPS activator, resulting in only 30% SMX removal in the same period. However, when 0.75 Cu.BVO and SPS were added to the system, complete 0.5 mg/L SMX degradation was obtained in 60 min. The synergy, as quantified by Equation (2), was 75%. Qualitatively similar results were recorded for the 3.0 Cu.BVO sample.

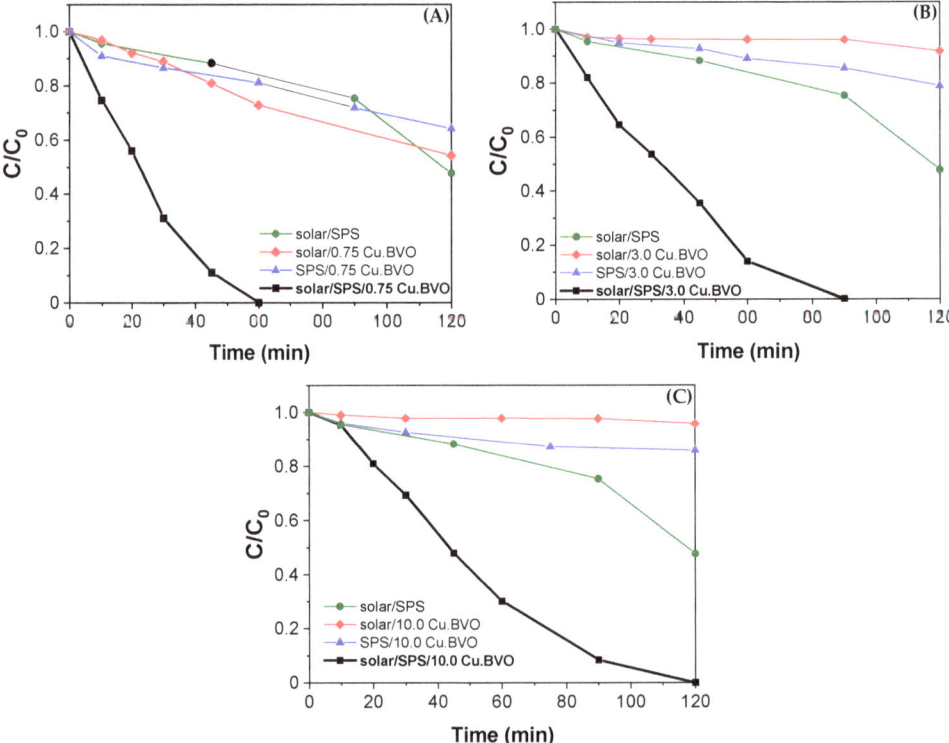

Figure 7. Removal of SMX using solar/SPS, solar/catalyst, SPS/catalyst, and solar/SPS/catalyst for (**A**) 0.75 Cu.BVO, (**B**) 3.0 Cu.BVO, and (**C**) 10.0 Cu.BVO in BW. Experimental conditions: 0.5 mg/L SMX, 100 mg/L SPS, and 500 mg/L catalyst.

Specifically, despite the higher amount of copper species on the surface of BiVO$_4$ that could lead to increased efficiency towards persulfate activation, SMX degradation was less than 20% in the SPS/3.0 Cu.BVO system, as shown in Figure 7B. However, the combination with simulated solar irradiation strongly enhanced SMX degradation, resulting in complete degradation after 120 min of irradiation. The degree of synergy was 61%. Considering the efficiency of the sample with the highest copper loading towards SMX degradation in BW (Figure 7C), it is observed that it was practically inactive in both systems examined (solar/10.0 Cu.BVO and solar/SPS/0.75 Cu.BVO), resulting in less than 10% SMX removal. In BW, all materials tested led to complete SMX degradation in the hybrid system, whereas SPS activation due to sunlight was lower than 40%. In addition, copper-promoted $BiVO_4$ samples showed low efficiency toward persulfate degradation.

Figure 8 outlines results obtained in BW in hybrid systems for all photocatalytic materials tested; SMX degradation was favored in the 0.75 Cu.BVO sample, but increasing copper loading slowed down the degradation kinetics. In contrast, the degree of synergy was similar in all cases (Figure 8B), implying that it is independent of the process's overall performance.

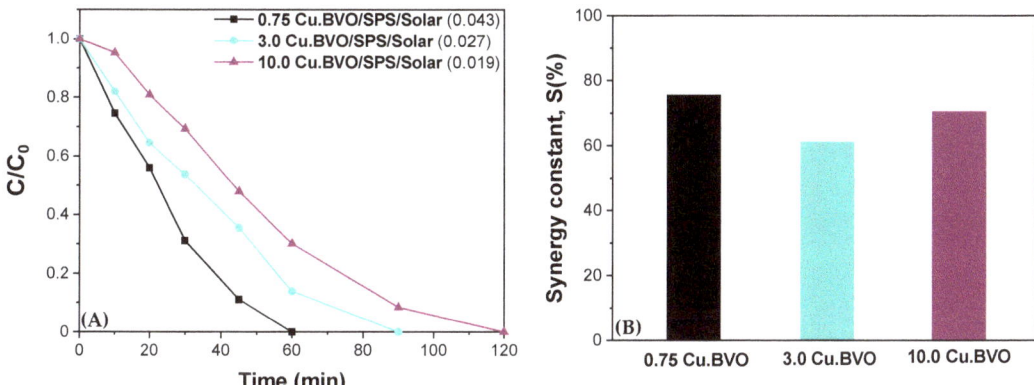

Figure 8. Effect of copper loading on (**A**) SMX removal and (**B**) *kapp* in solar/SPS/catalyst systems in BW.

Significantly lower yields were recorded in WW (Figure 9). It should be noted that in that case, 500 mg/L SPS were added to the photocatalytic reactor. Specifically, as shown in Figure 9A, practically no SMX degradation occurred in the solar/0.75 Cu.BVO system and less than 20% in the SPS/0.75 Cu.BVO. SMX degradation increased to 50% in the combined solar/SPS/0.75 Cu.BVO, resulting in 37% synergy. Synergistic rather than cumulative effects were also observed in WW for the solar/SPS/3.0 Cu.BVO system, as shown in Figure 9B and Table 2, with the degree of synergy increasing to 45%.

In agreement with results obtained in BW, 3.0 Cu.BVO showed very low efficiency towards the activation of SPS (SPS/3.0 Cu.BVO). Figure 9C shows results obtained for the 10.0 Cu.BVO catalyst; SPS cannot practically be activated either by sunlight or by 10.0 Cu.BVO in WW. However, SMX removal reaches 60% after 120 min of irradiation in the hybrid system. In general, the SMX degradation rate was lower in WW than in BW, as expected, due to the increased complexity of the water matrix. Interestingly, in that case (Figure 10) 3.0 CuBVO and 10.0 Cu.BVO showed similar catalytic activity in the combined system, whereas both the k_{app} and synergy degree were lower for 0.75 Cu.BVO.

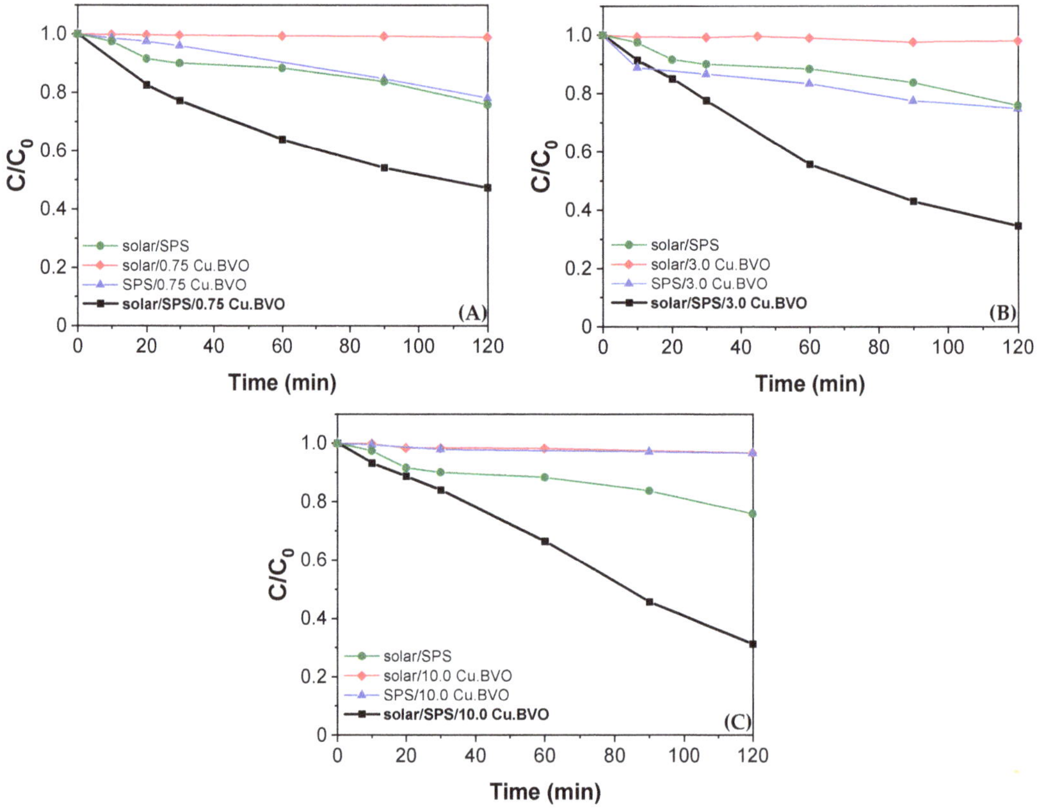

Figure 9. Removal of SMX using solar/SPS, solar/catalyst, SPS/catalyst, and solar/SPS/catalyst for (**A**) 0.75 Cu.BVO, (**B**) 3.0 Cu.BVO and (**C**) 10.0 Cu.BVO in WW. Experimental conditions: 0.5 mg/L SMX, 500 mg/L SPS, 500 mg/L catalyst.

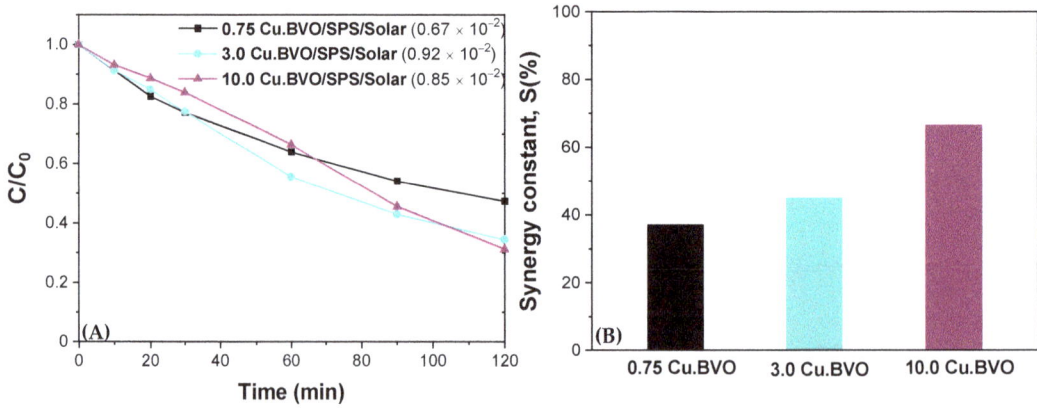

Figure 10. Effect of copper loading on (**A**) SMX removal and (**B**) k_{app} in solar/SPS/catalyst systems in WW.

Additional experiments using appropriate scavengers for different reactive species were performed to investigate the mechanism of SMX decomposition; results are shown in

Figure 11. More specifically, EDTA was used to bind photogenerated holes, tert butanol (which reacts mainly with hydroxyl radicals), and methanol (which reacts at a similar rate with hydroxyl radicals and sulfate radicals). From Figure 11, it is evident that the presence of t-butanol does not cause a significant inhibition. In contrast, SMX degradation was hindered in the presence of methanol; the inhibition becomes significant when EDTA is present in the solution, which indicates the dominant role of photogenerated holes.

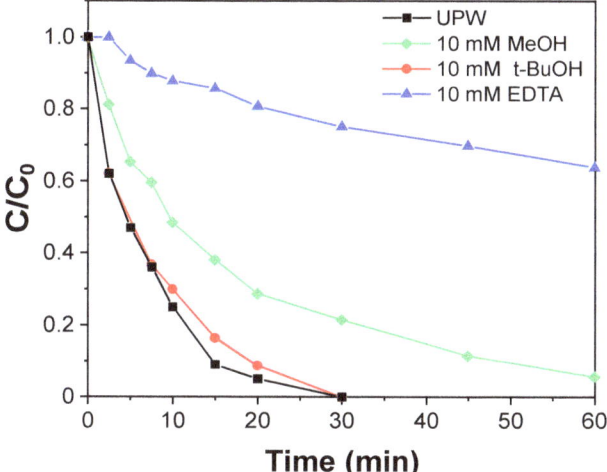

Figure 11. Effect of reactive species scavengers on SMX degradation. Experimental conditions: 0.5 mg/L SMX, 100 mg/L SPS, and 500 mg/L catalyst 3.0 Cu.BVO.

To shed light on the contribution of sulfate radicals, the consumption of SPS was measured in the presence and absence of the photocatalyst or solar irradiation. The consumption of SPS was significantly higher in the hybrid system than in the individual processes, as shown in Figure 12.

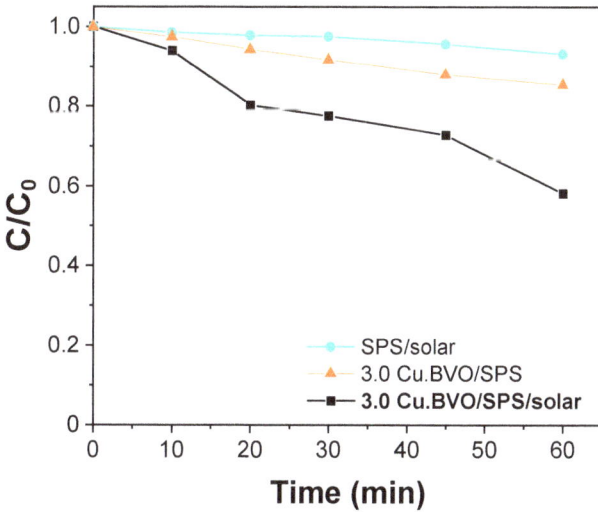

Figure 12. SPS consumption by solar/SPS, SPS/catalyst, and solar/SPS/catalyst for 3.0 Cu.BVO. Experimental conditions: 0.5 mg/L SMX, 100 mg/L SPS, and 500 mg/L catalyst.

For example, after 60 min the consumption of persulfate was 7%, 14%, and 4% for SPS/Solar, 3.0 Cu.BVO/SPS, and the hybrid 3.0 Cu.BVO/SPS/solar system, respectively.

Taking the aforementioned into consideration, the SMX degradation mechanism in the combined process can be described as follows: Under solar light irradiation, photoproduced pairs of electrons-holes are formed in copper-promoted $BiVO_4$ samples. SPS can now react with electrons in the semiconductor's conduction band, thus forming $SO_4^{\bullet-}$ which can participate in SMX degradation. In addition, the photogenerated holes in the valence band are now "free" to decompose SMX. Additional $SO_4^{\bullet-}$ are also available in the system due to SPS activation by sunlight. The proposed degradation mechanism is shown in Scheme 1.

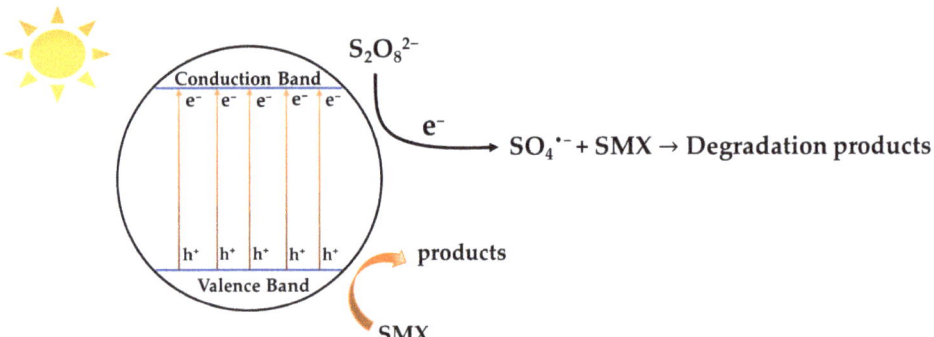

Scheme 1. Simplified SMX degradation mechanism in the solar/SPS/catalyst hybrid system.

3. Materials and Methods

3.1. Chemical and Water Matrices

Sulfamethoxazole ($C_{10}H_{11}N_3O_3S$, CAS: 723-46-6), sodium persulfate ($Na_2S_2O_8$, CAS: 7775-27-1), t-butanol ($C_4H_{10}O$, CAS: 75-65-0), methanol (CH_3OH, CAS: 67-56-1), humic acid (HA, CAS: 1415-93-6), sodium chloride (NaCl, CAS: 7647-14-5), sodium bicarbonate ($NaHCO_3$, CAS: 144-55-8), and acetonitrile (CH_3CN, CAS: 75-05-8, for HPLC analysis) were obtained from Sigma-Aldrich (St. Louis, MO, USA). Chemicals used for catalyst preparation can be found elsewhere [23].

Apart from ultrapure water (UPW, pH = 6.5), secondary effluent from the University of Patras campus wastewater treatment plant (WW) and commercially bottled water were used in degradation experiments. Their main characteristics are shown in Table 3.

Table 3. Real water matrices' characteristics.

Parameter	WW	BW
pH	8.0	7.4
Conductivity (μs/cm)	300	350
Sulfate (mg/L)	35	10
Chloride (mg/L)	75	8
Nitrate (mg/L)	60	10
Bicarbonate (mg/L)	180	240

3.2. Catalyst Preparation and Characterization

Copper-promoted $BiVO_4$ photocatalysts (0.75, 3.0, and 10.0 wt.% Cu) were prepared using a polyol-reduction method with ethylene glycol as a reductant [41]. The preparation method is described in detail in our previous work [23,32]. Photocatalysts thus prepared are denoted in the following as xCu.BVO, where x represents the copper loading (wt.%).

Photocatalysts were characterized through physisorption at the temperature of liquid nitrogen (77 K) (BET method) using a Micromeritics (Gemini III 2375, Norcross, GA, USA),

X-ray diffraction (XRD), (Brucker D8 Advance, Billerica, MA, USA), diffuse reflectance spectra (DRS) (Varian Cary 3E, Palo Alto, CA, USA) high resolution transmission electron microscopy (TEM) and scanning electron microscopy with an energy dispersive spectrometer (SEM/EDS) (JEOL 6300, Peabody, MA, USA). High-resolution transmission electron microscopy (HR-TEM) images were obtained using a JEOL JEM-2100 (Peabody, MA, USA) system operated at 200 kV (resolution: point 0.23 nm, lattice 0.14 nm). Details regarding the methodologies and procedures are available in former studies [42].

3.3. Experimental Procedure and Analytical Methods

In a typical experiment, an appropriate amount of a stock SMX solution was added to the photocatalytic reactor, which was filled with UPW, BW, or WW (the reactor's capacity is 120 mL). After adding the desired amount (typically 500 mg/L) of the photocatalyst, the system was stirred in the dark for 15 min to achieve the adsorption−desorption dynamic equilibrium. Next, the suspension was irradiated by simulated solar irradiation using a solar simulator (Oriel-LCS-100, 100 W Xe ozone-free lamp). The incident intensity in the photoreactor was calculated using chemical actinometry and was found to be equal to 7.6×10^{-7} einstein/(L.s). Sampling took place at fixed times; after filtration, analysis through high-performance liquid chromatography (HPLC) occurred.

To estimate reaction rates, SMX degradation was considered to follow a pseudo-first-order kinetic expression (Equation (1)) [43]:

$$-\frac{dC}{dt} = k_{app}C \Leftrightarrow \ln\frac{C_0}{C} = k_{app}t \quad (1)$$

where C and C_0 correspond to SMX concentration at time $t = t$ and $t = 0$, respectively, and k_{app} is the kinetic constant.

Persulfate consumption was measured using a Hach DR5000 (Loveland, CO, USA) spectrophotometer according to the method proposed by Liang et al. [44].

Leaching of copper in the liquid phase was measured via atomic absorption spectrometer (SHIMADZU AA-6800, Kyoto, Japan).

The degree of synergy, S, was quantified according to the following equation (Equation (2)) [45]:

$$S(\%) = \frac{k_{combined} - \sum_i^n k_i}{k_{combined}} \times 100 \quad (2)$$

$$S \begin{cases} > 0, \text{ synergistic effect} \\ = 0, \text{ cumulative effect} \\ < 0, \text{ antagonistic effect} \end{cases}$$

4. Conclusions

Summarizing, this work examines the possible synergy of the SPS/CuO$_x$.BiVO$_4$ system in the presence of simulated solar irradiation. According to the results, using an oxidant, such as persulfate, to improve the performance of the combined process seems promising, making the combined process more attractive for real aqueous matrices where the separate processes show significantly reduced performance.

In recent years, research has shifted to conditions that simulate real problems. Ideally, the evaluation of both catalytic materials and processes must be performed holistically by including parameters such as performance under different conditions, cost and reusability of the materials, toxicity, and environmental and energy costs.

In this light, and inspired by previous research, this work shows that the simultaneous use of more than one process seems to be a promising solution that overcomes some of the disadvantages of individual processes. However, it is not the de facto optimal solution proposed for all cases.

Since the performance of both materials and processes depends to a vast extent on the conditions, the evaluation of materials or systems must be performed according to the problem in question and in a range of experimental or operational conditions. Uncritical evaluation or adoption of optimal conditions from unrepresentative experiments can lead to incorrect conclusions.

The main conclusions extracted from the present study can be summarized as follows:
(1) The hybrid system, solar/SPS/CuO$_x$.BiVO$_4$, leads to significantly improved photocatalytic yields than the individual systems (solar/CuO$_x$.BiVO$_4$ and SPS/CuO$_x$.BiVO$_4$), considering SMX degradation in real water matrices.
(2) In real water matrices, the interaction between the photocatalytic-persulfate system and matrix components results in a synergistic rather than a cumulative effect, with a high degree of synergy. Using combined technologies seems to be preferred in very complex aqueous matrices where the competition for available oxidizing species and the catalyst surface are maximized.
(3) The complexity of the water matrix determines the best performing catalytic material; 0.75 Cu.BVO showed the highest efficiency in UPW and BW, whereas 3.0 Cu.BVO and 10.0 Cu.BVO were characterized by higher activity in WW.

Author Contributions: Conceptualization, A.P. and Z.F.; methodology, A.P., A.A.K. and Z.F.; formal analysis, K.K., A.P., A.A.K. and Z.F.; investigation, K.K. and A.P.; resources, A.P. and Z.F.; data curation, K.K., A.P., A.A.K. and Z.F.; writing—original draft preparation, K.K., A.P., A.A.K. and Z.F.; writing—review and editing, K.K., A.P., A.A.K. and Z.F.; visualization, K.K., A.P. and A.A.K.; supervision, A.P. and Z.F.; project administration, A.P. and Z.F. All authors have read and agreed to the published version of the manuscript.

Funding: This research was funded by the Hellenic Foundation for Research and Innovation (HFRI) and the General Secretariat for Research and Technology (GSRT). This work is part of the "2De4P: Development and Demonstration of a Photocatalytic Process for removing Pathogens and Pharmaceuticals from wastewaters" project, which was implemented under the action "H.F.R.I.–1st Call for Research Projects to Support Post-Doctoral Researchers", and funded by the HFRI and the GSRT.

Data Availability Statement: Not applicable.

Conflicts of Interest: The authors declare no conflict of interest.

References

1. Coha, M.; Farinelli, G.; Tiraferri, A.; Minella, M.; Vione, D. Advanced Oxidation Processes in the Removal of Organic Substances from Produced Water: Potential, Configurations, and Research Needs. *Chem. Eng. J.* **2021**, *414*, 128668. [CrossRef]
2. Petala, A.; Mantzavinos, D.; Frontistis, Z. Impact of Water Matrix on the Photocatalytic Removal of Pharmaceuticals by Visible Light Active Materials. *Curr. Opin. Green Sustain. Chem.* **2021**, *28*, 100445. [CrossRef]
3. Chen, G.; Yu, Y.; Liang, L.; Duan, X.; Li, R.; Lu, X.; Yan, B.; Li, N.; Wang, S. Remediation of Antibiotic Wastewater by Coupled Photocatalytic and Persulfate Oxidation System: A Critical Review. *J. Hazard. Mater.* **2021**, *408*, 124461. [CrossRef] [PubMed]
4. Wee, S.Y.; Aris, A.Z. Occurrence and Public-Perceived Risk of Endocrine Disrupting Compounds in Drinking Water. *Npj Clean Water* **2019**, *2*, 4. [CrossRef]
5. Dimitrakopoulou, D.; Rethemiotaki, I.; Frontistis, Z.; Xekoukoulotakis, N.P.; Venieri, D.; Mantzavinos, D. Degradation, Mineralization and Antibiotic Inactivation of Amoxicillin by UV-A/TiO$_2$ Photocatalysis. *J. Environ. Manag.* **2012**, *98*, 168–174. [CrossRef]
6. Loeb, S.K.; Alvarez, P.J.J.; Brame, J.A.; Cates, E.L.; Choi, W.; Crittenden, J.; Dionysiou, D.D.; Li, Q.; Li-Puma, G.; Quan, X.; et al. The Technology Horizon for Photocatalytic Water Treatment: Sunrise or Sunset? *Environ. Sci. Technol.* **2019**, *53*, 2937–2947. [CrossRef]
7. Spasiano, D.; Marotta, R.; Malato, S.; Fernandez-Ibañez, P.; Di Somma, I. Solar Photocatalysis: Materials, Reactors, Some Commercial, and Pre-Industrialized Applications. A Comprehensive Approach. *Appl. Catal. B Environ.* **2015**, *170–171*, 90–123. [CrossRef]
8. Wang, X.; Peng, Y.; Xie, T.; Zhang, Y.; Wang, Y.; Yang, H. Photocatalytic Removal of Sulfamethoxazole Using Yeast Biomass-Derived NixP/Biocarbon Composites in the Presence of Dye Sensitizer. *J. Environ. Chem. Eng.* **2022**, *10*, 107426. [CrossRef]
9. Evgenidou, E.; Chatzisalata, Z.; Tsevis, A.; Bourikas, K.; Torounidou, P.; Sergelidis, D.; Koltsakidou, A.; Lambropoulou, D.A. Photocatalytic Degradation of a Mixture of Eight Antibiotics Using Cu-Modified TiO2 Photocatalysts: Kinetics, Mineralization, Antimicrobial Activity Elimination and Disinfection. *J. Environ. Chem. Eng.* **2021**, *9*, 105295. [CrossRef]

10. Nguyen, V.-H.; Phan Thi, L.-A.; Chandana, P.S.; Do, H.-T.; Pham, T.-H.; Lee, T.; Nguyen, T.D.; Le Phuoc, C.; Huong, P.T. The Degradation of Paraben Preservatives: Recent Progress and Sustainable Approaches toward Photocatalysis. *Chemosphere* **2021**, *276*, 130163. [CrossRef]
11. Velegraki, T.; Hapeshi, E.; Fatta-Kassinos, D.; Poulios, I. Solar-Induced Heterogeneous Photocatalytic Degradation of Methyl-Paraben. *Appl. Catal. B Environ.* **2015**, *178*, 2–11. [CrossRef]
12. Wang, L.; Wang, K.; He, T.; Zhao, Y.; Song, H.; Wang, H. Graphitic Carbon Nitride-Based Photocatalytic Materials: Preparation Strategy and Application. *ACS Sustain. Chem. Eng.* **2020**, *8*, 16048–16085. [CrossRef]
13. Sendão, R.M.S.; Esteves da Silva, J.C.G.; Pinto da Silva, L. Photocatalytic Removal of Pharmaceutical Water Pollutants by TiO_2–Carbon Dots Nanocomposites: A Review. *Chemosphere* **2022**, *301*, 134731. [CrossRef] [PubMed]
14. Velempini, T.; Prabakaran, E.; Pillay, K. Recent Developments in the Use of Metal Oxides for Photocatalytic Degradation of Pharmaceutical Pollutants in Water—A Review. *Mater. Today Chem.* **2021**, *19*, 100380. [CrossRef]
15. Shanavas, S.; Haija, M.A.; Singh, D.P.; Ahamad, T.; Roopan, S.M.; Van Le, Q.; Acevedo, R.; Anbarasan, P.M. Development of High Efficient Co_3O_4/Bi_2O_3/RGO Nanocomposite for an Effective Photocatalytic Degradation of Pharmaceutical Molecules with Improved Interfacial Charge Transfer. *J. Environ. Chem. Eng.* **2022**, *10*, 107243. [CrossRef]
16. Luo, C.; Ma, J.; Jiang, J.; Liu, Y.; Song, Y.; Yang, Y.; Guan, Y.; Wu, D. Simulation and Comparative Study on the Oxidation Kinetics of Atrazine by UV/H_2O_2, UV/HSO_5^- and $UV/S_2O_8^{2-}$. *Water Res.* **2015**, *80*, 99–108. [CrossRef]
17. Guo, P.-C.; Qiu, H.-B.; Yang, C.-W.; Zhang, X.; Shao, X.-Y.; Lai, Y.-L.; Sheng, G.-P. Highly Efficient Removal and Detoxification of Phenolic Compounds Using Persulfate Activated by MnO_x@OMC: Synergistic Mechanism and Kinetic Analysis. *J. Hazard. Mater.* **2021**, *402*, 123846. [CrossRef]
18. Shukla, P.; Wang, S.; Singh, K.; Ang, H.M.; Tadé, M.O. Cobalt Exchanged Zeolites for Heterogeneous Catalytic Oxidation of Phenol in the Presence of Peroxymonosulphate. *Appl. Catal. B Environ.* **2010**, *99*, 163–169. [CrossRef]
19. Alexopoulou, C.; Petala, A.; Frontistis, Z.; Drivas, C.; Kennou, S.; Kondarides, D.I.; Mantzavinos, D. Copper Phosphide and Persulfate Salt: A Novel Catalytic System for the Degradation of Aqueous Phase Micro-Contaminants. *Appl. Catal. B Environ.* **2019**, *244*, 178–187. [CrossRef]
20. Manos, D.; Papadopoulou, F.; Margellou, A.; Petrakis, D.; Konstantinou, I. Heterogeneous Activation of Persulfate by $LaMO_3$ (M=Co, Fe, Cu, Mn, Ni) Perovskite Catalysts for the Degradation of Organic Compounds. *Catalysts* **2022**, *12*, 187. [CrossRef]
21. Rao, Y.F.; Zhang, Y.; Han, F.; Guo, H.; Huang, Y.; Li, R.; Qi, F.; Ma, J. Heterogeneous Activation of Peroxymonosulfate by $LaFeO_3$ for Diclofenac Degradation: DFT-Assisted Mechanistic Study and Degradation Pathways. *Chem. Eng. J.* **2018**, *352*, 601–611. [CrossRef]
22. Guo, H.; Zhou, X.; Zhang, Y.; Yao, Q.; Qian, Y.; Chu, H.; Chen, J. Carbamazepine Degradation by Heterogeneous Activation of Peroxymonosulfate with Lanthanum Cobaltite Perovskite: Performance, Mechanism and Toxicity. *J. Environ. Sci.* **2020**, *91*, 10–21. [CrossRef]
23. Kanigaridou, Y.; Petala, A.; Frontistis, Z.; Antonopoulou, M.; Solakidou, M.; Konstantinou, I.; Deligiannakis, Y.; Mantzavinos, D.; Kondarides, D.I. Solar Photocatalytic Degradation of Bisphenol A with $CuO_x/BiVO_4$: Insights into the Unexpectedly Favorable Effect of Bicarbonates. *Chem. Eng. J.* **2017**, *318*, 39–49. [CrossRef]
24. Lalas, K.; Petala, A.; Frontistis, Z.; Konstantinou, I.; Mantzavinos, D. Sulfamethoxazole Degradation by the CuO_x/Persulfate System. *Catal. Today* **2021**, *361*, 139–145. [CrossRef]
25. Zhang, T.; Chen, Y.; Wang, Y.; Le Roux, J.; Yang, Y.; Croué, J.P. Efficient peroxydisulfate activation process not relying on sulfate radical generation for water pollutant degradation. *Environ. Sci. Technol.* **2014**, *48*, 5868–5875. [CrossRef] [PubMed]
26. Lei, Y.; Chen, C.S.; Tu, Y.J.; Huang, Y.H.; Zhang, H. Heterogeneous degradation of organic pollutants by persulfate activated by $CuO-Fe_3O_4$: Mechanism, stability, and effects of pH and bicarbonate ions. *Environ. Sci. Technol.* **2015**, *49*, 6838–6845. [CrossRef] [PubMed]
27. Feng, C.; Wang, D.; Jin, B.; Jiao, Z. The enhanced photocatalytic properties of $BiOCl/BiVO_4$ p–n heterojunctions via plasmon resonance of metal Bi. *RSC Adv.* **2015**, *5*, 75947–75952. [CrossRef]
28. Hirota, K.; Komatsu, G.; Yamashita, M.; Takemura, H.; Yamaguchi, O. Formation, characterization and sintering of alkoxy-derived bismuth vanadate. *Mater. Res. Bull.* **1992**, *27*, 823–830. [CrossRef]
29. Sayama, K.; Nomura, A.; Zou, Z.; Abe, R.; Abe, Y.; Arakawa, H. Photoelectrochemical decomposition of water on nanocrystalline BiVO4 film electrodes under visible light. *Chem. Commun.* **2003**, *3*, 2908–2909. [CrossRef]
30. Park, Y.; Mc Donald, K.J.; Choi, K.S. Progress in bismuth vanadate photoanodes for use in solar water oxidation. *Chem. Soc. Rev.* **2013**, *42*, 2321–2337. [CrossRef]
31. Lu, Y.; Shang, H.; Shi, F.; Chao, C.; Zhang, X.; Zhang, B. Preparation and Efficient Visible Light-Induced Photocatalytic Activity of m-$BiVO_4$ with Different Morphologies. *J. Phys. Chem. Solids* **2015**, *85*, 44–50. [CrossRef]
32. Petala, A.; Bontemps, R.; Spartatouille, A.; Frontistis, Z.; Antonopoulou, M.; Konstantinou, I.; Kondarides, D.I.; Mantzavinos, D. Solar light-induced degradation of ethyl paraben with $CuO_x/BiVO_4$: Statistical evaluation of operating factors and transformation by-products. *Catal. Today* **2017**, *280*, 122–131. [CrossRef]
33. Aguilera-Ruiz, E.; García-Pérez, U.M.; De La Garza-Galván, M.; Zambrano-Robledo, P.; Bermúdez-Reyes, B.; Peral, J. Efficiency of $Cu_2O/BiVO_4$ Particles Prepared with a New Soft Procedure on the Degradation of Dyes under Visible-Light Irradiation. *Appl. Surf. Sci.* **2015**, *328*, 361–367. [CrossRef]

34. Xu, H.; Li, H.; Wu, C.; Chu, J.; Yan, Y.; Shu, H.; Gu, Z. Preparation, Characterization and Photocatalytic Properties of Cu-Loaded BiVO$_4$. *J. Hazard. Mater.* **2008**, *153*, 877–884. [CrossRef] [PubMed]
35. Jiang, H.-Q.; Endo, H.; Natori, H.; Nagai, M.; Kobayashi, K. Fabrication and Efficient Photocatalytic Degradation of Methylene Blue over CuO/BiVO$_4$ Composite under Visible-Light Irradiation. *Mater. Res. Bull.* **2009**, *44*, 700–706. [CrossRef]
36. Lado Ribeiro, A.R.; Moreira, N.F.F.; Li Puma, G.; Silva, A.M.T. Impact of Water Matrix on the Removal of Micropollutants by Advanced Oxidation Technologies. *Chem. Eng. J.* **2019**, *363*, 155–173. [CrossRef]
37. Ribeiro, R.S.; Frontistis, Z.; Mantzavinos, D.; Silva, A.M.T.; Faria, J.L.; Gomes, H.T. Screening of Heterogeneous Catalysts for the Activated Persulfate Oxidation of Sulfamethoxazole in Aqueous Matrices. Does the Matrix Affect the Selection of Catalyst? *J. Chem. Technol. Biotechnol.* **2019**, *94*, 2425–2432. [CrossRef]
38. Kemmou, L.; Frontistis, Z.; Vakros, J.; Manariotis, I.D.; Mantzavinos, D. Degradation of antibiotic sulfamethoxazole by biochar-activated persulfate: Factors affecting the activation and degradation processes. *Catal. Today* **2018**, *313*, 128–133. [CrossRef]
39. Yin, R.; Chen, Y.; He, S.; Li, W.; Zeng, L.; Guo, W.; Zhu, M. In situ photoreduction of structural Fe(III) in a metal–organic framework for peroxydisulfate activation and efficient removal of antibiotics in real wastewater. *J. Hazard. Mater.* **2020**, *388*, 121996. [CrossRef]
40. Song, Y.; Huang, L.; Zhang, X.; Zhang, H.; Wang, L.; Zhang, H.; Liu, Y. Synergistic effect of persulfate and g-C$_3$N$_4$ under simulated solar light irradiation: Implication for the degradation of sulfamethoxazole. *J. Hazard. Mater.* **2020**, *393*, 122379. [CrossRef]
41. Min, S.; Wang, F.; Jin, Z.; Xu, J. Cu$_2$O Nanoparticles Decorated BiVO$_4$ as an Effective Visible-Light-Driven p-n Heterojunction Photocatalyst for Methylene Blue Degradation. *Superlattices Microstruct.* **2014**, *74*, 294–307. [CrossRef]
42. Petala, A.; Tsikritzis, D.; Kollia, M.; Ladas, S.; Kennou, S.; Kondarides, D.I. Synthesis and Characterization of N-Doped TiO$_2$ Photocatalysts with Tunable Response to Solar Radiation. *Appl. Surf. Sci.* **2014**, *305*, 281–291. [CrossRef]
43. Abellán, M.N.; Giménez, J.; Esplugas, S. Photocatalytic Degradation of Antibiotics: The Case of Sulfamethoxazole and Trimethoprim. *Catal. Today* **2009**, *144*, 131–136. [CrossRef]
44. Liang, C.; Huang, C.F.; Mohanty, N.; Kurakalva, R.M. A rapid spectrophotometric determination of persulfate anion in ISCO. *Chemosphere* **2008**, *73*, 1540–1543. [CrossRef] [PubMed]
45. Metheniti, M.; Frontistis, Z.; Ribeiro, R.; Silva, A.; Faria, J.; Gomes, H.; Mantzavinos, D. Degradation of Propyl Paraben by Activated Persulfate Using Iron-Containing Magnetic Carbon Xerogels: Investigation of Water Matrix and Process Synergy Effects. *Environ. Sci. Pollut. Res.* **2018**, *25*, 34801–34810. [CrossRef]

MDPI
St. Alban-Anlage 66
4052 Basel
Switzerland
Tel. +41 61 683 77 34
Fax +41 61 302 89 18
www.mdpi.com

Catalysts Editorial Office
E-mail: catalysts@mdpi.com
www.mdpi.com/journal/catalysts

www.ingramcontent.com/pod-product-compliance
Lightning Source LLC
LaVergne TN
LVHW070449100526
838202LV00014B/1689